Progress in Inflammation Research

Forthcoming titles:

NPY Family of Peptides in Immune Disorders, Inflammation, Angiogenesis and Cancer, G.Z. Feuerstein, Z. Zukowska (Editors), 2005

Turning up the Heat on Pain: Vanilloid Receptors in Pain and Inflammation, A.B Malmberg, K.R. Bley (Editors), 2005

Sodium Channels, Pain, and Analgesia, K. Coward, M. Baker (Editors), 2005

Complement and Kidney Disease, P.F. Zipfel (Editor), 2005

(Already published titles see last page.)

Regulatory T Cells in Inflammation

Leonie S. Taams
Arne N. Akbar
Marca H. M. Wauben

Editors

Birkhäuser Verlag
Basel · Boston · Berlin

Editors

Leonie S. Taams
Infection & Immunity Research Group
Franklin-Wilkins Building
King's College London
150 Stamford Street
London SE1 9NN
UK

Arne N. Akbar
Department of Immunology and Molecular Pathology
Division of Infection and Immunity
University College London
46 Cleveland Street
London W1T 4JF
UK

Marca H. M. Wauben
Department of Immunohematology and
Blood Transfusion
E3-Q, Leiden University Medical Center
Albinusdreef 2, P.O. Box 9600
2300 RC Leiden
The Netherlands

Library of Congress Cataloging-in-Publication Data

Regulatory T cells in inflammation / Leonie Taams, Arne N. Akbar, Marca H.M. Wauben, editors.
p. ; cm. -- (Progress in inflammation research)
Includes bibliographical references and index.
ISBN 3-7643-7088-2 (alk. paper)
1. Inflammation. 2. T cells. I. Taams, Leonie. II. Akbar, Arne N. III. Wauben, Marca H. M. IV. PIR (Series)
[DNLM: 1. T-Lymphocytes--immunology. 2. Immunity, Cellular. 3. Immunotherapy--methods. QW 568 R3439 2005]
RB131.R44 2005
616'.0473--dc22

2004062666

Bibliographic information published by Die Deutsche Bibliothek
Die Deutsche Bibliothek lists this publication in the Deutsche Nationalbibliografie;
detailed bibliographic data is available in the internet at http://dnb.ddb.de

ISBN-10: 3-7643-7088-2 Birkhäuser Verlag, Basel – Boston – Berlin
ISBN-13: 978-3-7643-7088-6 Birkhäuser Verlag, Basel – Boston – Berlin

Printed on acid-free paper produced from chlorine-free pulp. TCF ∞
Cover design: Markus Etterich, Basel
Cover illustration: Cross-talk between antigen presenting cell and regulatory T cell. (By: E.N.M. Nolte'-t Hoen and M.H.M. Wauben)
Printed in Germany
ISBN-10: 3-7643-7088-2
ISBN-13: 978-3-7643-7088-6

9 8 7 6 5 4 3 2 1

www.birkhauser.ch

Contents

List of contributors

Luciano Adorini, BioXell, Via Olgettina 58, 20132 Milano, Italy;
e-mail: Luciano.Adorini@bioxell.com

Arne N. Akbar, Department of Immunology and Molecular Pathology, Division of Infection and Immunity, University College London, 46 Cleveland Street, London W1T 4JF, UK; e-mail: a.akbar@ucl.ac.uk

Ahmed Akl, Nuffield Department of Surgery, John Radcliffe Hospital, University of Oxford, Oxford OX3 9DU, United Kingdom; e-mail: alakl2001@yahoo.com

Clare Baecher-Allan, Laboratory of Molecular Immunology, Center for Neurologic Diseases, Brigham and Women's Hospital and Harvard Medical School, 77 Avenue Louis Pasteur, Boston, MA 02115, USA; e-mail: callan@rics.bwh.harvard.edu

Robert N. Barker, Department of Medicine and Therapeutics, Institute of Medical Sciences, University of Aberdeen, Foresterhill, Aberdeen AB25 2ZD, UK;
e-mail: r.n.barker@abdn.ac.uk

Johannes W. J. Bijlsma, Dept Rheumatology & Clinical Immunology, University Medical Center Utrecht, 3508 GA Utrecht, the Netherlands;
e-mail: j.w.j.bijlsma@azu.nl

Katie E. Birch, Department of Immunology and Molecular Pathology, Division of Infection and Immunity, University College London, 46 Cleveland Street, London W1T 4JF, UK; and Department of Dermatology, Royal Free Hospital, Pond Street, London NW3 2QG, UK; e-mail: Kebirch@aol.com

Sophie Candon, INSERM U580, Hôpital Necker, 161 Rue de Sèvres, 75743 Paris Cedex 15, France; e-mail: candon@necker.fr

Lucienne Chatenoud, INSERM U580, Hôpital Necker, 161 Rue de Sèvres, 75743 Paris Cedex 15, France; e-mail: chatenoud@necker.fr

Kenn C. Daniel, BioXell, Via Olgettina 58, 20132 Milano, Italy;
e-mail: kenn.daniel@bioxell.com

Nadia Giarratana, BioXell, Via Olgettina 58, 20132 Milano, Italy;
e-mail: nadigi@libero.it

Silvia Gregori, San Raffaele Telethon Institute for Gene Therapy (HSR-TIGET), Via Olgettina 58, 20132 Milano, Italy; e-mail: silvia.gregori@hsr.it

David A. Hafler, Laboratory of Molecular Immunology, Center for Neurologic Diseases, Brigham and Women's Hospital and Harvard Medical School, 77 Avenue Louis Pasteur, Boston, MA 02115, USA; e-mail: dhafler@rics.bwh.harvard.edu

Dirk Homann, University of Colorado Health Sciences Center, Barbara Davis Center for Childhood Diabetes, 4200 East 9th Avenue, Box 140, Denver, CO 80262, USA; e-mail: dhomann@scripps.edu

Eleanor M. Ling, Leukocyte Biology Section, Division of Biomedical Sciences and Allergy and Clinical Immunology, National Heart and Lung Institute, Sir Alexander Fleming Building, Imperial College London, Exhibition Road, London SW7 2AZ, UK

Kevin J. Maloy, Sir William Dunn School of Pathology, University of Oxford, South Parks Road, Oxford, OX1 3RE, UK; e-mail: kevin.maloy@path.ox.ac.uk

Esther N. M. Nolte-'t Hoen, Faculty of Biological Sciences, Imperial College London, South Kensington Campus, London SW7 2AZ, UK;
e-mail: e.nolte@imperial.ac.uk

Giuseppe Penna, BioXell, Via Olgettina 58, 20132 Milano, Italy;
e-mail: giuseppe.penna@bioxell.com

Fiona Powrie, Sir William Dunn School of Pathology, University of Oxford, South Parks Road, Oxford, OX1 3RE, UK; e-mail: fiona.powrie@path.ox.ac.uk

John R. Reed, Department of Immunology and Molecular Pathology, Division of Infection and Immunity, University College London, 46 Cleveland Street, London W1T 4JF, UK; e-mail: j.reed@rfc.ucl.ac.uk

Douglas S. Robinson, Leukocyte Biology Section, Division of Biomedical Sciences and Allergy and Clinical Immunology, National Heart and Lung Institute, Sir Alexander Fleming Building, Imperial College London, Exhibition Road, London SW7 2AZ, UK; e-mail: d.s.robinson@ic.ac.uk

Malcolm H.A. Rustin, Department of Dermatology, Royal Free Hospital, Pond Street, London NW3 2QG, UK

Noriko Sakaguchi, Department of Experimental Pathology, Institute for Frontier Medical Sciences, Kyoto University, Kyoto 606-8507, Japan; and Laboratory of Immunopathology, Research Center for Allergy and Immunology, The Institute for Physical and Chemical Research (RIKEN), Yokohama 230-0045, Japan;
e-mail: norikos@frontier.kyoto-u.ac.jp

Shimon Sakaguchi, Department of Experimental Pathology, Institute for Frontier Medical Sciences, Kyoto University, 53 Shogoin Kawahara-cho, Sakyo-ku, Kyoto 606-8507, Japan; and Laboratory of Immunopathology, Research Center for Allergy and Immunology, The Institute for Physical and Chemical Research (RIKEN), Yokohama 230-0045, Japan; and Core Research for Evolutional Science and Technology (CREST), Science and Technology Agency of Japan, Kawaguchi 332-0012, Japan; e-mail: shimon@frontier.kyoto-u.ac.jp

Leonie S. Taams, Infection & Immunity Research Group, Franklin-Wilkins Building, King's College London, 150 Stamford Street, London SE1 9NN, UK;
e-mail: leonie.taams@kcl.ac.uk

Jocea M. van Amelsfort, Dept Rheumatology & Clinical Immunology, University Medical Center Utrecht, 3508 GA Utrecht, the Netherlands;
e-mail: j.m.r.vanAmelsfort@azu.nl

Vissia Viglietta, Laboratory of Molecular Immunology, Center for Neurologic Diseases, Brigham and Women's Hospital and Harvard Medical School, 77 Avenue Louis Pasteur, Boston, MA 02115, USA; e-mail: vviglietta@rics.bwh.harvard.edu

Matthias G. von Herrath, La Jolla Institute for Allergy and Immunology, Division of Immune Regulation, 10355 Science Center Drive, La Jolla, San Diego, CA 92121, USA; e-mail: matthias@liai.org

Milica Vukmanovic-Stejic, Department of Immunology and Molecular Pathology, Division of Infection and Immunity, University College London, 46 Cleveland Street, London W1T 4JF, UK; e-mail: m.vukmanovic-stejic@ucl.ac.uk

Frank J. Ward, Department of Medicine and Therapeutics, Institute of Medical Sciences, University of Aberdeen, Foresterhill, Aberdeen AB25 2ZD, UK;
e-mail: f.j.ward@abdn.ac.uk

Marca H. M. Wauben, Department of Immunohematology and Blood Transfusion, E3-Q, Leiden University Medical Center, Albinusdreef 2, P.O. Box 9600, 2300 RC Leiden, The Netherlands; e-mail: M.H.M.Wauben@lumc.nl

Kathryn J. Wood, Nuffield Department of Surgery, John Radcliffe Hospital, University of Oxford, Oxford OX3 9DU, United Kingdom;
e-mail: kathryn.wood@nds.ox.ac.uk

Preface

Scientific interest in regulatory T cells has revived during the last decade. Initially described in the early seventies as suppressor T cells, the concept of suppressor/regulatory T cells went through turbulent times during the eighties when molecular analysis failed to identify putative suppressor genes. The constructive and elegant cellular experiments on regulatory T cells during the nineties, initiated by Shimon Sakaguchi and co-workers, however has brought these cells back into the limelight. Nowadays, regulatory T cells are regarded as essential components of the immune system, and several different subsets of regulatory T cells have been described. An important regulatory role has been attributed to the $CD4^{+}CD25^{+}$ T cells. They act by suppressing immune reactivity thereby maintaining or restoring the balance between immunity and tolerance. The aim of this book is to bring together recent developments and viewpoints in the field of $CD4^{+}CD25^{+}$ regulatory T cells and to discuss the potential use of regulatory T cells as target for immunotherapy of inflammatory diseases.

By linking data on regulatory T cells from experimental models with recent findings from the clinic, this topical book will be of interest to immunologists and other biomedical researchers as well as clinicians that are interested in regulation and manipulation of the immune response during (chronic) inflammatory disease.

October 2004

Leonie S. Taams
Arne N. Akbar
Marca H. M. Wauben

Part I
Origin, function and distribution of regulatory T cells

History of CD25⁺CD4⁺ regulatory T cells

Shimon Sakaguchi[1,2,3] *and Noriko Sakaguchi*[1,2]

[1]Department of Experimental Pathology, Institute for Frontier Medical Sciences, Kyoto University, 53 Shogoin Kawahara-cho, Sakyo-ku , Kyoto 606-8507, Japan; [2]Laboratory of Immunopathology, Research Center for Allergy and Immunology, The Institute for Physical and Chemical Research (RIKEN), Yokohama 230-0045, Japan; [3]Core Research for Evolutional Science and Technology (CREST), Science and Technology Agency of Japan, Kawaguchi 332-0012, Japan

Introduction

It is still a key issue in immunology to understand how the immune system discriminates between self and non-self, inhibiting autoimmune responses while allowing effective immune responses to non-self antigens. In addition to physical elimination of self-reactive T cells (clonal deletion) and their functional inactivation (clonal anergy), there is now substantial evidence that T-cell-mediated suppression of self-reactive T cells is also a key mechanism of immunologic self-tolerance [1–3]. Although the idea of T cells that negatively control immune responses is not a new one, for a long time it was controversial whether they actually constituted a definite cellular entity in the immune system [4]. In recent years, however, we have witnessed resurgent interest in suppressor or regulatory T (T_R) cells in many fields of basic and clinical immunology [5, 6]. This change is partly due to our new understanding that the normal immune system endogenously produces as its normal cellular constituent a T-cell subpopulation which is highly specified for suppressive function; abnormality of this population in number or function can indeed be a cause of immunological diseases – in particular autoimmune disease – and this naturally occurring T_R cell population can be exploited for induction of immunological tolerance (e.g., transplantation tolerance), negative control of pathological immune responses (e.g., allergy), and enhancement of host defense (e.g. tumor immunity and microbial immunity) [5]. In addition to these endogenous T_R cells, there are other types of T_R cell that can be induced by specific methods of antigenic stimulation *in vivo* or *in vivo* [7, 8]. In this chapter, we shall provide a brief historical sketch of how endogenous T_R cells were recognized and characterized, and discuss their roles in immunologic tolerance and immunoregulation.

Regulatory T Cells in Inflammation, edited by Leonie S. Taams, Arne N. Akbar and Marca H.M. Wauben

A historical note on suppressor or regulatory T cells

In 1970, Gershon and Kondo [9] made the seminal finding that T cells not only enhance but also dampen immune responses and that this down-regulation is mediated by T cells that are different from helper T cells. This T-cell population, called suppressor T cells, was intensively studied over the following years in various fields of immunology. The studies showed several types of suppressor T cell interacting in a cascade; some were antigen-specific and others were non-specific; some secreted antigen-specific suppressive factors and others non-specific ones. The phenotype of suppressor T cells was on the most part shown to be Lyt-1$^-$ Lyt-2, 3$^+$, corresponding to CD8$^+$, and they expressed the I-J molecule, which was supposed to be a key suppressor molecule intimately associated with their suppressive function [10]. The research, however, quite abruptly collapsed in the mid-1980s when scrutiny of the mouse MHC gene by molecular biology techniques showed no existence of the I-J region, which was assumed to encode the I-J molecule and locate within the MHC gene complex [11]. With this bewildering I-J episode as a turning point, immunologists' interest in suppressor T cells rapidly waned [4]. There are several reasons for this decline: e.g. failure in finding reliable markers for distinguishing suppressor T cells from other T cells, ambiguity in the molecular basis of suppression, and difficulty in preparing antigen-specific suppressor T-cell clones amenable to fine cellular and molecular analyses. In addition, approaches to immunologic tolerance with reliable molecular tools, such as T-cell receptor (TCR)-specific monoclonal antibodies and transgenic mice, in the late 1980s to the early 1990s unequivocally demonstrated clonal deletion, and also anergy, as key mechanisms of immunologic tolerance [12–14]. Molecular characterization of cytokines clearly revealed pleiotropism, cross-regulation, and redundancy in their function [15]. Collectively these findings generated a climate in which suppressor T cells played little meaningful part in immunologic tolerance. There were even doubts about suppressor T cells as a distinct cellular entity when suppressive phenomena were explicable by T cells secreting particular cytokines. Indeed, interleukin (IL)-10-secreting T cells produced *in vitro* by antigenic stimulation of T cells in the presence of IL-10- or transforming growth factor (TGF)-β-secreting T cells propagated from animals in oral tolerance did not encounter much resistance to being accepted by immunologists in the 1990s [7, 8].

In parallel with the suppressor T-cell research depicted above, there has been a different stream of study on T-cell-mediated suppression. A notable feature of the latter is that it examined from the beginning how the manipulation of the T-cell immune system breaks natural immunologic self-tolerance and causes autoimmune disease, rather than studying experimental immunologic tolerance induced by specific ways of antigen administration [1]. Nishizuka and Sakakura [16] showed in 1969 that neonatal thymectomy (NTx) of normal mice between day 2 and 4 after birth led to the destruction of ovaries, which was first supposed

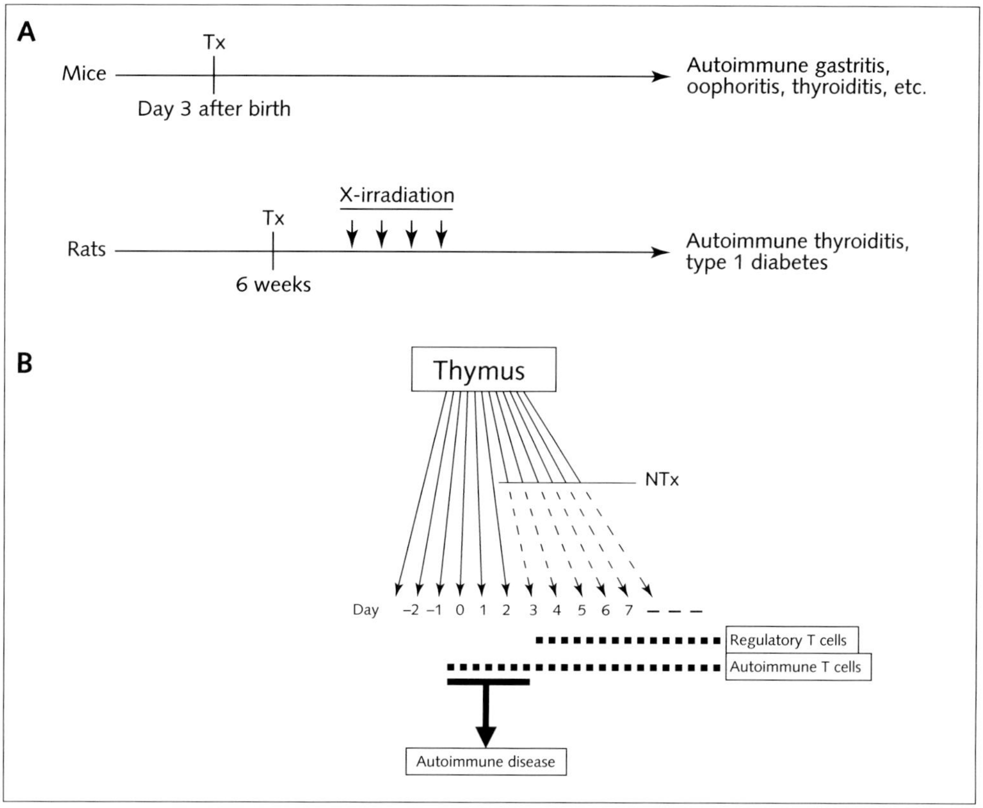

Figure 1
Induction of autoimmune disease by depleting natural T_R cells.
(A) Induction of autoimmune disease in animals by manipulating thymus/T cells. See text for details. Tx, thymectomy. (B). Ontogeny of autoimmune-preventive natural T_R cells. The normal thymus may start to release natural T_R cells around day 3 after birth. NTx on day 3, therefore, abrogates ontogenic development of natural T_R cells in the periphery, allowing self-reactive T cells that have migrated to the periphery before NTx to become activated and cause autoimmune disease.

to be due to deficiency of a certain ovary-tropic hormone secreted by the thymus, and hence was called "ovarian dysgenesis". This ovarian lesion later turned out to be of an autoimmune nature because subsequent investigation demonstrated that NTx also produced inflammatory tissue damage in other organs accompanying the appearance of tissue-specific autoantibodies in the circulation; e.g. thyroiditis, gastritis, orchitis, prostatitis, and sialadenitis of the salivary gland [17] (Fig. 1A). In 1973, Penhale et al. [18] reported that adult thymectomy (ATx) of rats followed by four sublethal doses of X-irradiation (2–2.5 Gray) every 2 weeks resulted in the development of autoimmune thyroiditis accompanied by anti-thyroglobulin autoantibodies [18]. They and others later showed that the same protocol can produce type 1 diabetes in other strains of rats [19, 20] (Fig. 1A).

A common element between autoimmune disease induction by NTx and by ATx and X-irradiation is thymectomy. A simple interpretation of how these treatments cause autoimmune disease would be that the normal thymus is producing a population of T cells having autoimmune-preventive activity; NTx shortly after birth may abrogate developmentally determined thymic production of autoimmune-preventive T cells, allowing those self-reactive T cells that have been produced before NTx to become activated and cause autoimmune disease because of the paucity of regulatory T cells in the periphery (Fig. 1B); likewise, ATx and X-irradiations may abrogate the thymic supply of regulatory T cells and reduce regulatory T cells in the periphery presumably because they may be relatively radiosensitive (see below). Supporting this notion, inoculation of normal T cells, especially $CD4^+$ T cells, to the treated mice or rats indeed prevented the development of autoimmune diseases [21, 22]. On the other hand, as in other experimental autoimmune diseases, $CD4^+$ helper T cells mediate these autoimmune diseases as helper T cells for autoantibody formation and effectors of cell-mediated immune destruction of the target organs/tissues [23, 24]. These findings led to the idea that there may exist two types of $CD4^+$ T cell in the normal immune system: one potentially capable of mediating autoimmune disease, the other dominantly suppressing it [25] (Fig. 1B).

Thymus-produced natural $CD25^+CD4^+$ T_R cells

The phenotype and function of natural $CD4^+$ T_R cells

The key issue was then to know how these two populations can be dissected out or differentiated from each other if they co-exist in the normal immune system, and to determine whether specific removal of the autoimmune-preventive $CD4^+$ T-cell population can break self-tolerance and cause autoimmune disease in otherwise normal animals. Attempts were made to separate the $CD4^+$ T-cell population in normal naïve mice into an autoimmune-inducing and an autoimmune-preventive population by the expression levels of cell-surface molecules [25–31] (Fig. 2). Our experiments showed in 1985 that, when splenic cell suspensions from normal BALB/c mice were depleted of $CD5^{high}CD4^+$ T cells *ex vivo* (by *in vitro* treatment of the cell suspensions with a mixture of anti-CD8 and anti-CD5 antibodies and complement) and the remaining cells transferred to congenitally T-cell-deficient BALB/c athymic nude mice, the nude mice spontaneously developed autoimmune disease in multiple

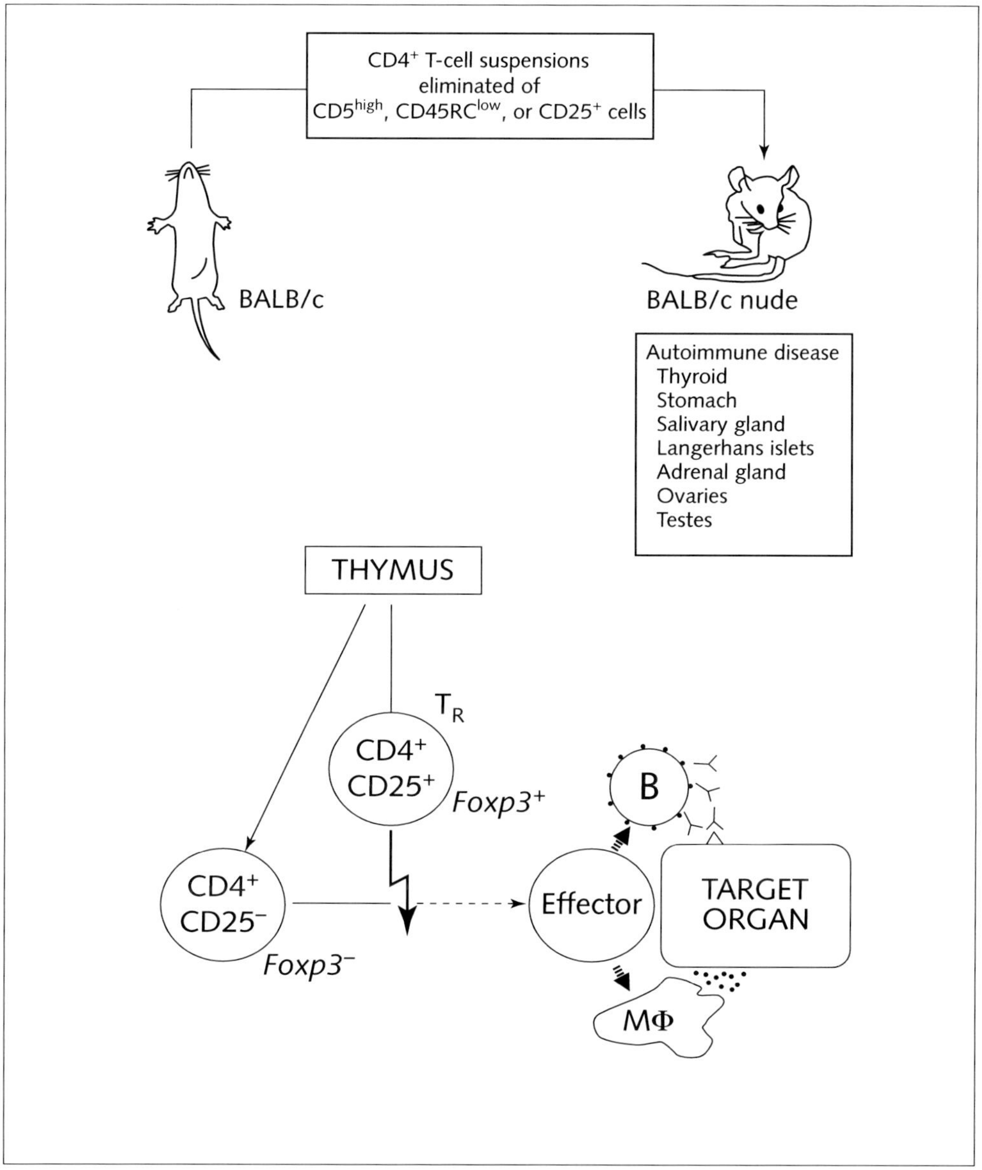

Figure 2
Induction of autoimmune disease by depleting Foxp3⁺CD25⁺CD4⁺ T_R cells.
Thymus-derived natural CD25⁺CD4⁺ T_R cells, which specifically express Foxp3, control T cells that mediate autoimmune disease, IBD, and allergy. Genetic defects of Foxp3 cause these immunological diseases by impairing the development or function of CD25⁺CD4⁺ T_R cells. MΦ, macrophage.

organs (stomach, thyroid, ovaries, or testes) a few months after cell transfer [25]. Co-transfer of normal untreated $CD4^+$ T cells with $CD5^{low}CD4^+$ T cells inhibited the autoimmunity. Likewise, transfer of $CD5^{low}CD4^+$ T cells from normal C3H mice to T-cell-depleted C3H mice produced autoimmune thyroiditis [26]. Powrie and Mason [28] subsequently reconstituted PVG athymic nude rats with splenic T cells depleted of $CD45RC^{low}CD4^+$ T cells, showing that the transferred $CD45RC^{high}$ $CD4^+$ T cells elicited graft-*versus*-host-disease-like systemic disease and autoimmune tissue damage in multiple organs including thyroid and islets of Langerhans. McKeever et al. [29] similarly showed that transfer of spleen cell suspensions depleted of $RT6.1^+$ T cells was able to produce type 1 diabetes mellitus and thyroiditis in histocompatible athymic nude rats. Powrie et al. [32] and Morrisey et al. [33] then independently showed that transfer of BALB/c $CD45RB^{high}CD4^+$ T cells to T-B-cell-deficient BALB/c SCID mice induced inflammatory bowel disease (IBD). These findings prompted us to search for a cell-surface molecule which would be more specific than CD5 or CD45RB in defining such autoimmunity/inflammation-preventive $CD4^+$ T cells. In 1995, the CD25 molecule (the IL-2 receptor α-chain) was found as a candidate since $CD25^+$ T cells, which constitute 5–10% of peripheral $CD4^+$ T cells and less than 1% of peripheral $CD8^+$ T cells in normal naive mice, are contained in the $CD5^{high}$ and $CD45RB^{low}$ fraction of $CD4^+$ T cells [30, 31]. Transfer of BALB/c splenic cell suspensions depleted of $CD25^+CD4^+$ T cells indeed produced in BALB/c athymic nude mice histologically and serologically evident autoimmune diseases at higher incidences and in a wider spectrum of organs (including stomach, thyroid, ovaries, adrenal glands, and islets of Langerhans) than the transfer of $CD5^{low}$ or $CD45RB^{high}$ T cells prepared from the same number of splenic cell suspensions. Co-transfer of a small number of $CD25^+CD4^+$ T cells with $CD25^-$ T cells inhibited the autoimmunity. Furthermore, transfer of $CD25^-CD4^+$ T cells alone sufficed to mediate the autoimmune disease by giving rise to $CD4^+$ helper T cells for humoral and cell-mediated autoimmunity, although the presence of $CD25^-CD8^+$ T cells also enhanced the autoimmune induction, presumably as a source of self-reactive cytotoxic T lymphocytes. Importantly, these experiments showed that removal of $CD25^+CD4^+$ T cells not only elicited autoimmune disease but also enhanced immune responses to non-self antigens including soluble proteins and allografts [30].

Subsequent *in vitro* characterization of $CD25^+CD4^+$ T cells in normal mice confirmed the CD25 molecule as a highly specific marker for natural T_R cells: $CD25^+CD4^+$ T cells are hyporesponsive to TCR stimulation but, upon stimulation, potently suppress the activation and proliferation of other T cells *in vitro*; this property was confined to the $CD25^+$ fraction of normal $CD4^+$ T cells [34–36]. Their TCR repertoire is as broad as other T cells but appears to have higher affinity for self-peptide/MHC ligands selecting them in the thymus; i.e., they are more self-reactive than other T cells but capable of recognizing a broad spectrum of self and non-self antigens [34, 37, 38]. Unlike *in vitro*, $CD25^+CD4^+$ T_R cells in the *in vivo* physiological state are more vigorously proliferating than $CD25^-CD4^+$ T cells, presumably by recognizing self-antigens [38, 39]. This may explain why $CD25^+CD4^+$ T_R cells are radiosensitive [18, 40].

Thymic production of natural $CD25^+CD4^+$ T_R cells

The above findings, such as the ability of NTx to cause autoimmune disease, requirement of ATx for autoimmune induction by radiation-induced autoimmune disease, and the activity of normal $CD4^+CD8^-$ thymocytes to prevent autoimmune disease in NTx mice, all indicate that the normal thymus produces $CD4^+$ regulatory T cells. As a direct demonstration of this, transfer of $CD4^+CD8^-$ mature thymocyte suspensions depleted of $CD25^+$ thymocytes produced various autoimmune diseases in syngeneic nude mice, as shown with the transfer of $CD25^-CD4^+$ spleen cells [41]. Thus, the normal thymus is continuously producing both pathogenic self-reactive $CD4^+$ T cells and functionally mature $CD25^+CD4^+$ T_R cells, and $CD25^+CD4^+$ T_R cells in the thymus and the periphery may constitute a distinct cellular lineage (Fig. 2). Indeed, as discussed below, both $CD25^+CD4^+CD8^-$ thymocytes and $CD25^+CD4^+$ T cells express *Foxp3*, a T_R-cell-specific transcription factor, and *Foxp3* deficiency abrogates the development of both populations [42–44]. In addition, both populations are functionally and phenotypically similar; for example, both are naturally anergic to *in vitro* TCR stimulation, exhibiting an equivalent *in vitro* suppressive activity, and constitutively expressing CTLA-4 and a high level of GITR (glucocorticoid-induced tumor necrosis factor (TNF) receptor-related protein) [45–49].

Ontogeny of natural $CD25^+CD4^+$ T_R cells

Ontogeny of $CD25^+CD4^+$ T cells also correlates well with that of natural T_R cells. $CD25^+CD4^+$ T cells become detectable in the periphery of normal mice from around day 3 after birth, rapidly increasing to the adult level (i.e. 5–10% of $CD4^+$ T cells) in 3 weeks [31]. NTx on day 3 substantially reduces peripheral $CD25^+CD4^+$ T cells; and the inoculation of $CD25^+CD4^+$ T cells from normal mice prevents the autoimmune development in NTx mice [31]. These results collectively indicate that the ontogenic time course of the thymic production and the peripheral migration of $CD25^+CD4^+$ T_R cells is developmentally determined (i.e. around day 3 after birth in mice); the abrogation of the thymic production of natural T_R cells from the very beginning of their ontogeny, therefore, results in their selective paucity in the periphery, leading to the activation of self-reactive T cells that have migrated to the periphery before NTx, producing autoimmune disease.

The role of IL-2 and CD25 for natural T_R cells

As discussed above, CD25 was found to be a good marker for operationally distinguishing endogenous T_R cells from other T cells in normal naïve animals. The following findings indicate that CD25 is also an indispensable molecule for the generation and maintenance of natural T_R cells. IL-2-deficient mice bear few CD25$^+$CD4$^+$ T cells, and spontaneously develop severe autoimmunity, although they develop a normal number of other T cells with a normal composition of CD4/CD8 subsets [50, 51]. CD25-deficient or CD122 (the IL-2 receptor β-chain)-deficient mice are also afflicted with similar autoimmunity [52–54]. Besides, *in vivo* neutralization of IL-2 by administration of anti-IL-2 monoclonal antibody substantially reduced CD25$^+$CD4$^+$ T cells, but not other T cells, in normal mice and consequently produced autoimmune disease [55], and R. Setoguchi et al., unpublished observations). Taken together, CD25 as a component of the high-affinity IL-2 receptor is a key functional molecule for natural CD4$^+$ T_R cells; IL-2 is a key growth/survival factor for them; and its expression, high-level expression in particular, is an excellent marker for natural CD4$^+$ T_R cells in mice and humans [36, 56].

Summary

To summarize, these findings on natural CD25$^+$CD4$^+$ T_R cells lead to the following notions (Fig. 2B). First, despite thymic negative selection, the normal immune system still harbors self-reactive T cells (CD4$^+$ T cells in particular) that are sufficiently pathogenic in TCR specificity and affinity to mediate various autoimmune diseases similar in immunopathology to their human counterparts, such as autoimmune gastritis/pernicious anemia, premature ovarian failure with autoimmune oophoritis, Hashimoto's thyroiditis, adrenalitis/Adison's disease, and insulitis/type 1 diabetes. Second, the activation/expansion of such self-reactive T cells is normally kept in check by a regulatory CD4$^+$ T-cell population, many if not all of which physiologically express CD25. Third, the normal thymus is continuously producing them, which is another key function of the thymus in self-tolerance, in addition to its role in positive and negative selection of T cells. Furthermore, elimination of this CD25$^+$CD4$^+$ regulatory population alone, without manipulating the target self-antigens, suffices to break natural self-tolerance and elicit chronic and destructive autoimmune diseases. The appearance of various disease-specific autoantibodies in the T_R-cell-depleted animals implies that the breakdown of this mode of T-cell self-tolerance and development of autoimmune CD4$^+$ helper T cells result in the breakdown of B-cell self-tolerance as well. Thus, one aspect of natural self-tolerance in T cells, and for that matter in B cells, is maintained by a regulatory subpopulation of CD4$^+$ T cells. Furthermore, such naturally present T_R cells also engage in the control of immune responses to non-self antigens as well.

IPEX (immune dysregulation, polyendocrinopathy, enteropathy, X-linked syndrome) as evidence for the key role of $CD25^+CD4^+$ natural T_R cells in natural self-tolerance in humans

The finding that thymectomy during a critical neonatal period induces autoimmune disease suggests that the generation of $CD4^+$ T_R cells may be developmentally controlled. Attempts were therefore made to produce autoimmune disease by genetically altering the normal developmental course of natural T_R cells [57]. An important clue to this developmental control of natural T_R cells came from the human disease called IPEX. IPEX was described in 1982 as an X-linked immunodeficiency syndrome associated with autoimmune disease in multiple endocrine organs (such as type 1 diabetes and thyroiditis), IBD, atopic dermatitis, and fatal infections [58]. The Scurfy strain of mice is an X-linked recessive mutant with lethality in hemizygous males within a month after birth, exhibiting hyperactivation of $CD4^+$ T cells and overproduction of proinflammatory cytokines [59]. The gene defective in Scurfy mice was identified and designated *Foxp3*, which encodes Scurfin, a new member of the forkhead/winged-helix family of transcription factors [60]. Subsequently, mutations of the human gene *FOXP3*, the ortholog of murine *Foxp3*, were found to be the cause of IPEX [61–63].

Recent studies have revealed the specific role of Foxp3 in the development and function of natural $CD25^+CD4^+$ T_R cells [42–44]. $CD25^+CD4^+$ peripheral T cells and $CD25^+CD4^+CD8^-$ thymocytes predominantly expressed *Foxp3* mRNA, whereas other thymocytes/T cells and B cells did not. Importantly, activation of $CD25^-CD4^+$ T cells, Th1 cells or Th2 cells failed to induce *Foxp3* expression [42, 43]. Furthermore, retroviral transduction of Foxp3 to $CD25^-CD4^+$ T cells converted them into $CD25^+CD4^+$ T_R-like cells with respect to phenotype and function [42, 43]. For example, *Foxp3* transduction induced expression of CD25, CTLA-4, and GITR, which are closely associated with the functions of natural T_R cells [42]. *Foxp3*-transduced $CD25^-CD4^+$ T cells were able to suppress proliferation of other T cells *in vitro* and the development of autoimmune disease and IBD *in vivo* [42]. Furthermore, in bone-marrow (BM) chimeric mice with a mixture of BM cells from wild-type and *Foxp3*-deficient mice, *Foxp3*-deficient BM cells failed to give rise to $CD25^+CD4^+$ T cells, while *Foxp3*-intact BM cells generated them [43], indicating an essential role of Foxp3 for the development of $CD25^+CD4^+$ T_R cells.

Thus, *Foxp3*/*FOXP3* appears to be a master control gene for the development and function of natural $CD25^+CD4^+$ T_R cells (Fig. 2B). Given that humans bear natural $CD25^+CD4^+$ T_R cells with a common phenotype and function to those found in rodents [64], it is most likely in IPEX that disruption of the *FOXP3* gene abrogates the development of the T_R cells, leading to hyperactivation of T cells reactive with self-antigens, commensal bacteria in the intestine, or innocuous environmental substances, thus causing autoimmune polyendocrinopathy, IBD, and allergy, respectively. This has several implications for self-tolerance and autoimmune/inflammatory disease in humans. First, thus far this is the clearest example that abnormality in naturally arising $CD25^+CD4^+$ T_R cells is a primary cause of human autoimmune disease and for that matter IBD and allergy. Second, the development of natural T_R cells is, at least in part, genetically and developmentally programmed, indicating that autoimmunity is in part a primary T-cell immunodeficiency. Third, females with hemizygous defects of the *FOXP3* gene illustrate that the mechanism of dominant self-tolerance is operating physiologically in humans. Because of random inactivation of the X chromosome (Lyonization) in individual T_R cells, some hemizygous females have a similar number of *FOXP3*-defective T_R cells and *FOXP3*-normal ones as a mosaic, but they are completely normal and do not show intermediate phenotypes [65]. This means that the normal T_R cells dominantly control self-reactive T cells in the presence of defective T_R cells. It also indicates that a partial reconstitution of IPEX patients with normal T_R cells (for example by BM transplantation) or *FOXP3*-transduced autologous T cells may suffice to control the disease dominantly.

Conclusion and perspective

A prominent feature of $CD25^+CD4^+$ T_R cells is that the majority, if not all, of them are endogenously produced by the normal thymus as a functionally distinct and mature subpopulation of T cells and persist in the periphery with stable function, and that their generation is, at least in part, developmentally controlled. Congenital deficiency of this population as in IPEX, therefore, results in serious impairment of self-tolerance and immunoregulation, leading to severe autoimmunity and allergy. Furthermore, as illustrated by IPEX, any genetic abnormality or environmental insult can be a cause or a predisposing factor of autoimmune disease if it would tip the balance between natural $CD25^+CD4^+$ T_R cells and self-reactive T cells towards the dominance of the latter [66]. On the other hand, their natural presence in the immune system as a phenotypically distinct population makes it a good target for designing ways to treat or prevent immunological diseases and to control pathological as well as physiological immune responses. In addition to this naturally arising "professional" T_R cell population there are several other types of T_R cell that can be induced from naïve T cells by antigenic stimulation under specialized conditions in the periphery. Although physiological roles for these inducible or "adaptive" T_R cells need to be fully established, they can still be exploited as a therapeutic tool. Furthermore, suppressive phenomena intensively studied in the 1970s and early 1980s can be re-interpreted from the vantage point of the present.

Acknowledgements
The authors thank Dr. Zoltan Fehervari for critical reading of the manuscript. This work was supported by grants-in-aid from the Ministry of Education, Science, Sports and Culture, the Ministry of Human Welfare, and the Science and Technology Agency of Japan.

References

1 Sakaguchi S (2000) Regulatory T cells: Key controllers of immunologic self-tolerance. *Cell* 101: 455–458
2 Shevach EM (2000) Regulatory T cells in autoimmunity. *Annu Rev Immunol* 18: 423–449
3 Maloy KJ, Powrie F (2001) Regulatory T cells in the control of immune pathology. *Nat Immunol* 2: 816–822
4 Bloom BR, Salgame P, Diamond B (1992) Revisiting and revising suppressor T cells. *Immunol Today* 13: 131–136
5 Sakaguchi S (2004) Naturally arising $CD4^+$ regulatory t cells for immunologic self-tolerance and negative control of immune responses. *Annu Rev Immunol* 22: 531–562
6 Baecher-Allan C, Hafler DA (2004) Suppressor T cells in human diseases. *J Exp Med* 200: 273–276
7 Chen Y, Kuchroo VK, Inobe J, Hafler DA, Weiner HL (1994) Regulatory T cell clones induced by oral tolerance: suppression of autoimmune encephalitis. *Science* 265: 1237–1240
8 Groux H, O'Garra A, Bigler M, Rouleau M, Antonenko S, de Vries JE, Roncarolo MG (1997) A $CD4^+$ T-cell subset inhibits antigen-specific T-cell responses and prevents colitis. *Nature* 389: 737–742
9 Gershon RK, Kondo K (1970) Cell interactions in the induction of tolerance: the role of thymic lymphocytes. *Immunology* 18: 723–737
10 Green DR, Flood PM, Gershon RK (1983) Immunoregulatory T-cell pathways. *Annu Rev Immunol* 1: 439–463
11 Kronenberg M, Steinmetz M, Kobori J, Kraig E, Kapp JA, Pierce CW, Sorensen CM, Suzuki G, Tada T, Hood L (1983) RNA transcripts for I-J polypeptides are apparently not encoded between the I-A and I-E subregions of the murine major histocompatibility complex. *Proc Natl Acad Sci USA* 80: 5704–5708
12 Kappler JW, Roehm N, Marrack P (1987) T cell tolerance by clonal elimination in the thymus. *Cell* 49: 273–280
13 Kisielow P, Bluthmann H, Staerz UD, Steinmetz M, von Boehmer H (1988) Tolerance in T-cell-receptor transgenic mice involves deletion of nonmature $CD4^+8^+$ thymocytes. *Nature* 333: 742–746
14 Goodnow CC, Cyster JG, Hartley SB, Bell SE, Cooke MP, Healy JI, Akkaraju S, Rath-

mell JC, Pogue SL, Shokat KP (1995) Self-tolerance checkpoints in B lymphocyte development. *Adv Immunol* 59: 279–368

15 O'Garra A, Murphy K (1994) Role of cytokines in determining T-lymphocyte function. *Curr Opin Immunol* 6: 458–466

16 Nishizuka Y, Sakakura T (1969) Thymus and reproduction: sex-linked dysgenesia of the gonad after neonatal thymectomy in mice. *Science* 166: 753–755

17 Kojima A, Prehn RT (1981) Genetic susceptibility to post-thymectomy autoimmune diseases in mice. *Immunogenetics* 14: 15–27

18 Penhale WJ, Farmer A, McKenna RP, Irvine WJ (1973) Spontaneous thyroiditis in thymectomized and irradiated Wistar rats. *Clin Exp Immunol* 15: 225–236

19 Penhale WJ, Stumbles PA, Huxtable CR, Sutherland RJ, Pethick DW (1990) Induction of diabetes in PVG/c strain rats by manipulation of the immune system. *Autoimmunity* 7: 169–179

20 Fowell D, Mason D (1993) Evidence that the T cell repertoire of normal rats contains cells with the potential to cause diabetes. Characterization of the $CD4^+$ T cell subset that inhibits this autoimmune potential. *J Exp Med* 177: 627–636

21 Sakaguchi S, Takahashi T, Nishizuka Y (1982) Study on cellular events in post-thymectomy autoimmune oophoritis in mice. II. Requirement of Lyt-1 cells in normal female mice for the prevention of oophoritis. *J Exp Med* 156: 1577–1586

22 Penhale WJ, Irvine WJ, Inglis JR, Farmer A (1976) Thyroiditis in T cell-depleted rats: suppression of the autoallergic response by reconstitution with normal lymphoid cells. *Clin Exp Immunol* 25: 6–16

23 Sakaguchi S, Takahashi T, Nishizuka Y (1982) Study on cellular events in postthymectomy autoimmune oophoritis in mice. I. Requirement of Lyt-1 effector cells for oocytes damage after adoptive transfer. *J Exp Med* 156: 1565–1576

24 Fukuma K, Sakaguchi S, Kuribayashi K, Chen WL, Morishita R, Sekita K, Uchino H, Masuda T (1988) Immunologic and clinical studies on murine experimental autoimmune gastritis induced by neonatal thymectomy. *Gastroenterology* 94: 274–283

25 Sakaguchi S, Fukuma K, Kuribayashi K, Masuda T 1985. Organ-specific autoimmune diseases induced in mice by elimination of T-cell subset. I. Evidence for the active participation of T cells in natural self-tolerance: deficit of a T-cell subset as a possible cause of autoimmune disease. *J Exp Med* 161: 72–87

26 Sugihara S, Izumi Y, Yoshioka T, Yagi H, Tsujimura T, Tarutani O, Kohno Y, Murakami S, Hamaoka T, Fujiwara H (1988) Autoimmune thyroiditis induced in mice depleted of particular T-cell subset. I. Requirement of Lyt-1^{dull} L3T4^{bright} normal T cells for the induction of thyroiditis. *J Immunol* 141: 105–113

27 Smith H, Lou YH, Lacy P, Tung, KSK (1992) Tolerance mechanism in experimental ovarian and gastric autoimmune disease. *J Immunol* 149: 2212–2218

28 Powrie F, Mason D (1990) OX-22^{high} $CD4^+$ T cells induce wasting disease with multiple organ pathology: prevention by OX-22^{low} subset. *J Exp Med* 172: 1701–1708

29 McKeever U, Mordes JP, Greiner DL, Appel MC, Rozing J, Handler ES, Rossini AA

(1990) Adoptive transfer of autoimmune diabetes and thyroiditis to athymic rats. *Proc Natl Acad Sci USA* 87: 7618–7622

30 Sakaguchi S, Sakaguchi N, Asano M, Itoh M, Toda M (1995) Immunologic tolerance maintained by activated T cells expressing IL-2 receptor α-chains (CD25): breakdown of a single mechanism of self-tolerance causes various autoimmune diseases. *J Immunol* 155: 1151–1164

31 Asano M, Toda M, Sakaguchi N, Sakaguchi S (1996) Autoimmune disease as a consequence of developmental abnormality of a T cell subpopulation. *J Exp Med* 184: 387–396

32 Powrie F, Leach MW, Mauze S, Caddle LB, Coffman RL (1993) Phenotypically distinct subsets of $CD4^+$ T cells induce or protect from chronic intestinal inflammation in C. B-17 scid mice. *Int Immunol* 5: 1461–1471

33 Morrissey PJ, Charrier K, Braddy S, Liggitt D, Watson JD (1993) $CD4^+$ T cells that express high levels of CD45RB induce wasting disease when transferred into congenic severe combined immunodeficient mice. Disease development is prevented by cotransfer of purified $CD4^+$ T cells. *J Exp Med* 178: 237–244

34 Takahashi T, Kuniyasu Y, Toda M, Sakaguchi N, Itoh M, Iwata M, Shimizu J, Sakaguchi S (1998) Immunologic self-tolerance maintained by $CD25^+CD4^+$ naturally anergic and suppressive T cells: induction of autoimmune disease by breaking their anergic/suppressive state. *Int Immunol* 10: 1969–1980

35 Thornton AM, Shevach EM (1998) $CD4^+$ $CD25^+$ immunoregulatory T cells suppress polyclonal T cell activation *in vitro* by inhibiting interleukin 2 production. *J Exp Med* 188: 287–296

36 Kuniyasu Y, Takahashi T, Itoh M, Shimizu J, Toda G, Sakaguchi S (2000) Naturally anergic and suppressive $CD25^+CD4^+$ T cells as a functionally and phenotypically distinct immunoregulatory T-cell subpopulation. *Int Immunol* 12: 1145–1155

37 Jordan MS, Boesteanu A, Reed AJ, Petrone AL, Holenbeck AE, Lerman MA, Naji A, Caton AJ (2001) Thymic selection of $CD4^+CD25^+$ regulatory T cells induced by an agonist self-peptide. *Nat Immunol* 2: 301–306

38 Sakaguchi S, Hori S, Fukui Y, Sasazuki T, Sakaguchi N, Takahashi T (2003) Thymic generation and selection of $CD25^+CD4^+$ regulatory T cells: implications of their broad repertoire and high self-reactivity for the maintenance of immunological self-tolerance. *Novartis Found Symp* 252: 6–16

39 Fisson S, Darrasse-Jeze G, Litvinova E, Septier F, Klatzmann D, Liblau R, Salomon BL (2003) Continuous activation of autoreactive $CD4^+CD25^+$ regulatory T cells in the steady state. *J Exp Med* 198: 737–746

40 Sakaguchi N, Miyai K, Sakaguchi S (1994) Ionizing radiation and autoimmunity. Induction of autoimmune disease in mice by high dose fractionated total lymphoid irradiation and its prevention by inoculating normal T cells. *J Immunol* 152: 2586–2595

41 Itoh M, Takahashi T, Sakaguchi N, Kuniyasu Y, Shimizu J, Otsuka F, Sakaguchi S (1999) Thymus and autoimmunity: Production of $CD25^+CD4^+$ naturally anergic and suppressive T cells as a key function of the thymus in maintaining immunologic self-tolerance.

J Immunol 162: 5317–5326
42 Hori S, Nomura T, Sakaguchi S (2003) Control of regulatory T cell development by the transcription factor Foxp3. *Science* 299: 1057–1061
43 Fontenot JD, Gavin MA, Rudensky AY (2003) Foxp3 programs the development and function of $CD4^+CD25^+$ regulatory T cells. *Nat Immunol* 4: 330–336.
44 Khattri R, Cox T, Yasayko SA, Ramsdell F (2003) An essential role for Scurfin in $CD4^+CD25^+$ T regulatory cells. *Nat Immunol* 4: 337–342
45 Takahashi T, Tagami T, Yamazaki S, Uede T, Shimizu J, Sakaguchi N, Mak TW, Sakaguchi S (2000) Immunologic self-tolerance maintained by $CD25^+CD4^+$ regulatory T cells constitutively expressing cytotoxic lymphocyte-associated antigen 4. *J Exp Med* 192: 303–310
46 Salomon B, Lenschow DJ, Rhee L, Ashourian N, Singh B, Sharpe A, Bluestone JA (2000) B7/CD28 costimulation is essential for the homeostasis of the $CD4^+CD25^+$ immunoregulatory T cells that control autoimmune diabetes. *Immunity* 12: 431–440
47 Read S, Malmstroem V, Powrie F (2000) CTLA-4 plays an essential role in the function of $CD25^+CD4^+$ regulatory cells which control intestinal inflammation. *J Exp Med* 192: 295–302
48 Shimizu J, Yamazaki S, Takahashi T, Ishida Y, Sakaguchi S (2002) Stimulation of $CD25^+CD4^+$ regulatory T cells through GITR breaks immunological self-tolerance. *Nat Immunol* 3: 135–142
49 McHugh RS, Whitters MJ, Piccirillo CA, Young DA, Shevach EM, Collins M, Byrne MC (2002) $CD4^+CD25^+$ immunoregulatory T cells: gene expression analysis reveals a functional role for the glucocorticoid-induced TNF receptor. *Immunity* 16: 311–323
50 Schorle H, Holtschke T, Hunig T, Schimpl A, Horak I (1991) Development and function of T cells in mice rendered interleukin-2 deficient by gene targeting. *Nature* 352: 621–624
51 Almeida AR, Legrand N, Papiernik M, Freitas AA (2002) Homeostasis of peripheral $CD4^+$ T cells: IL-2R alpha and IL-2 shape a population of regulatory cells that controls $CD4^+$ T cell numbers. *J Immunol* 169: 4850–4860
52 Willerford DM, Chen J, Ferry JA, Davidson L, Ma A, Alt FW (1995) Interleukin-2 receptor alpha chain regulates the size and content of the peripheral lymphoid compartment. *Immunity* 3: 521–530
53 Suzuki H, Zhou YW, Kato M, Mak TW, Nakashima I (1999) Normal regulatory alpha/beta T cells effectively eliminate abnormally activated T cells lacking the interleukin 2 receptor beta *in vivo*. *J Exp Med* 190: 1561–1572
54 Malek TR, Yu A, Vincek V, Scibelli P, Kong L (2002) CD4 regulatory T cells prevent lethal autoimmunity in IL-2Rbeta-deficient mice. Implications for the nonredundant function of IL-2. *Immunity* 17: 167–178
55 Murakami M, Sakamoto A, Bender J, Kappler J, Marrack P (2002) $CD25^+CD4^+$ T cells contribute to the control of memory $CD8^+$ T cells. *Proc Natl Acad Sci USA* 99: 8832–8837

56 Baecher-Allan C, Brown JA, Freeman GJ, Hafler DA (2001) $CD4^+CD25^{high}$ regulatory cells in human peripheral blood. *J Immunol* 167: 1245–1253
57 Sakaguchi S, Ermak TH, Toda M, Berg LJ, Ho W, Fazekas de St Groth B, Peterson PA, Sakaguchi N, Davis MM (1994) Induction of autoimmune disease in mice by germline alteration of the T cell receptor gene expression. *J Immunol* 152: 1471–1484
58 Powell BR, Buist NR, Stenzel P (1982) An X-linked syndrome of diarrhea, polyendocrinopathy, and fatal infection in infancy. *J Pediatr* 100: 731–737
59 Godfrey VL, Wilkinson JE, Russell LB (1991) X-linked lymphoreticular disease in the scurfy (sf) mutant mouse. *Am J Pathol* 138:1379–1387
60 Brunkow ME, Jeffery EW, Hjerrild KA, Paeper B, Clark LB, Yasayko SA, Wilkinson JE, Galas D, Ziegler SF, Ramsdell F (2001) Disruption of a new forkhead/winged-helix protein, scurfin, results in the fatal lymphoproliferative disorder of the scurfy mouse. *Nat Genet* 27: 68–73
61 Chatila TA, Blaeser F, Ho N, Lederman HM, Voulgaropoulos C, Helms C, Bowcock AM (2000) JM2, encoding a fork head-related protein, is mutated in X-linked autoimmunity-allergic disregulation syndrome. *J Clin Invest* 106: R75–R81
62 Wildin RS, Ramsdell F, Peake J, Faravelli F, Casanova JL, Buist N, Levy-Lahad E, Mazzella M, Goulet O, Perroni L et al (2001) X-linked neonatal diabetes mellitus, enteropathy and endocrinopathy syndrome is the human equivalent of mouse scurfy. *Nat Genet* 27: 18–20
63 Bennett CL, Christie J, Ramsdell F, Brunkow ME, Ferguson PJ, Whitesell L, Kelly TE, Saulsbury FT, Chance PF, Ochs HD (2001) The immune dysregulation, polyendocrinopathy, enteropathy, X-linked syndrome (IPEX) is caused by mutations of FOXP3. *Nat Genet* 27: 20–21
64 Shevach EM (2001) Certified professionals: $CD4^+CD25^+$ suppressor T cells. *J Exp Med* 193: F41–F46
65 Tommasini A, Ferrari S, Moratto D, Badolato R, Boniotto M, Pirulli D, Notarangelo LD, Andolina M (2002) X-chromosome inactivation analysis in a female carrier of FOXP3 mutation. *Clin Exp Immunol* 130: 127–130
66 Sakaguchi S, Sakaguchi N (1994) Thymus, T cells and autoimmunity: various causes but a common mechanism of autoimmune disease. In: A Coutinho, M Kazatchkine (eds): *Autoimmunity: Physiology and Disease*. Wiley-Liss, New York, 203–227

"Natural" and "induced" regulatory T cells – purpose and problems associated with an emerging distinction

Dirk Homann[1] *and Matthias G. von Herrath*[2]

[1]University of Colorado Health Sciences Center, Barbara Davis Center for Childhood Diabetes, 4200 East 9th Avenue, Box 140, Denver, CO 80262, USA; [2]La Jolla Institute for Allergy and Immunology, Division of Immune Regulation, 10355 Science Center Drive, La Jolla, San Diego, CA 92121, USA

What are regulatory T cells?

As judged by the sheer number of recent research articles, reviews and book chapters devoted to the topic of regulatory T cells (T_R) there appears to be no shortage of information about the role of T_R in health and disease. Yet with the proliferation of research endeavors, model systems employed and insights generated comes a host of new questions that challenge any simple classification of T_R according to phenotype or functionality. The current interest in T_R is grounded in a fundamental insight into the nature of the vertebrate immune system that is about as old as the science of modern immunology: immune responses, which are characterized by dynamic alterations of the physiological steady state, have to be balanced, curtailed, regulated or suppressed in order to preserve organismal homeostasis [1, 2]. This process involves an array of immune cell subtypes, effector molecules and tissue environments that control in concert the extent of protective immune responses, autoimmunity and transplantation immunity. Thus, immune regulation is not merely the property of isolated and committed "regulatory cells" but an integral part of the activity and functionality of the immune system. Nevertheless, the particularities of defined model systems have allowed the identification of several parameters that are preferentially associated with ability of immune regulation. In light of these considerations, we prefer the notion that regulatory cells are best defined as cells capable of suppressing other immune responses *in vitro* and/or *in vivo*. This capacity segregates to a variety of immune cell subsets that include in addition to $CD4^+$ and $CD8^+$ T cells, natural killer (NK) T (NKT) cells, NK cells, dendritic cells (DCs) and others as well as a wide array of factors and tissue microenvironments involved in regulatory cell differentiation and effector function. As a consequence of our rather broad conceptualization of regulatory cells, it appears to be more practical to characterize these cells, in particular T_R, at the level of suppressive functionality rather than phe-

Regulatory T Cells in Inflammation, edited by Leonie S. Taams, Arne N. Akbar and Marca H.M. Wauben

notype. Functional definitions might also be useful to allow for more rapid isolation and targeted use in therapeutic settings where the precise mechanisms invoked during immunosuppression will be of critical importance.

Natural and induced T_R

A perennial approach in our quest to make sense of the complex phenomena we encounter is the establishment of dichotomies, however fraught with shortcomings, inconsistencies and exceptions to the rule. The recent distinction between "natural" and "induced" T_R is a case in point [3, 4]. According to this classification, two basic categories of T_R can be distinguished that differ in origin, antigen specificity and preferred effector mechanisms: natural T_R (nT_R) are the product of potentially distinct thymic maturation processes, exhibit a defining $CD4^+CD25^+$ phenotype and are "naturally" poised to mediate immunosuppression. Induced T_R (iT_R), which are also thymus-derived, develop from naïve, $CD4^+CD25^-$ T cells in the periphery under conditions of "sub-optimal" antigen exposure such as altered costimulatory interactions and/or "tolerogenic" microenvironments. As helpful as this distinction is, recent work has demonstrated a considerable overlap between these classes of T_R and other T cell subpopulations with regulatory functions such as $CD8^+$ T cells, γδTCR T cells, Qa-1-restricted $CD8^+$ T cells and NKT cells remain unaccounted in this classification [5–8].

Thymic selection and antigen specificity

Perhaps the key element in the above taxonomy is the notion of a natural $CD4^+$ T-cell subset preferentially dedicated to immune regulation and suppression. Recent evidence suggests that nT_R arise in the thymus as a consequence of positive selection that involves interactions of "intermediate affinity" whereas conventional $CD4^+$ T cells are selected based on "low-affinity" interactions [9–11]. In most cases, however, the antigen specificity of nT_R remains enigmatic. It is well possible that particular thymic expression profiles of defined peripheral antigens dictate the preferential development of nT_R rather than classical naive $CD4^+$ T cells. This view suggests that it is not the antigen-specificity *per se* but rather the thymic "self shadow" [12] that determines the spectrum of nT_R *versus* naïve $CD4^+$ T cell specificity. Little information is available about the further requirements for the development of nT_R but costimulatory interactions including those between B7 and CD28, and CD40 and CD40L, as well as TRANCE and RANK appear to be of particular importance [13–16]. Nevertheless, the intriguing concept of a distinct thymic maturation process favoring the export of committed nT_R will require extensive experimentation to acquire convincing validation. Despite the absence of detailed information about the antigen specificity of most nT_R, the distribution of TCR repertories among murine CD25$^+$ and CD25$^-$ CD4$^+$ T cells is similar and both populations exhibit comparable TCR diversity [10, 17, 18]. A recent study has reported corresponding findings for human nT_R in thymus and peripheral blood [19], further supporting the notion that nT_R are generated in the thymus prior to migration to the periphery [20–22]. Interestingly, this study also observed reduced levels of CD3 expression among thymic and peripheral nT_R as well as decreased TCR excision circle (TREC) levels in peripheral CD4$^+$CD25$^+$ T cells which is consistent with T_R activation and the existence of active homeostatic mechanisms assuring the preservation of nT_R in the periphery [19].

A further question pertaining to the antigen specificity of T_R relates to the nature of antigens classified as "self" or "non-self" [23]. At present, this issue is not resolved and requires further study. However, some studies have indicated that reactivity to "self" antigens is found among nT_R [9–11] and it is conceivable that "self-reactivity" is also important for iT_R as it appears unlikely that immunosuppressive T_R constitute the first line of defense against pathogens. Indeed, many studies that analyzed the specific T-cell repertoire to foreign pathogens have not observed significant numbers of T_R. How can we reconcile these considerations with the recent discovery that T_R play an instrumental role in dampening immunity to pathogens [24]? In some of these studies, T_R activity was abolished by *in vivo* anti-CD25 depletion leaving the issue of nT_R specificity unaddressed. Rather than recognizing epitopes derived from pathogens, it is conceivable that T_R recognize autoantigens in the context of inflammatory alterations. We would favor this interpretation, since physiological, regulatory, self-reactivity would only then be augmented, if more autoantigens become presented during an inflammatory process. The consequence would be that T_R activity would only increase with a slight delay following the pathogen-specific effector response and in that way not compromise host defense. T_R, in turn, would be important for reducing the pathogen-specific response to limit immunopathology and ensure the return to organismal homeostasis (Fig. 1). In other studies on T_R activity in response to pathogens, TCR-transgenic T cells were utilized to facilitate identification of specific cells. While these reports demonstrate that T_R can in principle be reactive to "non-self" antigens, the relevance of these observations remains to be assessed under more physiological conditions.

In summary, autoantigens are likely the preferred targets for both nT_R and iT_R and physiological autoreactivity may provide a key function in the preservation of immune homeostasis. If T_R were to recognize a significant amount of foreign proteins, they would respond but yet not necessarily expand upon microbial challenge. This would flood the organism with immunosuppressive molecules such as interleukin (IL)-10 and transforming growth factor (TGF)-β and likely condition antigen-presenting cells, which impairs activation of other T cells. Such a scenario would be disadvantageous to the host defense response. If in contrast most T_R recognize autoantigens, they will become more important during chronic inflammato-

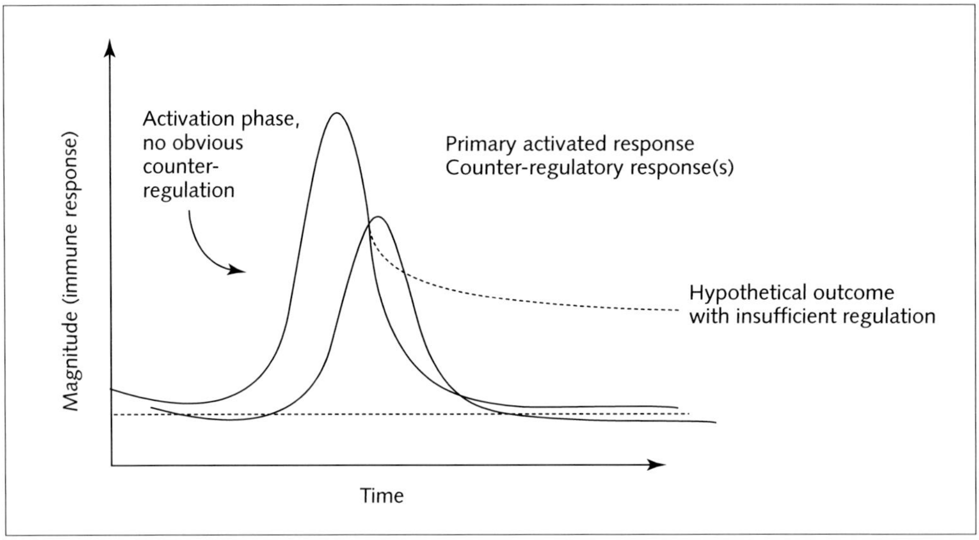

Figure 1
T_R induction to limit immunopathology but not host defense: counterbalance of immune responses by regulating to maintain baseline homeostasis.
Evidence suggests that most if not all iT_R and nT_R recognize autoantigens but not foreign antigens. This leads to their activation during chronic or prolonged immune responses but not during acute anti-viral immunity. The consequence is intact host defense yet limitation of immunopathology.

ry reactions. Indeed, most experimental evidence supports this argument; for example, T_R are capable of dampening chronic inflammation (for example, as present in long-term graft rejection or chronic type 1 diabetes), yet they are less efficient in limiting acute inflammatory reactions, such as primary graft rejection.

Phenotypic and functional properties

It is now well appreciated that phenotypic delineation of nT_R is complicated by the fact that many of the additional markers preferentially expressed by nT_R are also differentially regulated upon activation of "regular" naïve $CD4^+$ T cells. In addition to CD25, the IL-2 receptor α-subunit and a classical activation marker, regulatory T-cell activity has been segregated into cell subsets expressing $CD45RB^{low}$, DX5/CD49b, $CD62L^{high}$, glucocorticoid-induced tumor necrosis factor receptor (TNFR) family-related gene (GITR), OX40, CD103, CD38 or CTLA-4 (CD152) [25–32]. However, to what extent these markers participate in the predominantly contact-dependent activity of nT_R remains controversial and the topic of ongoing investigations. Functional distinctions appear more promising and the recent identification of Foxp3 as a signature transcription factor for development and function of nT_R in mice apparently bolstered the case for a defined $CD4^+$ T cell-subpopulation committed to immunosuppression [33–35]. Yet it is becoming increasingly clear that Foxp3 expression is neither necessary for T regulatory functions nor an exclusive hallmark nT_R: activation of human T cells, in contrast to mouse T cells, up-regulates expression of Foxp3 and induces regulatory activity in $CD4^+CD25^-$ T cells [36]. In the presence of TGF-β, Foxp3 can also be induced in naïve murine $CD4^+CD25^-$ T cells [37–39], while TGF-β in combination with IL-2 may also induce $CD8^+$ regulatory T cells [40]. On the other hand, IL-10-producing $CD4^+CD25^-$ regulatory T cells, shown to be comparable to nT_R in terms of low IL-2 expression, contact dependency and IL-10-independent activity mediated their suppressive function in the absence of Foxp3 expression [41] and transplantation tolerance, for example, is mediated by both $CD25^+$ and $CD25^-$ $CD4^+$ T cells alike [42].

iT_R comprise a heterogeneous group of T-cell subsets that arise as a consequence of the specific context of their generation [43]. Immunization with various antigens under different experimental conditions can give rise to conventional αβTCR $CD4^+$ (including T_H2 and T_H3 cells) or $CD8^+$ T cells as well as γδTCR $CD8^+$ T cells with regulatory activity. Here, it is the production of immunosuppressive cytokines such as TGF-β, IL-10 and IL-4 that conveys their regulatory properties and distinguishes them from the cytokine-independent nT_R. Thus, adaptive regulatory T cells represent a varied collection of T cells generated in the context of *bona fide* immune responses and may constitute a preferred target for immunomodulatory therapies.

The role of antigen in T_R development

Although T_R can often (but not always [44]) elaborate their immunosuppressive function in an antigen-non-specific fashion, a phenomenon termed "bystander suppression", "linked" or "infectious tolerance" [45–47], we have emphasized in the preceding paragraphs the importance of specific antigens in selection (nT_R) or induction (iT_R) of regulatory T cells. A number of recent studies has furthermore highlighted that the precise level of available antigen can skew immune responses to a regulatory phenotype. Evaluation of TCR transgenic T cells in *in vivo* environments expressing high or low levels of cognate antigen revealed an intrinsic tuning process in mature T cells that was associated with greater T-cell responsiveness in hosts with lower levels of antigen presentation and proliferative T-cell inhibition under conditions of higher antigen expression [48]. Similar findings were obtained in a model system where induction of IL-10-producing ovalbumin-specific regulatory T cells were generated in an inducible co-stimulator (ICOS)-dependent fashion in hosts transgenically expressing high but not low levels of ovalbumin [49]. The prolonged infusion of peptide antigens was also recently demonstrated as a means for transforming mature T cells into $CD4^+CD25^+$ T_R capable of immunosuppression upon antigen challenge [50]. However, similar to other antigen-specific TCR transgenic regulatory T cells generated in mice co-expressing the corresponding antigen, antigen persistence was not required to maintain phenotype and function of these cells [50, 51]. Furthermore, persisting pathogen antigens have been associated with increased numbers of T_R in both $CD25^+$ and $CD25^-$ CD4 compartments and clinical immunosuppression [52–54].

Based on these recent insights, we would like to argue that the extent and context of antigen presentation, both in the thymus and in the periphery, may indeed be the cardinal parameter that determines the nature of the specific immune response. Thus, immune regulation exerted by specific T-cell subpopulations is not a result of a separate T-cell lineage destined to perform immunosuppressive functions a priori but the ontological consequence of T-cell development, maturation and induction in the presence of varied patterns of antigen expression. Assuming that under physiological conditions there is a considerable overlap between individuals in regards to the distinct thymic patterns of self-antigen expression as well as to the presence of peripheral antigens, the repertoire and distribution of T_R in healthy individuals should be highly diverse and essentially similar. Due to differential "antigenic histories", differences will emerge with time to endow individuals with unique "signatures" of T_R specificity. Thus, the dynamics and diversities of T_R profiles should resemble those observed among "conventional" naïve, effector and/or memory T cells.

The role of homeostasis regulation in T_R function

We have recently emphasized that homeostatic imbalances are conducive for both T_R development and function [55]. In fact, a review of T_R function in type 1 diabetes demonstrated that the vast majority of experiments demonstrating a protective T_R effect *in vivo* were conducted in model systems that exhibit profound alterations of peripheral lymphocyte homeostasis (Tab. 1) [8, 13, 14, 29, 56–89]. These considerations are of particular importance since a central purpose of the study of T_R biology is their prospective use for various therapeutic applications. As based on T_R function in autoimmunity, it appears that effective nT_R and iT_R function *in vivo* requires specific TCR engagement as well as additional signals such as lymphopenia-induced proliferation/facilitated activation and/or inflammatory alterations. While the potential for active immune regulation by T_R, as suggested by Stockinger and colleagues [90, 91], may emerge only as an epiphenomenon of altered homeostatic conditions, it should be noted that the very same conditions may also promote autoimmunity in the first place.

As an example, we would like to amend our recent discussion of homeostatic alterations and T_R function in type 1 diabetes [55] with a recent publication that has proposed that autoimmunity in the non-obese diabetic (NOD) mouse arises preferentially as a consequence of homeostatic imbalances [92]. Although NOD mice are the preferred small-animal model for the study of type 1 diabetes, there is a surprising dearth of information about the composition and homeostatic characteristics of peripheral lymphocytes in this model [93]. While some studies have reported a numerical reduction of nT_R in NOD as compared to non-autoimmune mice [13, 14], another study has found essentially normal distribution of peripheral T-cell compartments and comparable proliferation of T cells transferred into irradiated NOD and B6 recipients [93]. A recent study by Sarvetnick's group [92] proposes that reduced T-cell numbers in the NOD mouse promote an exaggerated "homeostatic proliferation" and subsequent autoimmunity that can be averted by simple infusion of unfractioned T cells or purified $CD4^+$ T cells. Although this notion is clearly intriguing, additional studies will have to evaluate why comparable incidences of diabetes in different NOD colonies can be associated with reduced [92] or normal [93] peripheral T-cell counts. Furthermore the effects of IL-21, suggested to be important for the accelerated T-cell proliferation in NOD mice [92], may preferentially act under lymphopenic conditions as no impact on antigen-specific memory T-cell proliferation was noted in intact lymphatic compartments [94]. In summary, homeostatic imbalances may play a key role in both destructive autoimmunity and effective immune regulation, which together suggests a careful balance that needs to be maintained to effectively harness T_R for therapeutic purposes.

Therapeutic potential of T_R?

Beneficial auto/reactivity exerted by T_R has, in principle, therapeutic potential: there are several pathways by which autoreactive cells can suppress lymphocytes of differential specificity, most of which involve cytokines such as IL-10, IL-4 and TGF-β *in vivo*. This so-called phenomenon of "bystander suppression" or, if one T_R induces other T_R, also termed "infectious tolerance", can be instrumental in dampening local or systemic inflammatory processes [85, 95–97]. It follows that T_R can be useful (and necessary) in preventing autoimmunity, maintaining transplantation tolerance and possibly curbing some immune responses to infectious agents, thus reducing immunopathology. Selective and specific *in vitro* assays for T_R function will be required to translate findings from diverse experimental models to therapeutic applications in humans. This is of particular importance if one considers that T_R levels that dampen a given immune response might be set at different levels [85, 97] depending on the individual and the type of inflammatory reaction under observation. Thus T_R that exhibit beneficial potential for one disease might not do so in other diseases and their therapeutic augmentation will have to be tested and tracked at least for each therapeutic application and maybe on an individual basis.

Table 1 - Regulatory T cells and type 1 diabetes

Diabetes model	Treatment to induce direct protecion	Transferred cell populations	Recipient	Active regulatory T cell (functional profile; proposed mechanism)	Refs.
(A) Natural $CD4^+CD25^+$ regulatory T cells (nT_R)					
PVG.RT1 rat (thymectomized/ irradiated)	Not done	$CD4^+CD45RC^{lo}RT6^+$	PVG.RT1 rat (thymectomized/ irradiated)	$CD4^+CD45RC^{lo}RT6^+T_R$ (IL-2/IL-4/B cell help)	[56]
PVG.RT1 rat (thymectomized/ irradiated)	Not done	$CD4^+CD62L^+$	PVG.RT1 rat (thymectomized/ irradiated)	$CD4^+CD62L^+T_R$ (B cell help)	[57]
NOD	Not done	Diabetogenic NOD T cells & CD4 T cells from prediabetic NOD mice	Irradiated NOD	CD4 T_R	[58]
NOD	Not done	Diabetogenic NOD T cells & T cells from male NOD mice	Irradiated NOD	CD4 T_R	[59]
NOD	Not done	Diabetogenic NOD T cells & $CD4^+CD45RB^{lo}$ T cells from prediabetic NOD mice	NOD.SCID	$CD4^+CD45RB^{lo}$ T_R (TH2)	[60]
NOD	Not done	Diabetogenic NOD T cells & $CD4^+CD62L^+$ T cells from prediabetic NOD mice	NOD.SCID	$CD4^+CD62L^+$ T_R	[29]
NOD	Not done	Ddiabetogenic NOD T cells & $CD4^+CD62L^+$ T cells from prediabetic NOD mice	NOD.SCID	$CD4^+CD62L^+$ T_R	[61]

NOD & NOD.$CD28^{-/-}$	Not done	Diabetogenic NOD.$CD28^{-/-}$ & $CD4^+CD25^+$ T cells from prediabetic NOD mice	NOD.SCID	$CD4^+CD25^+$ T_R (CTLA-4; CTLA-4-dependent immune suppression)	[14]
NOD & NOD.$CD28^{-/-}$	Not done	8×10^6 $CD4^+CD25^+$ T cells from prediabetic NOD	NOD.$CD28^{-/-}$	$CD4^+CD25^+$ T_R (delay of blood glucose elavation)	[14]
NOD	Not done	Diabetogenic NOD T cells & $CD4^+CD25^+CD62L^+CCR7^+$ T cells from prediabetic NOD mice	NOD.SCID	$CD4^+CD25^+CD62L^+CCR7^+$ T_R (differential chemokine receptor expression)	[62]
8.3NOD.$RAG2^{-/-}$. $CD154^{-/-}$	Not done	$CD4^+CD25^+CD154^{-/-}$	8.3NOD.$RAG2^{-/-}$. $CD154^{-/-}$	$CD4^+CD25^+CD154^-$ T_R (CTLA-4; suppression of dendritic cell maturation)	[63]
Tet-TNF-α/CD80	Not done	$CD4^+CD25^+$ from Tet-TNF-α/CD80 mice expressing TNF-α d0-25	Tet-TNF-α/CD80 mice expressing TNF-α d0-28	$CD4^+CD25^+$ T_R (TRANCE-RANK-dependent development; suppression of of intraislet differentiation of diabetogenic $CD8^+$ T cells)	[13]
Tet-TNF-α/CD80	Not done	$CD4^+CD25^+$ from Tet-TNF-α/CD80 mice expressing TNF-α d0-25	Tet-TNF-α/CD80 mice expressing TNF-α d0-28	$CD4^+CD25^+$ T_R (TGF-β; direct (CD8) and/or indirect (APC) TGF-β-dependent suppression)	[64] [64]
(B) Induced regulatory T cells (iT_R)					
NOD	Insulin (oral)	Diabetogenic NOD T cells & spleen cells from treated NOD mice	Irradiated NOD	iT_R (bystander suppression)	[65]

Table 1 (continued)

Diabetes model	Treatment to induce direct protecion	Transferred cell populations	Recipient	Active regulatory T cell (functional profile; proposed mechanism)	Refs.
NOD	Insulin (oral)	Diabetogenic NOD T cells & $CD4^+$ T cells from treated mice	Irradiated NOD	$CD4^+$ iT_R (blockade of diabetogenic T cells)	[67]
NOD	Insulin (oral)	Not done	Not done	$CD4^+$ iT_R (IL-4/IL-10/ TGF-β/prostaglandin E)	[69]
NOD	Insulin, insulin B-chain(IFA)	Diabetogenic NOD T cells & T cells from treated mice	Irradiated NOD	$CD4^+$ & $CD8^+$ T_R	[70]
NOD	Insulin (aerosol)	Diabetogenic NOD T cells & γδTCR $CD8^+$ T cells from treated mice	Irradiated NOD	γδTCR $CD8^+$ iT_R (IL-4/IL-10)	[8]
NOD	Diabetic NOD mouse	Autoreactive $CD4^+$ T cell clone	Young NOD (4–20 weeks)	Specific $CD4^+$ T_R (TGF-β/IL-10/IFN-γ; inhibition of activation and migration of effector T cells)	[72, 73]
NOD	Insulin B-chain (oral)	Diabetogenic NOD T cells & T cells from treated mice	Irradiated NOD	Specific $CD4^+$ iT_R (IL-4/IL-10/ TGF-β; bystander suppression)	[74]
NOD	CTB-insulin (oral)	Diabetogenic NOD T cells & T cells from treated NOD mice	Irradiated NOD	Specific iT_R (selective homing)	[68]
NOD	Schistosome egg ag & insulin B-chain (oral)*	Not done	Not done	Specific $CD4^+$ iT_R (IL-10/IL-4/TGF-β)	[76]

NOD	Insulin B-chain insulin B10-24 GAD524-543 (oral, intranasal)	Diabetogenic NOD T cells & specific $CD4^+$ T cell lines	Irradiated NOD, NOD.SCID, neonatal NOD	Specific $CD4^+$ iT_R (IL-10/TGF-β; bystander suppression)	[75]
NOD	Insulin (oral)	Diabetogenic NOD T cells & T cells from treated mice	NOD.SCID	Specific $CD4^+$ iT_R (target/IL-2 competition with diabetogenic T cells)	[66]
NOD	Iinsulin B chain (i.m. DNA vaccination)	Not done	Not done	iTREG (IL-4)	[77]
NOD	Proinsulin B24-C36 (intranasal)	Diabetogenic NOD T cells & $CD4^+$ T cells from treated mice	Irradiated NOD	Specific $CD4^+$ iT_R	[71]
NOD	Insulin B9-23 peptide (foot pad/IFA)	Diabetogenic NOD T cells & cultured $CD4^+$ T cells from treated mice	NOD.SCID	Specific $CD4^+$ iT_R (TGF-β/TNF-α; TGF-β-dependent suppression of APC activation)	[78]
NOD	GAD65 peptides (intranasal)	Diabetogenic NOD T cells & $CD4^+$ T cells from treated mice	NOD.SCID	Specific $CD4^+$ iT_R (IL-5; TH1 to TH2 shift, APC modulation, bystander suppression)	[79]
NOD	GAD65 (i.v.)	Diabetogenic NOD T cells & $CD4^+$ T cells from treated mice	Irradiated NOD	specific $CD4^+$ T_R (IL-4; modulation of local milieu conducive for further T_R development)	[80]
NOD	GAD65/IL-4 (i.m. DNA vaccination)	Diabetogenic NOD T cells & $CD4^+$ T cells from treated mice	Neonatal NOD	Specific $CD4^+$ iT_R (IL-4)	[81]

Table 1 (continued)

Diabetes model	Treatment to induce direct protecion	Transferred cell populations	Recipient	Active regulatory T cell (functional profile; proposed mechanism)	Refs.
NOD	GAD 206 & 221 peptides (i.p./IFA)	Diabetogenic NOD T cells & GAD-specific $CD4^+$ T cells	NOD.SCID	Specific $CD4^+$ iT_R (IL-4/IL-10 & IFN-γ/IL-10; inhibition of migration)	[82]
NOD, NOD.$CD28^{-/-}$	Anti-CD3	Diabetogenic NOD T cells & $CD4^+CD25^+$ T cells from treated mice	NOD.SCID	$CD4^+CD25^+$ iT_R (TGF-β; anti-CD3-induced TGF-β^+ TREG)	[83]
NOD.TCR$\alpha^{-/-}$	Not done	*In vitro* activated BDC2.5 TCRtg $CD4^+$ T cells & dimeric mimotope:I-A^{g7} chimera treatment (i.v.)	NOD.TCR$\alpha^{-/-}$	Specific $CD4^+$ iT_R (IL-10; bystander suppression in conjunction with clonal deletion of pathogenic TCRtg T cells)	[89]
RIP-LCMV	Insulin (oral)	Not done	Not done	iT_R (IL-4/IL-10/TGF-β; Th1 to Th2 shift, bystander suppression)	[87]
RIP-LCMV	Insulin (oral)	Insulin B-chain-specific $CD4^+$ T cell lines	LCMV-infected RIP-LCMV	Specific $CD4^+$ iT_R (IL-4/IL-10/IFN-γ; local IL-4-dependent APC modulation)	[85, 86]
RIP-LCMV	Insulin B-chain (i.m. DNA vaccination)	$CD8^+$ T cell-depleted cells from treated mice	LCMV-infected RIP-LCMV	non-$CD8^+$ iT_R (IL-4; bystander suppression)	[84]

Influenza HA double transgenic	Dimeric $HA_{110-120}$:I-E^d chimera (i.v.)	HA-specific TCR transgenic $CD4^+$ T cells & *in vitro* dimeric chimera-treated pancreatic $CD4^+$ T cells	HA-transgenic $RAG2^{-/-}$	Specific pancreatic $CD4^+$ iT_R (IL-10; altered trafficking and differentiation of pathogenic $CD4^+$ CTL; anergy induction)	[88]

Table 1 lists animal models for diabetes in which presence or induction of active regulatory T cells has been demonstrated. Adoptive-transfer experiments (columns 3 and 4) are performed in most studies to determine the phenotype/function of putative regulatory T cells. Reports are listed according to animal model (shaded/unshaded background), year of publication and immunizing antigen.

APC: *antigen-presenting cell.*

CTB: *choleratoxin B subunit. Conjugation of antigens to CTB facilitates uptake of orally administered antigens.*

CTL: *cytotoxic T lymphocyte.*

GAD: *glutamic acid decarboxylase.*

Influenza HA double transgenic: *transgenic mice that express influenza hemagglutinin (HA) under control of the rat insulin promoter as well as HA110-120-specific TCR transgenic $CD4^+$ T cells.*

IFA: *incomplete Freund's adjuvant.*

LCMV: *lymphocytic choriomeningitis virus.*

NOD: *non-obese diabetic mice.*

NOD.SCID: *NOD mice with severe combined immunodeficiency. As NOD.SCID mice are devoid of T cells, induction of diabetes requires adoptive transfer of diabetogenic T cells.*

Irradiated NOD/neonatal NOD: *similar to NOD.SCID mice, irradiated and neonatal NOD mice exhibit a pronounced lymphopenia.*

8.3NOD.$Rag2^{-/-}$.$CD154^{-/-}$ mice: *these mice contain diabetogenic TCR transgenic $CD8^+$ T cells in the absence of other T cells and are deficient for CD154 (CD40L).*

RIP-LCMV: *transgenic mice expressing LCMV proteins under control of the rat insulin promoter. Diabetes is induced in RIP-LCMV mice after infection with LCMV. LCMV-specific T cells control the virus and destroy β-cells expressing the LCMV transgene.*

TCRtg: *T-cell receptor transgenic.*

Tet-TNF-α/CD80 mice: *these mice exhibit tetracyclin controlled induction of TNF-α as well as β-cell-specific expression of CD80. Duration of TNF-α expression determines diabetes induction or prevention.*

Th1, Th2, Th3, $CD4^+$ helper T cell subsets: *subsets are classified according to signature cytokines they produce. Th1, IFN-γ; Th2, IL-4; Th3, TGF-β.*

* *Effect of immunization on diabetes incidence was not studied*

References

1 Ehrlich P, Morgenroth J (1901) Zytotoxine als Antikörper. *Berl Klin Wochenschr* 38: 251–260

2 Römer P, Gebb H (1912) *Albrecht von Graefes Arch Ophtalmol* 81: 367–387

3 Bluestone JA, Abbas AK (2003) Natural *versus* adaptive regulatory T cells. *Nat Rev Immunol* 3: 253–257

4 Jonuleit H, Schmitt E (2003) The regulatory T cell family: distinct subsets and their interrelations. *J Immunol* 171: 6323–6327

5 Hu D, Ikizawa K, Lu L, Sanchirico ME, Shinohara ML, Cantor H (2004) Analysis of regulatory CD8 T cells in Qa-1-deficient mice. *Nat Immunol* 5: 516–523

6 Hammond KJ, Kronenberg M (2003) Natural killer T cells: natural or unnatural regulators of autoimmunity? *Curr Opin Immunol* 15: 683–689

7 Chess L, Jiang H (2004) Resurrecting $CD8^+$ suppressor T cells. *Nat Immunol* 5: 469–471

8 Harrison LC, Dempsey-Collier M, Kramer DR, Takahashi K (1996) Aerosol insulin induces regulatory CD8 gamma delta T cells that prevent murine insulin-dependent diabetes. *J Exp Med* 184: 2167–2174

9 Jordan MS, Boesteanu A, Reed AJ, Petrone AL, Holenbeck AE, Lerman MA, Naji A, Caton AJ (2001) Thymic selection of $CD4^+CD25^+$ regulatory T cells induced by an agonist self-peptide. *Nat Immunol* 2: 301–306

10 Pacholczyk R, Kraj P, Ignatowicz L (2002) Peptide specificity of thymic selection of $CD4^+CD25^+$ T cells. *J Immunol* 168: 613–620

11 Suto A, Nakajima H, Ikeda K, Kubo S, Nakayama T, Taniguchi M, Saito Y, Iwamoto I (2002) CD4(+)CD25(+) T-cell development is regulated by at least 2 distinct mechanisms. *Blood* 99: 555–560

12 Anderson MS, Venanzi ES, Klein L, Chen Z, Berzins SP, Turley SJ, von Boehmer H, Bronson R, Dierich A, Benoist C et al (2002) Projection of an immunological self shadow within the thymus by the aire protein. *Science* 298: 1395–1401

13 Green EA, Choi Y, Flavell RA (2002) Pancreatic lymph node-derived CD4(+)CD25(+) Treg cells: highly potent regulators of diabetes that require TRANCE-RANK signals. *Immunity* 16: 183–191

14 Salomon B, Lenschow DJ, Rhee L, Ashourian N, Singh B, Sharpe A, Bluestone JA (2000) B7/CD28 costimulation is essential for the homeostasis of the $CD4^+CD25^+$ immunoregulatory T cells that control autoimmune diabetes. *Immunity* 12: 431–440

15 Kumanogoh A, Wang X, Lee I, Watanabe C, Kamanaka M, Shi W, Yoshida K, Sato T, Habu S, Itoh M et al (2001) Increased T cell autoreactivity in the absence of CD40-CD40 ligand interactions: a role of CD40 in regulatory T cell development. *J Immunol* 166: 353–360

16 Tang Q, Henriksen KJ, Boden EK, Tooley AJ, Ye J, Subudhi SK, Zheng XX, Strom TB, Bluestone JA (2003) Cutting edge: CD28 controls peripheral homeostasis of $CD4^+CD25^+$ regulatory T cells. *J Immunol* 171: 3348–3352

17 Romagnoli P, Hudrisier D, van Meerwijk JP (2002) Preferential recognition of self antigens despite normal thymic deletion of CD4(+)CD25(+) regulatory T cells. *J Immunol* 168: 1644–1648
18 Taams L, Vukmanovic-Stejic M, Salmon M, Akbar A (2003) Immune regulation by $CD4^+CD25^+$ regulatory T cells: implications for transplantation tolerance. *Transpl Immunol* 11: 277–285
19 Kasow KA, Chen X, Knowles J, Wichlan D, Handgretinger R, Riberdy JM (2004) Human $CD4^+CD25^+$ regulatory T cells share equally complex and comparable repertoires with $CD4^+CD25^-$ counterparts. *J Immunol* 172: 6123–6128
20 Curotto de Lafaille MA, Lafaille JJ (2002) CD4(+) regulatory T cells in autoimmunity and allergy. *Curr Opin Immunol* 14: 771–778
21 Bensinger SJ, Bandeira A, Jordan MS, Caton AJ, Laufer TM (2001) Major histocompatibility complex class II-positive cortical epithelium mediates the selection of CD4(+)25(+) immunoregulatory T cells. *J Exp Med* 194: 427–438
22 Papiernik M, de Moraes ML, Pontoux C, Vasseur F, Penit C (1998) Regulatory CD4 T cells: expression of IL-2R alpha chain, resistance to clonal deletion and IL-2 dependency. *Int Immunol* 10: 371–378
23 Cohn M (2004) Whither T-suppressors: if they didn't exist would we have to invent them? *Cell Immunol* 227: 81–92
24 Mittrucker HW, Kaufmann SH (2004) Mini-review: regulatory T cells and infection: suppression revisited. *Eur J Immunol* 34: 306–312
25 Read S, Mauze S, Asseman C, Bean A, Coffman R, Powrie F (1998) $CD38^+$ CD45RB(low) $CD4^+$ T cells: a population of T cells with immune regulatory activities *in vitro*. *Eur J Immunol* 28: 3435–3447
26 Lehmann J, Huehn J, de la Rosa M, Maszyna F, Kretschmer U, Krenn V, Brunner M, Scheffold A, Hamann A (2002) Expression of the integrin alpha Ebeta 7 identifies unique subsets of $CD25^+$ as well as CD25- regulatory T cells. *Proc Natl Acad Sci USA* 99: 13031–13036
27 McHugh RS, Whitters MJ, Piccirillo CA, Young DA, Shevach EM, Collins M, Byrne MC (2002) CD4(+)CD25(+) immunoregulatory T cells: gene expression analysis reveals a functional role for the glucocorticoid-induced TNF receptor. *Immunity* 16: 311–323
28 Shimizu J, Yamazaki S, Takahashi T, Ishida Y, Sakaguchi S (2002) Stimulation of CD25(+)CD4(+) regulatory T cells through GITR breaks immunological self-tolerance. *Nat Immunol* 3: 135–142
29 Herbelin A, Gombert JM, Lepault F, Bach JF, Chatenoud L (1998) Mature mainstream TCR alpha $beta^+CD4^+$ thymocytes expressing L-selectin mediate "active tolerance" in the nonobese diabetic mouse. *J Immunol* 161: 2620–2628
30 Read S, Malmstrom V, Powrie F (2000) Cytotoxic T lymphocyte-associated antigen 4 plays an essential role in the function of CD25(+)CD4(+) regulatory cells that control intestinal inflammation. *J Exp Med* 192: 295–302
31 Singh B, Read S, Asseman C, Malmstrom V, Mottet C, Stephens LA, Stepankova R,

Tlaskalova H, Powrie F (2001) Control of intestinal inflammation by regulatory T cells. *Immunol Rev* 182: 190–200
32 Gonzalez A, Andre-Schmutz I, Carnaud C, Mathis D, Benoist C (2001) Damage control, rather than unresponsiveness, effected by protective $DX5^+$ T cells in autoimmune diabetes. *Nat Immunol* 2: 1117–1125
33 Khattri R, Cox T, Yasayko SA, Ramsdell F (2003) An essential role for Scurfin in $CD4^+CD25^+$ T regulatory cells. *Nat Immunol* 4: 337–342
34 Fontenot JD, Gavin MA, Rudensky AY (2003) Foxp3 programs the development and function of $CD4^+CD25^+$ regulatory T cells. *Nat Immunol* 4: 330–336
35 Hori S, Nomura T, Sakaguchi S (2003) Control of regulatory T cell development by the transcription factor Foxp3. *Science* 299: 1057–1061
36 Walker MR, Kasprowicz DJ, Gersuk VH, Benard A, Van Landeghen M, Buckner JH, Ziegler SF (2003) Induction of FoxP3 and acquisition of T regulatory activity by stimulated human $CD4^+$CD25- T cells. *J Clin Invest* 112: 1437–1443
37 Fantini MC, Becker C, Monteleone G, Pallone F, Galle PR, Neurath MF (2004) Cutting edge: TGF-beta induces a regulatory phenotype in $CD4^+CD25^-$ T cells through Foxp3 induction and down-regulation of Smad7. *J Immunol* 172: 5149–5153
38 Chen W, Jin W, Hardegen N, Lei KJ, Li L, Marinos N, McGrady G, Wahl SM (2003) Conversion of peripheral $CD4^+CD25^-$ naive T cells to $CD4^+CD25^+$ regulatory T cells by TGF-beta induction of transcription factor Foxp3. *J Exp Med* 198: 1875–1886
39 Zheng SG, Wang JH, Gray JD, Soucier H, Horwitz DA (2004) Natural and induced $CD4^+CD25^+$ cells educate $CD4^+CD25^-$ cells to develop suppressive activity: the role of IL-2, TGF-beta, and IL-10. *J Immunol* 172: 5213–5221
40 Zheng SG, Wang JH, Koss MN, Quismorio F Jr, Gray JD, Horwitz DA (2004) $CD4^+$ and $CD8^+$ regulatory T cells generated *ex vivo* with IL-2 and TGF-beta suppress a stimulatory graft-*versus*-host disease with a lupus-like syndrome. *J Immunol* 172: 1531–1539
41 Vieira PL, Christensen JR, Minaee S, O'Neill EJ, Barrat FJ, Boonstra A, Barthlott T, Stockinger B, Wraith DC, O'Garra A (2004) IL-10-secreting regulatory T cells do not express Foxp3 but have comparable regulatory function to naturally occurring $CD4^+CD25^+$ regulatory T cells. *J Immunol* 172: 5986–5993
42 Graca L, Thompson S, Lin CY, Adams E, Cobbold SP, Waldmann H (2002) Both CD4(+)CD25(+) and CD4(+)CD25(–) regulatory cells mediate dominant transplantation tolerance. *J Immunol* 168: 5558–5565
43 von Herrath MG, Harrison LC (2003) Antigen-induced regulatory T cells in autoimmunity. *Nat Rev Immunol* 3: 223–232
44 Tanchot C, Vasseur F, Pontoux C, Garcia C, Sarukhan A (2004) Immune regulation by self-reactive T cells is antigen specific. *J Immunol* 172: 4285–4291
45 Tian J, Olcott AP, Hanssen LR, Zekzer D, Middleton B, Kaufman DL (1998) Infectious Th1 and Th2 autoimmunity in diabetes-prone mice. *Immunol Rev* 164: 119–127
46 Cobbold S, Waldmann H (1998) Infectious tolerance. *Curr Opin Immunol* 10: 518–524
47 Weiner HL, Friedman A, Miller A, Khoury SJ, al-Sabbagh A, Santos L, Sayegh M,

Nussenblatt RB, Trentham DE, Hafler DA (1994) Oral tolerance: immunologic mechanisms and treatment of animal and human organ-specific autoimmune diseases by oral administration of autoantigens. *Annu Rev Immunol* 12: 809–837
48 Singh NJ, Schwartz RH (2003) The strength of persistent antigenic stimulation modulates adaptive tolerance in peripheral $CD4^+$ T cells. *J Exp Med* 198: 1107–1117
49 Kohyama M, Sugahara D, Sugiyama S, Yagita H, Okumura K, Hozumi N (2004) Inducible costimulator-dependent IL-10 production by regulatory T cells specific for self-antigen. *Proc Natl Acad Sci USA* 101: 4192–4197
50 Apostolou I, Von Boehmer H (2004) *In vivo* instruction of suppressor commitment in naive T cells. *J Exp Med* 199: 1401–1408
51 Klein L, Khazaie K, von Boehmer H (2003) *In vivo* dynamics of antigen-specific regulatory T cells not predicted from behavior *in vitro*. *Proc Natl Acad Sci USA* 100: 8886–8891
52 Iwashiro M, Messer RJ, Peterson KE, Stromnes IM, Sugie T, Hasenkrug KJ (2001) Immunosuppression by $CD4^+$ regulatory T cells induced by chronic retroviral infection. *Proc Natl Acad Sci USA* 98: 9226–9230
53 Hasenkrug KJ (2003) $CD4^+$ regulatory T cells in chronic viral infection. *Novartis Found Symp* 252: 194–199; discussion 199–210
54 Dittmer U, He H, Messer RJ, Schimmer S, Olbrich AR, Ohlen C, Greenberg PD, Stromnes IM, Iwashiro M, Sakaguchi S et al (2004) Functional impairment of CD8(+) T cells by regulatory T cells during persistent retroviral infection. *Immunity* 20: 293–303
55 Homann D, von Herrath MG (2004) Regulatory T cells and type 1 diabetes. *Clin Immunol* 112: 202–209
56 Fowell D, Mason D (1993) Evidence that the T cell repertoire of normal rats contains cells with the potential to cause diabetes. Characterization of the $CD4^+$ T cell subset that inhibits this autoimmune potential. *J Exp Med* 177: 627–636
57 Seddon B, Saoudi A, Nicholson M, Mason D (1996) $CD4^+CD8^-$ thymocytes that express L-selectin protect rats from diabetes upon adoptive transfer. *Eur J Immunol* 26: 2702–2708
58 Boitard C, Yasunami R, Dardenne M, Bach JF (1989) T cell-mediated inhibition of the transfer of autoimmune diabetes in NOD mice. *J Exp Med* 169: 1669–1680
59 Hutchings PR, Cooke A (1990) The transfer of autoimmune diabetes in NOD mice can be inhibited or accelerated by distinct cell populations present in normal splenocytes taken from young males. *J Autoimmun* 3: 175–185
60 Shimada A, Rohane P, Fathman CG, Charlton B (1996) Pathogenic and protective roles of CD45RB(low) $CD4^+$ cells correlate with cytokine profiles in the spontaneously autoimmune diabetic mouse. *Diabetes* 45: 71–78
61 Lepault F, Gagnerault MC (2000) Characterization of peripheral regulatory $CD4^+$ T cells that prevent diabetes onset in nonobese diabetic mice. *J Immunol* 164: 240–247
62 Szanya V, Ermann J, Taylor C, Holness C, Fathman CG (2002) The subpopulation of

CD4$^+$CD25$^+$ splenocytes that delays adoptive transfer of diabetes expresses L-selectin and high levels of CCR7. *J Immunol* 169: 2461–2465
63 Serra P, Amrani A, Yamanouchi J, Han B, Thiessen S, Utsugi T, Verdaguer J, Santamaria P (2003) CD40 ligation releases immature dendritic cells from the control of regulatory CD4$^+$CD25$^+$ T cells. *Immunity* 19: 877–889
64 Green EA, Gorelik L, McGregor CM, Tran EH, Flavell RA (2003) CD4$^+$CD25$^+$ T regulatory cells control anti-islet CD8$^+$ T cells through TGF-beta-TGF-beta receptor interactions in type 1 diabetes. *Proc Natl Acad Sci USA* 100: 10878–10883
65 Zhang ZJ, Davidson L, Eisenbarth G, Weiner HL (1991) Suppression of diabetes in nonobese diabetic mice by oral administration of porcine insulin. *Proc Natl Acad Sci USA* 88: 10252–10256
66 Bergerot I, Arreaza GA, Cameron MJ, Burdick MD, Strieter RM, Chensue SW, Chakrabarti S, Delovitch TL (1999) Insulin B-chain reactive CD4$^+$ regulatory T-cells induced by oral insulin treatment protect from type 1 diabetes by blocking the cytokine secretion and pancreatic infiltration of diabetogenic effector T-cells. *Diabetes* 48: 1720–1729
67 Bergerot I, Fabien N, Maguer V, Thivolet C (1994) Oral administration of human insulin to NOD mice generates CD4$^+$ T cells that suppress adoptive transfer of diabetes. *J Autoimmun* 7: 655–663
68 Bergerot I, Ploix C, Petersen J, Moulin V, Rask C, Fabien N, Lindblad M, Mayer A, Czerkinsky C, Holmgren J et al (1997) A cholera toxoid-insulin conjugate as an oral vaccine against spontaneous autoimmune diabetes. *Proc Natl Acad Sci USA* 94: 4610–4614
69 Hancock WW, Polanski M, Zhang J, Blogg N, Weiner HL (1995) Suppression of insulitis in non-obese diabetic (NOD) mice by oral insulin administration is associated with selective expression of interleukin-4 and -10, transforming growth factor-beta, and prostaglandin-E. *Am J Pathol* 147: 1193–1199
70 Muir A, Peck A, Clare-Salzler M, Song YH, Cornelius J, Luchetta R, Krischer J, Maclaren N (1995) Insulin immunization of nonobese diabetic mice induces a protective insulitis characterized by diminished intraislet interferon-gamma transcription. *J Clin Invest* 95: 628–634
71 Martinez NR, Augstein P, Moustakas AK, Papadopoulos GK, Gregori S, Adorini L, Jackson DC, Harrison LC (2003) Disabling an integral CTL epitope allows suppression of autoimmune diabetes by intranasal proinsulin peptide. *J Clin Invest* 111: 1365–1371
72 Han HS, Jun HS, Utsugi T, Yoon JW (1996) A new type of CD4$^+$ suppressor T cell completely prevents spontaneous autoimmune diabetes and recurrent diabetes in syngeneic islet-transplanted NOD mice. *J Autoimmun* 9: 331–339
73 Han HS, Jun HS, Utsugi T, Yoon JW (1997) Molecular role of TGF-beta, secreted from a new type of CD4$^+$ suppressor T cell, NY4.2, in the prevention of autoimmune IDDM in NOD mice. *J Autoimmun* 10: 299–307
74 Polanski M, Melican NS, Zhang J, Weiner HL (1997) Oral administration of the immunodominant B-chain of insulin reduces diabetes in a co-transfer model of diabetes

in the NOD mouse and is associated with a switch from Th1 to Th2 cytokines. *J Autoimmun* 10: 339–346

75 Maron R, Melican NS, Weiner HL (1999) Regulatory Th2-type T cell lines against insulin and GAD peptides derived from orally- and nasally-treated NOD mice suppress diabetes. *J Autoimmun* 12: 251–258

76 Maron R, Palanivel V, Weiner HL, Harn DA (1998) Oral administration of schistosome egg antigens and insulin B-chain generates and enhances Th2-type responses in NOD mice. *Clin Immunol Immunopathol* 87: 85–92

77 Bot A, Smith D, Bot S, Hughes A, Wolfe T, Wang L, Woods C, von Herrath M (2001) Plasmid vaccination with insulin B chain prevents autoimmune diabetes in nonobese diabetic mice. *J Immunol* 167: 2950–2955

78 Mukherjee R, Chaturvedi P, Qin HY, Singh B (2003) $CD4^+CD25^+$ regulatory T cells generated in response to insulin B:9-23 peptide prevent adoptive transfer of diabetes by diabetogenic T cells. *J Autoimmun* 21: 221–237

79 Tian J, Clare-Salzler M, Herschenfeld A, Middleton B, Newman D, Mueller R, Arita S, Evans C, Atkinson MA, Mullen Y et al. (1996) Modulating autoimmune responses to GAD inhibits disease progression and prolongs islet graft survival in diabetes-prone mice. *Nat Med* 2: 1348–1353

80 Tisch R, Liblau RS, Yang XD, Liblau P, McDevitt HO (1998) Induction of GAD65-specific regulatory T-cells inhibits ongoing autoimmune diabetes in nonobese diabetic mice. *Diabetes* 47: 894–899

81 Tisch R, Wang B, Weaver DJ, Liu B, Bui T, Arthos J, Serreze DV (2001) Antigen-specific mediated suppression of beta cell autoimmunity by plasmid DNA vaccination. *J Immunol* 166: 2122–2132

82 Chen C, Lee WH, Yun P, Snow P, Liu CP (2003) Induction of autoantigen-specific Th2 and Tr1 regulatory T cells and modulation of autoimmune diabetes. *J Immunol* 171: 733–744

83 Belghith M, Bluestone JA, Barriot S, Megret J, Bach JF, Chatenoud L (2003) TGF-beta-dependent mechanisms mediate restoration of self-tolerance induced by antibodies to CD3 in overt autoimmune diabetes. *Nat Med* 9: 1202–1208

84 Coon B, An LL, Whitton JL, von Herrath MG (1999) DNA immunization to prevent autoimmune diabetes. *J Clin Invest* 104: 189–194

85 Homann D, Holz A, Bot A, Coon B, Wolfe T, Petersen J, Dyrberg TP, Grusby MJ, von Herrath MG (1999) Autoreactive $CD4^+$ T cells protect from autoimmune diabetes *via* bystander suppression using the IL-4/Stat6 pathway. *Immunity* 11: 463–472

86 Homann D, Dyrberg T, Petersen J, Oldstone MB, von Herrath MG (1999) Insulin in oral immune "tolerance": a one-amino acid change in the B chain makes the difference. *J Immunol* 163: 1833–1838

87 von Herrath MG, Dyrberg T, Oldstone MB (1996) Oral insulin treatment suppresses virus-induced antigen-specific destruction of beta cells and prevents autoimmune diabetes in transgenic mice. *J Clin Invest* 98: 1324–1331

88 Casares S, Hurtado A, McEvoy RC, Sarukhan A, von Boehmer H, Brumeanu TD (2002)

Down-regulation of diabetogenic CD4+ T cells by a soluble dimeric peptide-MHC class II chimera. *Nat Immunol* 3: 383–391

89 Masteller EL, Warner MR, Ferlin W, Judkowski V, Wilson D, Glaichenhaus N, Bluestone JA (2003) Peptide-MHC class II dimers as therapeutics to modulate antigen-specific T cell responses in autoimmune diabetes. *J Immunol* 171: 5587–5595

90 Stockinger B, Barthlott T, Kassiotis G (2001) T cell regulation: a special job or everyone's responsibility? *Nat Immunol* 2: 757–758

91 Barthlott T, Kassiotis G, Stockinger B (2003) T cell regulation as a side effect of homeostasis and competition. *J Exp Med* 197: 451–460

92 King C, Ilic A, Koelsch K, Sarvetnick N (2004) Homeostatic expansion of T cells during immune insufficiency generates autoimmunity. *Cell* 117: 265–277

93 Berzins SP, Venanzi ES, Benoist C, Mathis D (2003) T-cell compartments of prediabetic NOD mice. *Diabetes* 52: 327–334

94 Lenz DC, Kurz SK, Lemmens E, Schoenberger SP, Sprent J, Oldstone MBA, Homann D (2004) IL-7 regulates basal homeostatic proliferation of antiviral CD4+ T cell memory. *Proc Natl Acad Sci USA* 101: 9357–9362

95 Alpan O, Bachelder E, Isil E, Arnheiter H, Matzinger P (2004) 'Educated' dendritic cells act as messengers from memory to naive T helper cells. *Nat Immunol* 5: 615–622

96 von Herrath M, Homann D (2004) Tolerance tag team. *Nat Med* 10: 585–587

97 von Herrath M, Homann D (2003) Introducing baselines for therapeutic use of regulatory T cells and cytokines in autoimmunity. *Trends Immunol* 24: 540–545

The role of interleukin-10 in regulatory-T-cell suppression: reconciling the discrepancies

Robert N. Barker and Frank J. Ward

Department of Medicine and Therapeutics, Institute of Medical Sciences, University of Aberdeen, Foresterhill, Aberdeen AB25 2ZD, UK

Introduction

There is compelling evidence from animal models that various forms of $CD4^+$ regulatory T (T_R) cell play important roles in immunological tolerance and in the control of immune pathology [1–3]. It has now been established that T_R cells of similar types also exist in humans [4–8], and the manipulation of their activity *in vivo* therefore holds out the prospect of safe, effective treatments for many diseases associated with either pathogenic immune responses [9–11], or a failure of specific immunity [12–15]. However, before treatments of this type can become a reality, it will be necessary to answer fundamental questions about T_R cells. These issues include determining T_R-cell specificities and activation requirements, and, the subject of this review, resolving discrepancies in the evidence as to their means of suppression.

In most experimental systems, the mechanisms by which T_R cells inhibit responses are antigen non-specific, even where their activation is specific. There is evidence for two major modes of suppression, the secretion of inhibitory cytokines, and ill-defined mechanisms dependent on direct physical contact between T_R cells and their targets. Controversy surrounds the question as to which of these major mechanisms is the more important for T_R function. The cytokines to which regulatory properties are most often ascribed are interleukin (IL)-10 and transforming growth factor (TGF)-β, and this chapter will focus on the conflicting views surrounding the roles of IL-10 in T_R-mediated suppression.

Properties of IL-10

IL-10 is a pleiotropic cytokine with receptors on various immune and non-immune cells [16]. Despite the current interest in IL-10 as a T_R-cell mediator, it is important to recognize that IL-10 has both immunoregulatory and immunostimulatory properties. Indeed, soon after its initial identification as cytokine-synthesis inhibitory factor (CSIF), a murine Th2 cell product that suppressed Th1 cytokine responses [17],

Regulatory T Cells in Inflammation, edited by Leonie S. Taams, Arne N. Akbar and Marca H.M. Wauben

IL-10 was also described as a B-cell-derived thymocyte growth-promoting factor [18].

The immunostimulatory effects of IL-10 are now known to be diverse [16]. They include up-regulation of MHC class II expression by resting B cells [19]; the enhancement of the proliferation of activated B cells and their differentiation to produce antibody [20], particularly IgG4 [21], the augmentation of thymocyte proliferation [18], co-stimulation of $CD8^+$ T-cell expansion and cytotoxicity [22, 23] and mast cell stimulation [24].

In addition, *via* effects on other cell types, IL-10 is a potent anti-inflammatory agent [16]. In particular, IL-10 strongly inhibits the activation of monocytes [25], macrophages [26] and dendritic cells [27], suppressing the secretion of inflammatory mediators and reducing the effectiveness of these cells as antigen-presenting cells (APCs) for $CD4^+$ Th cell responses. IL-10 also directly inhibits secretion of inflammatory cytokines by Th1 cells [17, 28], and suppresses inflammatory functions of granulocytes [29], natural killer (NK) cells and endothelial cells [30].

These divergent properties of IL-10 help to explain effects *in vivo* that are at odds with its commonly perceived role as a suppressive cytokine. For example, although IL-10 does reduce pathology in some models of inflammatory disease, such as rheumatoid arthritis [31], inflammatory bowel disease [32], uveitis [33], hepatitis [34] and brain injury [35], administration of IL-10 can exacerbate particular examples of graft-*versus*-host disease (GVHD) [36] and of diabetes in non-obese diabetic (NOD) mice [37].

The cellular sources of IL-10 are also varied [16]. Cells that produce IL-10 include not only certain $CD4^+$ cell subsets [4, 17], but also $CD8^+$ cells with a Th2-like cytokine-secretion pattern [38], monocyte/macrophages [25], B cells [18, 19], particularly B1 B cells that express CD5 [39], and epithelial cells. [40]. Regarding $CD4^+$ T-cell sources, as mentioned above, IL-10 was initially recognized as a murine Th2 cytokine [4], but in humans it is not a distinguishing feature of Th2 responses, and can also be produced by Th1 and Th0 cells [41].

Following the recent resurgence of interest in active cell-mediated immunoregulation, and the description of various $CD4^+$ T_R subsets [1, 2], there is now considerable focus on IL-10 as a major mediator of T_R activity. However, there is a divergence of opinion as to how important cytokines such as IL-10 are to T_R function. This lack of consensus reflects a number of factors, including the discovery of different T_R forms, variation between models and the difficulties in relating *in vitro* to *in vivo* results.

Major forms of T_R cell

The current, widely accepted paradigm is that there are two major forms of T_R cell, defined as "induced" and "natural" [1, 2, 42]. Antigen-induced T_R cells are held to be generated in the periphery following chronic exposure to antigen, or naïve T-cell activation under tolerogenic conditions, and to be part of a mechanism to limit pathology resulting from excessive or prolonged responses to foreign agents. In contrast, natural T_R cells are believed to differentiate into suppressor cells during thymic development and to play a role in self-tolerance. The classification of T_R cells into these two major forms is relevant to the role of IL-10 in suppression since there is wide acceptance that induced T_R cell activity is dependent on secreted cytokine [1–3], whereas in many experiments natural T_R inhibition is apparently mediated *via* direct cell–cell contact [2, 42]. However, despite the popularity of the classification, doubt remains as to whether this distinction between the T_R forms is complete [43], particularly in humans, where studies have inevitably been more limited than in mice.

Antigen-induced T_R

Mosmann and Coffman [44] originally classified CD4$^+$ Th cell clones into different functional types on the basis of the cytokines that they secrete, and it became clear that at least some forms of immunoregulation could be attributed to mutual antagonism between these helper subpopulations. Initially, attention focused on Th1 and Th2 cells, which produce the respective signature cytokines interferon (IFN)-γ and IL-4 [44]. For example, many models of autoimmune disease are driven by helper responses that are strongly dominated by the Th1 subset, and inducing a corresponding Th2 bias can prevent or ameliorate the pathology [45, 46]. However, further, T_R cell subpopulations with different patterns of cytokine production, and important roles in immunoregulation and tolerance, have now been defined.

The first such subset identified with the potential for immunoregulation *via* secreted cytokine was termed Th3 [47], and is important for immunity at mucosal sites. The principal response of these cells is to secrete TGF-β, which acts as a switch factor for B cells to make IgA antibodies to protect the mucosae, but this cytokine also has potent immunosuppressive properties. Thus, Th3 cells can inhibit Th1 inflammatory responses, and mediate some forms of oral tolerance. This ability has been exploited to prevent or treat experimental models of autoimmune pathology [47, 48].

The next T_R phenotype to be described were CD4$^+$ T cells that secrete predominantly IL-10 [1]. This subset, defined as T_R1 by Groux et al. [4], were originally produced by repeated stimulation of murine or human T cells *in vitro* in the presence of high levels of IL-10, and it is now known that IFN-α enhances this differentiation [49]. Such T_R cells were demonstrated to have potent immunosuppressive effects *in vitro*, inhibiting Th1 cells, and to possess anti-inflammatory properties *in vivo* [1, 4]. T_R1 cells are anergic, failing to proliferate in response to antigen, but can be expanded non-specifically with cytokines such as IL-2 and IL-15 [50].

Although the T_R1 response is characterized by IL-10 secretion in the absence of IL-4, there is some production of TGF-β, IFN-γ and IL-5 [1, 4]. Similar cells can be produced by other manoeuvres *in vitro*, for example by biasing naïve T cells towards an IL-10-secreting phenotype by stimulation in the presence of the immunosuppressive compounds dexamethasone and vitamin D_3 [51] or antibody to CD46 [52]. The question as to whether T_R1 cells can arise *in vivo* has been addressed, with the demonstration in mouse models that chronic T-cell receptor (TCR) stimulation can generate anergic, IL-10-secreting T cells. Thus, specific T cells from double transgenic mice expressing TCR and corresponding viral antigen [53], or from mice repeatedly stimulated with bacterial superantigen [54] or peptide [55], can adopt an anergic, IL-10-secreting phenotype. In humans, there is also evidence that T_R1 cells can arise *in vivo*, since the T-cell response to Epstein-Barr virus (EBV) latent membrane protein 1 is strongly biased towards IL-10 secretion, presumably reflecting priming in the presence of BCRF1, the IL-10 homologue expressed by the virus [12]. It has also become clear that the type of APC can determine whether T cells are biased towards IL-10 production, since immature dendritic cells can induce such responses *in vitro* and *in vivo*, and many of the manoeuvres known to generate T_R1 cells first condition APCs, which then go on to deviate the T-cell pool towards a regulatory phenotype [56]. Evidence is growing that the differentiation of anergic or regulatory T cells from naïve precursors is directed by dendritic cells that either inherently secrete IL-10, or are stimulated to produce the cytokine in response to microbial products [33, 56]. Furthermore, the propensity of dendritic cells in the airways to produce IL-10, which can result from interactions with T cells *via* an inducible co-stimulator [57], helps to explain the induction of T_R cells at these mucosal surfaces, and the maintenance of tolerance in the respiratory tract [58, 59].

Although IL-10 is clearly a major mediator of antigen-induced T_R-cell suppressive activity *in vitro* and *in vivo*, there are a number of apparent anomalies to reconcile. First, as mentioned earlier, the suppressive roles of T_R1 cells are not necessarily replicated by administration of IL-10 in a number of models of immune pathology, and indeed can exacerbate diseases such as GVHD [36]. The view that T cells secreting IL-10 are necessarily suppressive is also challenged by the observation that anti-tumour immunity can be mediated by such a population [60]. Furthermore, when IL-10-secreting T cells do inhibit, it should not be assumed that the effect is due solely to the cytokine response. Thus, inhibition in a TCR transgenic model in which IL-10-secreting T_R cells are induced by chronic peptide stimulation, the *in vivo* suppression mediated by these cells was IL-10 dependent, whereas their *in vitro* effects were not [61]. Indeed, IL-10-secreting T_R cells induced either by such stimulation *in vivo*, or by priming in the presence of dexamethasone and vitamin D_3 *in vitro*, can inhibit responses *in vitro* exclusively *via* contact-dependent mechanisms [62]. This observation illustrates that the properties of T_R1-like cells can depend on the means by which they are obtained, as summarized in Table 1. These apparent discrepancies highlight the need to identify the precise sites of T_R activity and the cellular targets for IL-10. For example, if T_R IL-10 predominantly mediates its suppressive properties *via* an effect on APCs in the microenvironment of secondary lymphoid organs, then this may not be replicated by systemic administration or the cytokine *in vivo*, or by tissue culture of single-cell suspensions *in vitro*. Another factor contributing to the difficulties in establishing the mechanism of action of T_R-cell populations is the growing evidence of inter-relationships between cytokine-dependent and -independent suppressors. Thus, regulatory cytokines, including IL-10 and TGF-β, can help *in vitro* [63, 64] and *in vivo* [65] to recruit "natural" $CD25^+$ T_R cells that inhibit *via* cell–cell contact; and *in vitro* suppression by the T_R population induced with chronic superantigen exposure *in vivo* [54] switches from contact to IL-10/TGF-β dependence if the cells are generated in *rag-2*$^{-/-}$ mice lacking "natural" $CD25^+$ T_R cells [66].

Progress is being made in establishing the roles of cells with a T_R1 phenotype in human infectious, neoplastic, autoimmune and atopic diseases. As mentioned above, IL-10-secreting T cells with a T_R1 phenotype dominate the response to particular EBV latent proteins [12], and are likely to play an important role in immune evasion, the establishment of persistent infection and the survival of EBV-related tumours [14]. Cells of a similar phenotype specific for core protein are detectable in patients with hepatitis C virus, and the suppressive effects of the IL-10 that they secrete may contribute both to chronicity of infection and to protection from immune pathology that leads to cirrhosis [13, 67]. Furthermore, IL-10 and TGF-β responses are also associated with hyporesponsiveness to antigens from the filarial nematode *Onchocerca volvulus* [68]. T_R cells, such as the mixed IL-10-secreting and $CD25^+$ populations that infiltrate neoplastic lymph nodes in Hodgkin's lymphoma [14], have also been identified in cancer patients and are thought to represent a novel means of tumour immune escape. Conversely, evidence is accumulating that IL-10-secreting regulatory populations may play a protective role in human autoimmune diseases. For example, T_R1-type cells specific for naturally processed epitopes on the dominant red-cell autoantigens are present in peripheral blood and spleen of autoimmune haemolytic anaemia (AIHA) patients, and these cells are capable of inhibiting the Th1 effector responses *in vitro* by secretion of IL-10 [9]. This T_R activity correlates with periods of remission, consistent with the view that the autoimmune disease perpetuates due to an imbalance between pathogenic and regulatory responses [9]. Similarly, in Goodpasture's disease, resolution of autoimmune glomerular inflammation is coincident with a dramatic deviation of the $CD4^+$ T-cell response against the dominant antigen, α3(IV)NC1, from Th1 to IL-10 [10]. There are also reports that IL-10 T_R responses can prevent atopy, since patients de-sensitized to bee venom show an IL-10 response to the allergen [69], and individuals exposed to high levels of the cat allergen FelD1 develop an IL-10 response, associated with a reduced risk of cat allergy [70]. One noteworthy point regarding the relevance of secreted cytokine to induced T_R function is that, in all these examples of human disease, the regulatory activity appears to be largely IL-10-dependent. How-

Table 1 - Properties of TR1 cell clones

Induction method	Host	Specificity	Phenotype	Cytokine-mediated suppression?	Contact-dependent suppression?
a) Artificially induced clones					
Generation of OVA TCR transgenic T_R1 from naïve T cells with IL-10/IL-4 in presence of IL-2, anti-CD3 and irradiated APCs [4].	DO11-10 mouse (H-2^d) (*in vitro*)	OVA peptide 323–339	IL-10^{++} IFN-γ^+ IL-5^+ IL-4^- IL-2^- $CD25^+$ CTLA-4^+ TGF-β^{low}	Yes	Yes
Repeated nasal dosing of Tg4 TCR transgenic mouse, specific for MBP peptide Ac1-9, with high-affinity Ac1-9[4Y] analogue [61].	Tg4 mouse B10.PL (H-2^u)	MBP peptide Ac1-9	IL-10^{++} IL-2^- IL-5^- IL-4^- TGF-β^- $CD38^+$ $CD25^+$ CTLA-4^+ ICOS-1^+	*In vivo*, yes *In vitro*, no	Yes
Immunization with pathogen *Bordetella pertussis* followed by cloning of specific T_R in the presence of antigen, IL-2, IL-10 (PRN only) and feeder cells [59].	Balb/c (H-2^d)	Filamentous haemagglutinin, Pertactin (PRN)	IL-10^{++} IFN-$\gamma^{+/-}$ IL-5^+ IL-4^- IL-2^- TGF-β^+ $CCR5^+$	Yes	No
Conditioning of naïve human $CD4^+$ T cells with IL-10 and/or IFN-α in presence of anti-CD3 stimulation, IL-2/IL-15, and murine L cells expressing hCD32, hCD58, and hCD80 [103].	Human naïve $CD4^+$ T cells (*in vitro*)	Non-specific	IL-10^{++} IFN-$\gamma^{+/-}$ IL-$2^{-/low}$	Yes	Yes
Generation of OVA-specific T_R1 T cells from naïve $CD4^+$ T cells incubated with OVA323-339, and APCs in presence of vitamin D_3 (40 nM) and dexamethasone (10 nM) [51].	DO11-10	OVA peptide 323-339	IL-10^{+++} IFN-γ^- IL-5^- IL-4^- IL-10 secreting T_R proliferate strongly in response to $OVA_{323\text{-}339}$	Yes	Yes

Table 1 (continued)

Induction method	**Host**	**Specificity**	**Phenotype**	**Cytokine-mediated suppression?**	**Contact-dependent suppression?**
$CD4^+CD25^-$ cells converted to IL-10-secreting T_R upon incubation with equal numbers of $CD4^+CD25^+$ T cells for 48 h in presence of anti-CD3/CD28 antibodies and allogeneic dendrictic cells [95].	Healthy volunteer donors (*in vitro*)	Non-specific	$IL\text{-}10^{++}$ $IFN\text{-}\gamma^{low}$ $IL\text{-}5^-$ $IL\text{-}4^-$ $IL\text{-}2^-$	Yes	No
b) Clones expanded ex vivo					
Derivation method					
T_R cell lines and clones derived from PBMC of individuals exposed to the filaria *Onchocerca volvulus* (OV) by expansion with OV antigen, IL-2 and APCs [68].	Human OV-infected individuals	Saline-soluble extracts from OV	$IL\text{-}10^{++}$ $IL\text{-}5^{low}$ $IL\text{-}2^-$ $TGF\text{-}\beta^{++}$	Yes	Unknown but factor additional to IL-10/TGF-β contributes to suppression
PBMC from patients infected with hepatitis C virus (HCV) incubated with HCV core fragment derived peptides for 7 days followed by expansion phase in presence of IL-2, antigen and APCs. Clones derived by limiting dilution [67].	Human HCV-infected individuals	HCV-derived dominant core peptides	$IL\text{-}10^{++}$ $IFN\text{-}\gamma^+$ $IL\text{-}5^{low}$ $IL\text{-}4^-$ Most $TGF\text{-}\beta^-$	Yes	Unknown
Expansion of T_R1 T cells from PBMC of patients with AIHA that secrete IL-10 in response to peptides representing the RhD red-blood-cell autoantigen *ex vivo* with IL-15 and T-cell expander beads ([9] and R.N. Barker, F.J. Ward, unpublished observations)	Human AIHA patients	RhD peptides	$IL\text{-}10^{++}$ $IFN\text{-}\gamma^+$ $IL\text{-}4^-$ $CD25^+$ $FoxP3^+$ $CTLA\text{-}4^+$	Yes	Additional factor contributes to suppression

ever, in the different disease categories, the effects of IL-10 may not be mediated *via* the same mechanisms: suppression of cytotoxic responses in viral infection and neoplasia [12–14, 67]; inhibition of Th1 driven antibody production or inflammation in autoimmune disease [9,10]; and contributing to an isotype switch away from IgE in allergy [69, 70]. Overall, these studies encourage the view that the ability to manipulate IL-10 T_R responses to antigens of pathological relevance will open the door to novel forms of therapy for a wide range of human diseases.

Natural T_R cells

The second major form of T_R cells are those that constitutively express high levels of the activation marker CD25, and which are believed to arise as a "natural" suppressive population in the thymus [1, 2, 71–73]. They also constitutively express glucocorticoid-induced tumour necrosis factor (TNF) receptor family-related gene (GITR) [74, 75], CTLA-4 [72] and other activation markers, but it has not proved possible to identify any unique surface molecules on these cells that can distinguish them from $CD25^+$-activated T cells with no regulatory potential. However, the transcription factor Foxp3 has recently been shown to be characteristic of this suppressor population, and to lead to differentiation into the $CD25^+$ T_R phenotype [76]. Unlike antigen-induced regulatory populations, one of the major limitations in studying $CD25^+$ T_R cells has been ignorance of their specificities, and this difficulty has hampered efforts to elucidate their mode of inhibition.

Whereas it is clear that inhibitory cytokines such as IL-10 can be important in the regulation mediated by antigen-induced T_R cells, such as the T_R 1 subset [1], there are major discrepancies in the evidence as to their relevance to suppression by $CD25^+$ T_R cells [77]. There is no single explanation for these apparent anomalies, and a number of factors that may be responsible will be considered below. There may be alternative modes of suppression by $CD25^+$ T_R cells when comparing *in vivo* and *in vitro* experimental systems or different models of pathology *in vivo*; other cell types may replace T_R IL-10 production; the interpretation of data can be complicated by inter-relationships between induced and natural T_R cells; the $CD25^+$ subset may not be homogenous; and it is not clear that the human and rodent T_R subpopulations studied are precise equivalents.

In vivo versus in vitro

The reported differences between the mode of action of $CD25^+$ T_R cells *in vitro* and *in vivo* are summarized in Table 2. Initial work *in vitro* demonstrated that the murine $CD25^+$ T_R population suppressed the proliferation of naïve T cells by inhibiting IL-2 production, and that this effect was independent of T_R cytokine secretion [71]. Instead, suppression was ascribed to undefined mechanisms dependent on physical contact between effector-cell and T_R-cell populations. Despite one report that TGF-β bound to the $CD25^+$ T_R cell membrane, rather than the secreted form, mediated the inhibition [78], the lack of dependence on any of the recognized T_R cytokines *in vitro* was formally demonstrated by the effectiveness of suppression by $CD25^+$ T_R cells from genetically manipulated mice that cannot produce IL-4, IL-10 or TGF-β [71, 79].

In contrast, suppression *in vivo* is cytokine dependent in one of the first examples of disease shown to be controlled by T_R activity, a murine model of colitis [80–82]. The transfer of predominantly naïve $CD4^+CD45RB^{high}$ T cells from normal mice into lymphopenic recipients causes a Th1-mediated colitis, which can be prevented by co-transfer of regulatory cells within the $CD4^+CD45RB^{low}$ fraction [80]. These T_R cells are predominantly $CD25^+$, and their suppression *in vivo* is mediated by secreted cytokine, since inhibition can be blocked by antibody to IL-10 receptor or TGF-β, and $CD4^+CD45RB^{low}$ cells from IL-$10^{-/-}$ mice lack regulatory activity [81–83]. As noted above, recombinant IL-10 is also protective against colitis [32]. Furthermore, in another example of experimental colitis, where Th1-mediated inflammation is induced by trinitrobenzene sulphonic acid, protection transferred by T_R cells from tolerized mice is mediated primarily by TGF-β *in vivo*, with IL-10 necessary as a secondary factor that facilitates TGF-β production [84]. The control by $CD25^+$ T_R cells of a superantigen-induced cytokine burst *in vivo* is also dependent on the ability to secrete IL-10 [85].

The notion that suppression by $CD25^+$ T_R cells is cytokine-independent *in vitro*, but may be cytokine-dependent under some conditions *in vivo*, has received further support. In studies of immune evasion by Leishmania major, Belkaid et al. [86] identified $CD25^+$ T cells in skin lesions from persistently infected mice that secreted IL-10 in response to antigen, but which suppressed naïve T-cell proliferation *in vitro* in a cytokine-independent manner. Transfer experiments revealed that these T_R cells also mediated immune suppression *in vivo*, and resulted in persistence of *Leishmania major* infection. However, a comparison of the effects of transferring $CD25^+$ cells derived from IL-$10^{-/-}$ and wild-type mice revealed that T_R secretion of IL-10 did not alter the response to early infection, but that this cytokine did mediate suppression later in the disease [86].

One inference from the discrepancies between these studies is that *in vitro* experiments alone should be interpreted with caution. Clearly, the complex architecture of secondary lymphoid organs is destroyed in tissue culture of cell suspensions, and so subtle effects of cytokines such as IL-10 on APCs seen *in vivo* may not be recapitulated *in vitro*. However, the situation is not quite so simple, since IL-10 dependency of $CD25^+$ T_R cells *in vivo* is not consistent between different experimental models.

Table 2 - Models in vitro versus in vivo *that test the relevance of IL-10/TGF-β for $CD4^+CD25^+$ T_R function*

Disease model	T_R cell (source)	Target of suppression	Defining experiments	Cytokine required for suppression?	$CD4^+CD25^+$ phenotype
Multisystemic autoimmune disease mediated by $CD4^+$ T cells following neonate thymectomy at Day3 (d3Tx). *In vitro* model [71].	$CD4^+CD25^+$ (Balb/c) (C57BL IL-10 KO) (Balb/c IL-4 KO)	Balb/c $CD4^+CD25^-$ T cells *In vitro*	1. Separation of $CD4^+CD25^+$ T_R from $CD4^+CD25^-$ T cells in transwells prevents suppressor function. 2. Anti-IL-10 blockade did not inhibit suppressor function. 3. $CD4^+CD25^+$ T_R derived from IL-10 or IL-4 KO mice retained a capacity for suppression.	*In vitro*, no	$GITR^+$ $CTLA\text{-}4^+$ $CD45RB^{low}$ $CD69^+$ $CD62L^{low}$ $IL\text{-}10^+$ $IFN\text{-}\gamma^-$ $IL\text{-}2^-$ $IL\text{-}4^-$
Intestinal inflammation triggered upon infection with the bacterium *Helicobacter hepaticus*. *In vivo* model [88].	$CD4^+CD45RB^{low}$ $CD25^+$ (129SvEv)	$129SvEvRAG^{-/-}$ mice reconstituted with $CD4^+CD45RB^{lo}CD25^-$ or $CD4^+CD45RB^{high}$ T cells and infected with *H. hepaticus*	1. Reconstituted $129SvEvRAG^{-/-}$ mice treated during infection with anti-IL-10 receptor or anti-TGF-β antibodies abolished the regulatory potential of $CD4^+CD45RB^{low}CD25^+$ T_R.	*In vivo*, yes	$IL\text{-}10^{++}$ $TGF\text{-}\beta^+$
Model of allogeneic skin graft rejection. CBA ($H\text{-}2^k$) mice received grafts from C57BL/10 ($H\text{-}2^b$) mice. *In vivo* model [104].	$CD4^+CD25^+$ (CBA mice pre-treated with anti-CD4 & B10 blood)	$CD4^+CD45RB^{high}$ T cells isolated from naïve CBA mice & transfused into T-cell-deficient CBA mice	1. Antibody blockade of IL-10 receptor abolished suppressive function of $CD4^+CD25^+$ T_R. 2. Blockade of CTLA-4 also abolished T_R function.	*In vivo*, yes	$IL\text{-}10^+$ $CTLA\text{-}4^+$ $CD122^+$

Table 2 (continued)

Disease model	**T_R cell (source)**	**Target of suppression**	**Defining experiments**	**Cytokine required for suppression?**	**$CD4^+CD25^+$ phenotype**
Induction of EAE disease with high-affinity Ac1-9[4Y] peptide in transgenic mouse Tg4 expressing TCR specific for wild type MBP peptide Ac1-9. T_R induced by repeated nasal dosing with Ac1-9[4Y]. *In vivo versus in vitro* discrepancy of T_R function [61].	$CD4^+$ (Tg4-derived peptide induced tolerant cells (PItol))	*In vitro*: naïve Tg4 $CD4^+$ cells stimulated with Ac1-9 *In vivo*: CFSE labelled splenocytes from naïve Tg4 mice transferred i.v. to Tg4 naïve or PItol recipients	1. *In vitro*: IL-10 receptor and TGF-β antibody blockade did not affect PItol mediated suppression. 2. Separation of naïve from PItol $CD4^+$ T cells (transwells) abolished suppression. 3. *In vivo*: anti-IL-10 receptor antibody ransferred to host *in vivo* annulled PItol mediated tolerance.	*In vitro*, no *In vivo*, yes	$IL\text{-}10^{++}$ $CD25^+$ $CD38^+$ $CTLA\text{-}4^{++}$ $ICOS\text{-}1^+$ $IL\text{-}2^-$ $IL\text{-}4^-$ $IL\text{-}5^-$ $TGF\text{-}\beta^-$ $IFN\text{-}\gamma^{low}$
Human donor derived $CD25^-$ and $CD25^+$ thymocytes used in PBMC mixing experiments *in vitro* [105].	$CD4^+CD25^+$ thymocytes	$CD4^+CD25^-$-derived responses to allogeneic PBMC	1. Blockade of CTLA-4 and membrane-bound TGF-β abolished T_R activity. 2. Separation of T_R from $CD4^+CD25^-$ in transwells prevents suppression mediated by soluble TGF-β.	*In vitro*, no	$IL\text{-}2^-$ $IL\text{-}4^-$ $IL\text{-}10^+$ $IFN\text{-}\gamma^-$ $CTLA\text{-}4^{++}$
Human-donor PBMC-derived $CD25^-$ and $CD25^+$ T cell clones mixed with $CD4^+$ T cells (*in vitro* model) [6].	$CD4^+CD25^+$ T cell clones	$CD4^+$ T cells activated with anti-CD3 antibody	1. IL-10-suppressive capacity retained in $CD25^-$ $CD4^+$ T cell population. 2. Partial reversal of T_R suppression by TGF-β-neutralizing antibody.	*In vitro*, partial (TGF-β only)	$CTLA\text{-}4^{++}$ $GITR^+$ $IL\text{-}2^-$ $IL\text{-}4^-$ $IL\text{-}10^-$ $IFN\text{-}\gamma^-$ $TGF\text{-}\beta^+$

Table 2 (continued)

Disease model	**T_R cell (source)**	**Target of suppression**	**Defining experiments**	**Cytokine required for suppression?**	**$CD4^+CD25^+$ phenotype**
$CD4^+$ $CD25^-$ $CD25^+$ T-cell mixing experiments in the presence or absence of recombinant latency-associated peptide for TGF-β1 (rLAP) (*in vitro* and *in vivo*) [106].	$CD4^+CD25^+$ (Balb/c) (TGF-$\beta 1^-$ deficient mice) $CD4^+CD25^+$ human	*In vitro*: murine or human $CD4^+CD25^-$ T cells stimulated with anti-CD3 antibody *In vivo*: $CD4^+CD45RB^{high}$ (Balb/c)	1. Mouse *in vitro*: incubation of rLAP reversed suppressive potential of $CD4^+CD25^+$ T_R in a dose dependent manner. 2. Mouse *in vitro*: $CD4^+CD25^+$ T_R from TGF-β1-deficient mice inhibit T-cell responses suggesting that TGF-β1 is not required. 3. Mouse *in vivo*: $CD4^+CD25^+$ T_R from TGF-$\beta 1^-$-deficient mice not able to protect $CD4^+CD45RB^{low}$-mediated colitis in SCID model whereas wild type $CD4^+CD25^+$ T_R did suppress disease. 4. Human *in vitro*: rLAP inhibited $CD25^+$ T_R function.	*In vitro*, partial *In vivo*, yes	$CD103^+$ (α_E integrin) LAP^{++}

KO, knockout; EAE, experimented autoimmune encephalomyelitis.

Different mediators of pathology

As already described, *in vivo* T_R suppression in experimentally transferred colitis appears to be IL-10-dependent [81, 82]. However, in a transfer model of a different disease, Suri-Payer and Cantor [87] demonstrated that $CD25^+$ T cells from wild-type and IL-10-deficient mice were equally capable of suppressing autoimmune gastritis *in vivo*. In the same report, it was confirmed that T_R cell cytokine secretion was required for maximal protection from transferred colitis [87]. To explain this discrepancy, it could be argued that pathology in the colitis models is more robust than autoimmune inflammation, being driven by potent innate inflammatory stimuli from enteric bacteria. Such potent inflammation may only be controlled by T_R cells that can deploy multiple suppressive mechanisms, including IL-10 secretion, which could be redundant in less-aggressive models. This explanation is supported by the finding that $CD25^+$ T_R suppression of T-independent colitis triggered by *Helicobacter hepaticus* is dependent on IL-10 [88], whereas the secretion of this cytokine is not is necessary to control disease in the absence of this potent inflammatory stimulus [89]. In addition to innate stimuli, adaptive immune factors may determine the dependence of $CD25^+$ T_R cells on IL-10, including the characteristics of the inflammatory T cells that are being suppressed. One hypothesis is that autoimmune diseases such as Suri-Payer's gastritis model [87] are driven by fewer T-cell clones, of lower avidity, than those that recognize complex foreign bacterial antigens, and that these "weaker" responses are more susceptible to T_R cells lacking secreted cytokine from their suppressive armoury. Whereas this possibility has not been directly addressed, there is clear evidence that the antigen experience of T cells can determine their susceptibility to regulation in the absence of T_R IL-10. Thus, the transfer models of colitis reveal that $CD25^+$ T_R cells can inhibit inflammation driven by T cells with either naïve or memory phenotypes, but T_R IL-10 secretion is absolutely required for suppression only in the latter case [90]. Although it is clear that different pathological processes do vary in their sensitivity to T_R IL-10, in some cases the factors that determine the mode of suppression are very subtle, and include the contribution to cytokine secretion by other cell types.

Replacement of T_R IL-10 production by other cells

The requirement for T_R IL-10 in suppression *in vivo* can be influenced by the presence or absence of other cell types. For example, in transfer models of immune pathology, $CD25^+$ T_R cells are effective in protecting $RAG2^{-/-}$ hosts from wasting disease and in controlling peripheral lymphocyte expansion only if they are capable of secreting IL-10 [91], whereas in $CD3\varepsilon^{-/-}$ mice regulation of wasting, but not lymphocyte numbers, is dependent of IL-10 production by the T_R cells [92]. In addition to reinforcing the conclusion that different manifestations of pathology can vary in their susceptibility to T_R cytokine, these results demonstrate that the cellular milieu in which T_R cells operate may determine which suppressive mechanisms are relevant. Since B-cells are absent in RAG2$^{-/-}$ hosts, but present in CD3ε$^{-/-}$ mice, one straightforward interpretation of the data is that B-cell IL-10 can restore the suppressive ability of T_R cells that are deficient in this cytokine. The notion that B-cell cytokine can play a key role in controlling immune pathology is supported by the finding that recovery from experimental allergic encephalomyelitis is dependent on autoreactive B cells secreting IL-10 [93]. Further, in a study of T_R responses in grass-pollen allergy, atopic individuals responding to specific allergen immunotherapy had higher levels of IL-10 in local mucosa than non-responders, provided by both T cells and macrophages [94]. Taken together, these results raise the possibility that T_R cells can synergize with other IL-10-secreting cells, which complement T_R activity and, under some circumstances, substitute for T_R IL-10.

Inter-relationships between different T_R cell subsets

Another explanation for the lack of consensus regarding the mode of suppression by CD25$^+$ T_R cells is that this natural regulatory population co-operates with induced T_R subsets that produce TGF-β and/or IL-10. Some such inter-relationships have been reported, as already mentioned. Thus, human CD4$^+$CD25$^+$ T_R cells can induce, *via* contact-dependent mechanisms *in vitro*, CD4$^+$CD25$^-$ cells to attain an anergic, IL-10- [95] or TGF-β- [63] secreting T_R phenotype. Conversely, co-stimulation *in vitro* of murine naïve CD4$^+$CD25$^-$ T cells *via* the TCR and with TGF-β induces *Foxp3* and converts them into CD25$^+$ T_R cells capable of suppression both *in vitro* and *in vivo* [96]. A similar effect of this cytokine *in vivo* was also demonstrated by the ability of a transient pulse of TGF-β in the pancreatic islets during the priming of autoimmune murine diabetes to expand Foxp3-expressing CD4$^+$CD25$^+$ T_R cells in the target tissue and protect against disease [65]. Some of the inter-relationships of induction and expansion between different T_R forms may be complex and cyclical [64], and presumably reflect the need to generate from a limited T_R pool long-lasting and effective suppression of potentially damaging responses. A more radical view would be that CD25$^+$ and induced T_R are different forms of the same cell type, with the former state representing a "memory" or resting T_R phenotype [56]. Although there is little direct evidence to sustain this argument, it has recently been demonstrated that the phenotype of cloned IL-10 secreting autoreactive T_R cells from a patient with AIHA can deviate between two forms (R.N. Barker and F.J. Ward, unpublished observations). The clones express high levels of CD25 when expanding non-specifically, but this natural T_R marker is down-regulated, and the cells secrete IL-10, after specific stimulation with autoantigenic peptide. Furthermore, Foxp3 expression is increased after stimulation with the cognate autoantigen, but not upon non-specific expansion. These results raise the possibility that apparently distinct T_R1 and CD25$^+$ cells may be interconvertible, and that this property may not be revealed in the many studies of suppression that have been carried out without knowledge of T_R specificity. Further, circumstantial evidence for a relationship between T_R1 and $CD25^+$ T_R cells in humans comes from work on Goodpasture's disease. Different studies have shown that IL-10-secreting T cells specific for the glomerular antigen are virtually undetectable in patients with active disease, but dominate the response after resolution of inflammation [10], and that this clinical change is associated with recovery of $CD25^+$ T_R cell activity in peripheral blood [97]. This parallelism suggests a close relationship between T_R1 and $CD25^+$ T_R cells in human autoimmune disease.

Heterogeneity of the natural T_R-cell population

In some cases, discrepancies in the means of suppression by $CD25^+$ T_R cells may reflect, not inter-relationships with induced subsets, but heterogeneity within the natural population [98]. Evidence is emerging that differences in integrin expression can identify functionally distinct subsets that exist within the $CD4^+CD25^+$ population. Thus, in mice, the integrin $\alpha_E\beta_7$ (CD103) is a marker for a potently suppressive $CD25^+$ T_R subpopulation [99]. Compared to the $CD103^-CD25^+$ T_R subset, such $CD103^+CD25^+$ cells are highly enriched in gut-associated lymphoid tissue, secrete IL-10, and preferentially control inflammatory bowel disease, although both phenotypes inhibit *in vitro* through contact-dependent mechanisms [100]. Human $CD25^+$ T_R cells are also heterogeneous with respect to integrin expression, with the $\alpha_4\beta_1{}^+$ and $\alpha_4\beta_7{}^+$ $CD4^+CD25^+$ subpopulations respectively able to induce TGF-β- or IL-10-secreting T_R cells [98].

Murine *versus* human $CD25^+$ T_R cells

One potential source of confusion when studying T_R responses in such unrelated species as mice and humans is that the phenotypes and functions of particular subsets may not be precisely conserved. The induced and natural T_R populations do appear to have counterparts in both species, and studies of human $CD25^+$ T_R cells have consistently shown that, like their murine equivalents, their suppression *in vitro* is mediated exclusively by contact [5–8]. However, studies of human $CD25^+$ T_R cells have been limited to *in vitro* or *ex vivo* work, and so it is not yet clear whether they may have alternative or additional properties *in vivo*. In addition, it is possible that human $CD25^+$ T_R cells may not conform to the "natural" epithet, since they have short telomeres, consistent with a highly differentiated phenotype and suggestive of generation by repeated antigenic stimulation in the periphery [8]. It is attractive to argue that long-lived species such as humans need continually to induce peripheral $CD25^+$ T_R cells, and cannot rely on the thymic, natural source of such cells identified in mice.

Conclusion

Overall, there are three extreme interpretations of the discrepant data regarding the cytokine dependency of T_R cells.

First, one could argue that T_R activity is very complex, with T_R cells being drawn from many disparate induced and natural subsets, each with diverse and overlapping regulatory properties, and with effector mechanisms varying in relevance depending on the subset activated, disease and experimental system. Fresh discrepancies can be resolved by invoking further subsets, or the abilities of subpopulations to induce one another.

An alternative would be to subscribe to the view of William of Ockham in the 14th century: "Pluralitas non est ponenda sine neccesitate", which translates as "entities should not be multiplied unnecessarily". Thus, T_R activity may be more simple than appears at first sight, with, for example, contact-dependent "natural" and cytokine-dependent "induced" T_R subsets representing different states of the same cell type, which may be dependent on time since activation or age.

A third, radical possibility is that much of the apparent suppressor activity of T_R cells is artefactual, a view encouraged by the poor past record of immunologists in characterizing immunoregulatory phenomena [101]. Indeed, recent experiments on the transfer of wasting disease to lymphopenic recipients indicates that some apparent manifestations of regulatory activity can be attributed, not to a particular T_R subset, but to normal competition for shared resources [102]. Thus, T cells with high potential for homeostatic expansion, but no known regulatory characteristics, inhibited wasting disease transferred by naïve T cells into lymphopenic hosts [102]. Such phenomena cannot explain all of the many examples of apparent T_R suppression, but do illustrate the need for caution in attributing every inhibitory effect to T_R activity.

The three interpretations above are extreme, but are clearly not mutually exclusive, and illustrate the challenge posed to immunologists in building a comprehensive paradigm of T_R function. However, resolving which mechanisms of T_R action are relevant to human diseases where immune responses play a central role remains an important goal, because of their relevance to the development of new therapies.

References

1 Roncarolo MG, Levings MK (2000) The role of different subsets of T regulatory cells in controlling autoimmunity. *Curr Opin Immunol* 12: 676–683

2 Shevach EM, McHugh RS, Piccirillo CA, Thornton AM (2001) Control of T-cell activation by $CD4^+$ $CD25^+$ suppressor T cells. *Immunol Rev* 182: 58–67

3 Weiner HL (2001) Induction and mechanism of action of transforming growth factor-beta-secreting Th3 regulatory cells. *Immunol Rev* 182: 207–214

4 Groux H, O'Garra A, Bigler M, Rouleau M, Antonenko S, de Vries JE, Roncarolo GG (1997) A CD4$^+$ T-cell subset inhibits antigen-specific T-cell responses and prevents colitis. *Nature* 389: 737–741

5 Jonuleit H, Schmitt E, Stassen M, Tuettenberg A, Knop J, Enk AH (2001) Identification and functional characterisation of human CD4$^+$CD25$^+$ T cells with regulatory properties isolated from peripheral blood. *J Exp Med* 193: 1285–1294

6 Levings MK, Sangregorio R, Roncarolo MG (2001) Human CD25$^+$CD4$^+$ T regulatory cells suppress naïve and memory T cell proliferation and can be expanded *in vitro* without loss of function. *J Exp Med* 193: 1295–1301

7 Dieckmann D, Plottner H, Berchtold S, Berger T, Schuler G (2001) *Ex vivo* isolation and characterisation of CD4$^+$CD25$^+$ T cells with regulatory properties from human blood. *J Exp Med* 193: 1303–1310

8 Taams LS, Vukmanovic-Stejic M, Smith J, Dunne PJ, Fletcher JM, Plunkett FJ, Ebeling SB, Lombardi G, Rustin MH, Bijlsma JW et al. (2002) Antigen-specific T cell suppression by human CD4$^+$CD25$^+$ regulatory T cells. *Eur J Immunol* 32: 1621–1630

9 Hall AM, Ward FJ, Vickers MA, Stott LM, Urbaniak SJ, Barker RN (2002) Interleukin-10-mediated regulatory T-cell responses to epitopes on a human red blood cell autoantigen. *Blood* 100: 4529–4536

10 Cairns LS, Phelps RG, Bowie L, Hall AM, Saweirs WW, Rees AJ, Barker RN (2003) The fine specificity and cytokine profile of T-helper cells responsive to the alpha3 chain of type IV collagen in Goodpasture's disease. *J Am Soc Nephrol* 14: 2801–2812

11 Ling EM, Smith T, Nguyen XD, Pridgeon C, Dallman M, Arbery J, Carr VA, Robinson DS (2004) Relation of CD4$^+$CD25$^+$ regulatory T-cell suppression of allergen-driven T-cell activation to atopic status and expression of allergic disease. *Lancet* 363: 608–615

12 Marshall NA, Vickers MA, Barker RN (2003) Regulatory T cells secreting IL-10 dominate the immune response to EBV latent membrane protein 1. *J Immunol* 170: 6183–6189

13 McGuirk P, Mills KH (2002) Pathogen-specific regulatory T cells provoke a shift in the Th1/Th2 paradigm in immunity to infectious diseases. *Trends Immunol* 23: 450–455

14 Marshall NA, Christie LE, Munro LR, Culligan DJ, Johnston PW, Barker RN, Vickers MA (2004) Immunosuppressive regulatory T cells are abundant in the reactive lymphocytes of Hodgkin lymphoma. *Blood* 103: 1755–1762

15 Wang HY, Lee DA, Peng G, Guo Z, Li Y, Kiniwa Y, Shevach EM, Wang RF (2004) Tumor-specific human CD4$^+$ regulatory T cells and their ligands: implications for immunotherapy. *Immunity* 20: 107–118

16 Moore KW, de Waal Malefyt R, Coffman RL, O'Garra A (2001) Interleukin-10 and the interleukin-10 receptor. *Annu Rev Immunol* 19: 683–765

17 Fiorentino DF, Bond MW, Mosmann TR (1989) Two types of mouse T helper cell. IV. Th2 clones secrete a factor that inhibits cytokine production by Th1 clones. *J Exp Med* 170: 2081–2095

18 Suda T, O'Garra A, MacNeil I, Fischer M, Bond MW, Zlotnik A (1990) Identification

of a novel thymocyte growth-promoting factor derived from B cell lymphomas. *Cell Immunol* 129: 228–240

19 Go NF, Castle BE, Barrett R, Kastelein R, Dang W, Mosmann TR, Moore KW, Howard M (1990) Interleukin 10, a novel B cell stimulatory factor: unresponsiveness of X chromosome-linked immunodeficiency B cells. *J Exp Med* 172: 1625–1631

20 Rousset F, Garcia E, Defrance T, Peronne C, Vezzio N, Hsu DH, Kastelein R, Moore KW, Banchereau J (1992) Interleukin 10 is a potent growth and differentiation factor for activated human B lymphocytes. *Proc Natl Acad Sci USA* 89: 1890–1893

21 Jeannin P, Lecoanet S, Delneste Y, Gauchat JF, Bonnefoy JY (1998) IgE *versus* IgG4 production can be differentially regulated by IL-10. *J Immunol* 160: 3555–3561

22 Chen WF, Zlotnik A (1991) IL-10: a novel cytotoxic T cell differentiation factor. *J Immunol* 147: 528–534

23 Groux H, Bigler M, de Vries JE, Roncarolo MG (1998) Inhibitory and stimulatory effects of IL-10 on human CD8$^+$ T cells. *J Immunol* 160: 3188–3193

24 Thompson-Snipes L, Dhar V, Bond MW, Mosmann TR, Moore KW, Rennick DM (1991) Interleukin 10: a novel stimulatory factor for mast cells and their progenitors. *J Exp Med* 173: 507–510

25 de Waal Malefyt R, Abrams J, Bennett B, Figdor CG, de Vries JE (1991) Interleukin 10 (IL-10) inhibits cytokine synthesis by human monocytes: an autoregulatory role of IL-10 produced by monocytes. *J Exp Med* 174: 1209–1220

26 Bogdan C, Vodovotz Y, Nathan C (1991) Macrophage deactivation by interleukin 10. *J Exp Med* 174: 1549–1555

27 Peguet-Navarro J, Moulon C, Caux C, Dalbiez-Gauthier C, Banchereau J, Schmitt D (1994) Interleukin-10 inhibits the primary allogeneic T cell response to human epidermal Langerhans cells. *Eur J Immunol* 24: 884–891

28 de Waal Malefyt R, Yssel H, de Vries JE (1993) Direct effects of IL-10 on subsets of human CD4$^+$ T cell clones and resting T cells. Specific inhibition of IL-2 production and proliferation. *J Immunol* 150: 4754–4765

29 Cassatella MA, Meda L, Bonora S, Ceska M, Constantin G (1993) Interleukin 10 (IL-10) inhibits the release of proinflammatory cytokines from human polymorphonuclear leukocytes. Evidence for an autocrine role of tumor necrosis factor and IL-1 beta in mediating the production of IL-8 triggered by lipopolysaccharide. *J Exp Med* 178: 2207–2211

30 Groux H, Cottrez F (2003) The complex role of interleukin-10 in autoimmunity. *J Autoimmun* 20: 281–285

31 Kasama T, Strieter RM, Lukacs NW, Lincoln PM, Burdick MD, Kunkel SL (1995) Interleukin-10 expression and chemokine regulation during the evolution of murine type II collagen-induced arthritis. *J Clin Invest* 95: 2868–2876

32 Powrie F, Coffman RL (1993) Inhibition of cell-mediated immunity by IL4 and IL10. *Res Immunol* 144: 639–643

33 Rosenbaum JT, Angell E (1995) Paradoxical effects of IL-10 in endotoxin-induced uveitis. *J Immunol* 155: 4090–4094

34 Arai T, Hiromatsu K, Kobayashi N, Takano M, Ishida H, Nimura Y, Yoshikai Y (1995)

IL-10 is involved in the protective effect of dibutyryl cyclic adenosine monophosphate on endotoxin-induced inflammatory liver injury. *J Immunol* 155: 5743–5749

35 Woiciechowsky C, Asadullah K, Nestler D, Eberhardt B, Platzer C, Schoning B, Glockner F, Lanksch WR, Volk HD, Docke WD (1998) Sympathetic activation triggers systemic interleukin-10 release in immunodepression induced by brain injury. *Nat Med* 4: 808–813

36 Blazar BR, Taylor PA, Smith S, Vallera DA (1995) Interleukin-10 administration decreases survival in murine recipients of major histocompatibility complex disparate donor bone marrow grafts. *Blood* 85: 842–851

37 Moritani M, Yoshimoto K, Tashiro F, Hashimoto C, Miyazaki J, Ii S, Kudo E, Iwahana H, Hayashi Y, Sano T et al. (1994) Transgenic expression of IL-10 in pancreatic islet A cells accelerates autoimmune insulitis and diabetes in non-obese diabetic mice. *Int Immunol* 6: 1927–1936

38 Sad S, Marcotte R, Mosmann TR (1995) Cytokine-induced differentiation of precursor mouse $CD8^+$ T cells into cytotoxic $CD8^+$ T cells secreting Th1 or Th2 cytokines. *Immunity* 2: 271–279

39 O'Garra A, Chang R, Go N, Hastings R, Haughton G, Howard M (1992) Ly-1 B (B-1) cells are the main source of B cell-derived interleukin 10. *Eur J Immunol* 22: 711–717

40 Rivas JM, Ullrich SE (1992) Systemic suppression of delayed-type hypersensitivity by supernatants from UV-irradiated keratinocytes. An essential role for keratinocyte-derived IL-10. *J Immunol* 149: 3865–3871

41 Yssel H, De Waal Malefyt R, Roncarolo MG, Abrams JS, Lahesmaa R, Spits H, de Vries JE (1992) IL-10 is produced by subsets of human $CD4^+$ T cell clones and peripheral blood T cells. *J Immunol* 149: 2378–2384

42 Shevach EM (2002) $CD4^+$ $CD25^+$ suppressor T cells: more questions than answers. *Nat Rev Immunol* 2: 389–400

43 Wraith DC (2003) Role of interleukin-10 in the induction and function of natural and antigen-induced regulatory T cells. *J Autoimmun* 20: 273–275

44 Mosmann TR, Coffman RL (1989) TH1 and TH2 cells: different patterns of lymphokine secretion lead to different functional properties. *Annu Rev Immunol* 7: 145–173

45 Elson CJ, Barker RN (2000) Helper T cells in antibody-mediated, organ-specific autoimmunity. *Curr Opin Immunol* 12: 664–669

46 Liblau RS, Singer SM, McDevitt HO (1995) Th1 and Th2 $CD4^+$ T cells in the pathogenesis of organ-specific autoimmune diseases. *Immunol Today* 16: 34–38

47 Weiner HL (2001) Induction and mechanism of action of transforming growth factor-beta-secreting Th3 regulatory cells. *Immunol Rev* 182: 207–214

48 Chen Y, Kuchroo VK, Inobe J, Hafler DA, Weiner HL (1994) Regulatory T cell clones induced by oral tolerance: suppression of autoimmune encephalomyelitis. *Science* 265: 1237–1240

49 McRae BL, Semnani RT, Hayes MP, van Seventer GA (1998) Type I IFNs inhibit human dendritic cell IL-12 production and Th1 cell development. *J Immunol* 160: 4298–4304

50 Bacchetta R, Sartirana C, Levings MK, Bordignon C, Narula S, Roncarolo MG (2002) Growth and expansion of human T regulatory type 1 cells are independent from TCR activation but require exogenous cytokines. *Eur J Immunol* 32: 2237–2245
51 Barrat FJ, Cua DJ, Boonstra A, Richards DF, Crain C, Savelkoul HF, de Waal-Malefyt R, Coffman RL, Hawrylowicz CM, O'Garra A (2002) In vitro generation of interleukin 10-producing regulatory CD4(+) T cells is induced by immunosuppressive drugs and inhibited by T helper type 1 (Th1)- and Th2-inducing cytokines. *J Exp Med* 195: 603–616
52 Kemper C, Chan AC, Green JM, Brett KA, Murphy KM, Atkinson JP (2003) Activation of human CD4+ cells with CD3 and CD46 induces a T-regulatory cell 1 phenotype. *Nature* 421: 388–392
53 Buer J, Lanoue A, Franzke A, Garcia C, von Boehmer H, Sarukhan A (1998) Interleukin 10 secretion and impaired effector function of major histocompatibility complex class II-restricted T cells anergized *in vivo*. *J Exp Med* 187: 177–183
54 Sundstedt A, Hoiden I, Rosendahl A, Kalland T, van Rooijen N, Dohlsten M (1997) Immunoregulatory role of IL-10 during superantigen-induced hyporesponsiveness *in vivo*. *J Immunol* 158: 180–186
55 Burkhart C, Liu GY, Anderton SM, Metzler B, Wraith DC (1999) Peptide-induced T cell regulation of experimental autoimmune encephalomyelitis: a role for IL-10. *Int Immunol* 11: 1625–1634
56 Jonuleit H, Schmitt E, Steinbrink K, Enk AH (2001) Dendritic cells as a tool to induce anergic and regulatory T cells. *Trends Immunol* 22: 394–40
57 Akbari O, Freeman GJ, Meyer EH, Greenfield EA, Chang TT, Sharpe AH, Berry G, DeKruyff RH, Umetsu DT (2002) Antigen-specific regulatory T cells develop *via* the ICOS-ICOS-ligand pathway and inhibit allergen-induced airway hyperreactivity. *Nat Med* 8: 1024–1032
58 Akbari O, DeKruyff RH, Umetsu DT (2001) Pulmonary dendritic cells producing IL-10 mediate tolerance induced by respiratory exposure to antigen. *Nat Immunol* 2: 725–731
59 McGuirk P, McCann C, Mills KH (2002) Pathogen-specific T regulatory 1 cells induced in the respiratory tract by a bacterial molecule that stimulates interleukin 10 production by dendritic cells: a novel strategy for evasion of protective T helper type 1 responses by *Bordetella pertussis*. *J Exp Med* 195: 221–231
60 Segal BM, Glass DD, Shevach EM (2002) Cutting Edge: IL-10-producing CD4+ T cells mediate tumor rejection. *J Immunol* 168: 1–4
61 Sundstedt A, O'Neill EJ, Nicolson KS, Wraith DC (2003) Role for IL-10 in suppression mediated by peptide-induced regulatory T cells *in vivo*. *J Immunol* 170: 1240–1248
62 Vieira PL, Christensen JR, Minaee S, O'Neill EJ, Barrat FJ, Boonstra A, Barthlott T, Stockinger B, Wraith DC, O'Garra A (2004) IL-10-secreting regulatory T cells do not express Foxp3 but have comparable regulatory function to naturally occurring CD4(+)CD25(+) regulatory T cells. *J Immunol* 172: 5986–5993
63 Jonuleit H, Schmitt E, Kakirman H, Stassen M, Knop J, Enk AH (2002) Infectious tolerance: human CD25(+) regulatory T cells convey suppressor activity to conventional CD4(+) T helper cells. *J Exp Med* 196: 255–260

64 Zheng SG, Wang JH, Gray JD, Soucier H, Horwitz DA (2004) Natural and induced $CD4^+CD25^+$ cells educate $CD4^+CD25^-$ cells to develop suppressive activity: the role of IL-2, TGF-beta, and IL-10. *J Immunol* 172: 5213–5221

65 Peng Y, Laouar Y, Li MO, Green EA, Flavell RA (2004) TGF-beta regulates *in vivo* expansion of Foxp3-expressing $CD4^+CD25^+$ regulatory T cells responsible for protection against diabetes. *Proc Natl Acad Sci USA* 101: 4572–4577

66 Grundstrom S, Cederbom L, Sundstedt A, Scheipers P, Ivars F (2003) Superantigen-induced regulatory T cells display different suppressive functions in the presence or absence of natural $CD4^+CD25^+$ regulatory T cells *in vivo*. *J Immunol* 170: 5008–5017

67 MacDonald AJ, Duffy M, Brady MT, McKiernan S, Hall W, Hegarty J, Curry M, Mills KH (2002) CD4 T helper type 1 and regulatory T cells induced against the same epitopes on the core protein in hepatitis C virus-infected persons. *J Infect Dis* 185: 720–727

68 Doetze A, Satoguina J, Burchard G, Rau T, Loliger C, Fleischer B, Hoerauf A (2000) Antigen-specific cellular hyporesponsiveness in a chronic human helminth infection is mediated by Th3/Tr1-type cytokines IL-10 and transforming growth factor but not by a Th1 to Th2 shift. *Int Immunol* 12: 623–630

69 Akdis CA, Blesken T, Akdis M, Wüthrich B, Blaser K (1998) Role of IL-10 in specific immunotherapy. *J Clin Invest* 102: 98–106

70 Reefer AJ, Carneiro RM, Custis NJ, Platts-Mills TA, Sung SS, Hammer J, Woodfolk JA (2004) A role for IL-10-mediated HLA-DR7-restricted T cell-dependent events in development of the modified Th2 response to cat allergen. *J Immunol* 172: 2763–2772

71 Thornton AM, Shevach EM (1998) $CD4^+CD25^+$ immunoregulatory T cells suppress polyclonal T cell activation *in vitro* by inhibiting interleukin 2 production. *J Exp Med* 188: 287–296

72 Takahashi T, Tagami T, Yamazaki S, Uede T, Shimizu J, Sakaguchi N, Mak TW, Sakaguchi S (2000) Immunologic self-tolerance maintained by CD25(+)CD4(+) regulatory T cells constitutively expressing cytotoxic T lymphocyte-associated antigen 4. *J Exp Med* 192: 303–310

73 Sakaguchi S, Sakaguchi N, Asano M, Itoh M, Toda M (1995) Immunologic self-tolerance maintained by activated T cells expressing IL-2 receptor alpha-chains (CD25). Breakdown of a single mechanism of self-tolerance causes various autoimmune diseases. *J Immunol* 155: 1151–1164

74 McHugh RS, Whitters MJ, Piccirillo CA, Young DA, Shevach EM, Collins M, Byrne MC (2002) CD4(+)CD25(+) immunoregulatory T cells: gene expression analysis reveals a functional role for the glucocorticoid-induced TNF receptor. *Immunity* 16: 311–323

75 Shimizu J, Yamazaki S, Takahashi T, Ishida Y, Sakaguchi S (2002) Stimulation of CD25(+)CD4(+) regulatory T cells through GITR breaks immunological self-tolerance. *Nat Immunol* 3: 135–142

76 Hori S, Nomura T, Sakaguchi S (2003) Control of regulatory T cell development by the transcription factor Foxp3. *Science* 299: 1057–1061

77 Wraith DC (2003) Role of interleukin-10 in the induction and function of natural and antigen-induced regulatory T cells. *J Autoimmun* 20: 273–275

78 Nakamura K, Kitani A, Strober W (2001) Cell contact-dependent immunosuppression by CD4(+)CD25(+) regulatory T cells is mediated by cell surface-bound transforming growth factor beta. *J Exp Med* 194: 629–644
79 Piccirillo CA, Letterio JJ, Thornton AM, McHugh RS, Mamura M, Mizuhara H, Shevach EM (2002) CD4(+)CD25(+) regulatory T cells can mediate suppressor function in the absence of transforming growth factor beta1 production and responsiveness. *J Exp Med* 196: 237–246
80 Powrie F, Leach MW, Mauze S, Caddle LB, Coffman RL (1993) Phenotypically distinct subsets of CD4$^+$ T cells induce or protect from chronic intestinal inflammation in C. B-17 scid mice. *Int Immunol* 5: 1461–1471
81 Powrie F, Carlino J, Leach MW, Mauze S, Coffman RL (1996) A critical role for transforming growth factor-beta but not interleukin 4 in the suppression of T helper type 1-mediated colitis by CD45RB(low) CD4$^+$ T cells. *J Exp Med* 183: 2669–2674
82 Asseman C, Mauze S, Leach MW, Coffman RL, Powrie F (1999) An essential role for interleukin 10 in the function of regulatory T cells that inhibit intestinal inflammation. *J Exp Med* 190: 995–1004
83 Liu, H, Hu, B, Xu, D and Liew FY (2003) CD4$^+$CD25$^+$ regulatory T cells cure murine colitis: the role of IL-10, TGF-beta, and CTLA-4. *J Immunol* 171: 5012–5017
84 Fuss IJ, Boirivant M, Lacy B, Strober W (2002) The interrelated roles of TGF-beta and IL-10 in the regulation of experimental colitis. *J Immunol* 168: 900–908
85 Pontoux C, Banz A, Papiernik M (2002) Natural CD4 CD25(+) regulatory T cells control the burst of superantigen-induced cytokine production: the role of IL-10. *Int Immunol* 14: 233–239
86 Belkaid Y, Piccirillo CA, Mendez S, Shevach EM, Sacks DL (2002) CD4$^+$CD25$^+$ regulatory T cells control *Leishmania major* persistence and immunity. *Nature* 420: 502–507
87 Suri-Payer E, Cantor H (2001) Differential cytokine requirements for regulation of autoimmune gastritis and colitis by CD4(+)CD25(+) T cells. J Autoimmun 16: 115–123
88 Maloy KJ, Salaun L, Cahill R, Dougan G, Saunders NJ, Powrie F (2003) CD4$^+$CD25$^+$ T(R) cells suppress innate immune pathology through cytokine-dependent mechanisms. *J Exp Med* 197: 111–119
89 Annacker O, Asseman C, Read S, Powrie F (2003) Interleukin-10 in the regulation of T cell-induced colitis. *J Autoimmun* 20: 277–279
90 Asseman C, Read S, Powrie F (2003) Colitogenic Th1 cells are present in the antigen-experienced T cell pool in normal mice: control by CD4$^+$ regulatory T cells and IL-10. *J Immunol* 171: 971–978
91 Annacker O, Pimenta-Araujo R, Burlen-Defranoux O, Barbosa TC, Cumano A, Bandeira A (2001) CD25$^+$ CD4$^+$ T cells regulate the expansion of peripheral CD4 T cells through the production of IL-10. *J Immunol* 166: 3008–3018
92 Almeida AR, Legrand N, Papiernik M, Freitas AA (2002) Homeostasis of peripheral CD4$^+$ T cells: IL-2R alpha and IL-2 shape a population of regulatory cells that controls CD4$^+$ T cell numbers. *J Immunol* 169: 4850–4860

93 Fillatreau S, Sweenie CH, McGeachy MJ, Gray D, Anderton SM (2002) B cells regulate autoimmunity by provision of IL-10. *Nat Immunol* 3: 944–950

94 Nouri-aria KT, Wachholz PA, Francis JN, Jacobson MR, Walker SM, Wilcock LK, Staple SQ, Aalberse RC, Till SJ, Durham SR (2004) Grass pollen immunotherapy induces mucosal and peripheral IL-10 responses and blocking IgG activity. *J Immunol* 172: 3252–3259

95 Dieckmann D, Bruett CH, Ploettner H, Lutz MB, Schuler G (2002) Human CD4(+)CD25(+) regulatory, contact-dependent T cells induce interleukin 10-producing, contact-independent type 1-like regulatory T cells. *J Exp Med* 196: 247–253

96 Chen W, Jin W, Hardegen N, Lei KJ, Li L, Marinos N, McGrady G, Wahl SM (2003) Conversion of peripheral $CD4^+CD25$- naive T cells to $CD4^+CD25^+$ regulatory T cells by TGF-beta induction of transcription factor Foxp3. *J Exp Med* 198: 1875–1886

97 Salama AD, Chaudhry AN, Holthaus KA, Mosley K, Kalluri R, Sayegh MH, Lechler RI, Pusey CD, Lightstone L (2003) Regulation by $CD25^+$ lymphocytes of autoantigen-specific T-cell responses in Goodpasture's (anti-GBM) disease. *Kidney Int* 64: 1685–1694

98 Jonuleit H, Schmitt E (2003) The regulatory T cell family: distinct subsets and their interrelations. *J Immunol* 171: 6323–6327

99 Lehmann J, Huehn J, de la Rosa M, Maszyna F, Kretschmer U, Krenn V, Brunner M, Scheffold A, Hamann A (2002) Expression of the integrin alpha E beta 7 identifies unique subsets of $CD25^+$ as well as $CD25^-$ regulatory T cells. *Proc Natl Acad Sci USA* 99: 13031–13036

100 Banz A, Peixoto A, Pontoux C, Cordier C, Rocha B, Papiernik M (2003) A unique subpopulation of $CD4^+$ regulatory T cells controls wasting disease, IL-10 secretion and T cell homeostasis. *Eur J Immunol* 33: 2419–2428

101 Moller G (1988) Do suppressor T cells exist? *Scand J Immunol* 27: 247–250

102 Barthlott T, Kassiotis G, Stockinger B (2003) T cell regulation as a side effect of homeostasis and competition. *J Exp Med* 197: 451–460

103 Levings MK, Sangregorio R, Galbiati F, Squadrone S, de Waal Malefyt R, Roncarolo MG (2001) IFN-alpha and IL-10 induce the differentiation of human type 1 T regulatory cells. *J Immunol* 166: 5530–5539

104 Kingsley CI, Karim M, Bushell AR, Wood KJ (2002) $CD25^+CD4^+$ regulatory T cells prevent graft rejection: CTLA-4 and IL-10-dependent immunoregulation of alloresponses. *J Immunol* 168: 1080–1086

105 Annunziato F, Cosmi L, Liotta F, Lazzeri E, Manetti R, Vanini V, Romagnani P, Maggi E, Romagnani S (2002) Phenotype, localization, and mechanism of suppression of CD4(+)CD25(+) human thymocytes. *J Exp Med* 196: 379–387

106 Nakamura K, Kitani A, Fuss I, Pedersen A, Harada N, Nawata H, Strober W (2004) TGF-beta1 plays an important role in the mechanism of $CD4^+CD25^+$ regulatory T cell activity in both humans and mice. *J Immunol* 172: 834–842

Activation and distribution of regulatory T cells in naïve and antigen-stimulated immune systems

Esther N.M. Nolte-'t Hoen[1] *and Marca H.M. Wauben*[2]

[1]Faculty of Biological Sciences, Imperial College London, South Kensington Campus, London SW7 2AZ, UK; [2]Department of Immunohematology and Blood Transfusion, E3-Q, Leiden University Medical Center, Albinusdreef 2, P.O. Box 9600, 2300 RC Leiden, The Netherlands

Introduction

Several T-cell subsets capable of inhibiting the activation of other T cells have now been characterized, and are referred to as "regulatory T cells". The best-described regulatory-T-cell population consists of a subpopulation of $CD4^+$ T cells expressing the interleukin-2 receptor α-chain (CD25). These $CD4^+CD25^+$ T cells could both control autoreactive T cells *in vivo* [1, 2] and inhibit proliferation of $CD4^+CD25^-$ T cells *in vitro* [3–5]. Mouse, rat and human studies indicate that regulatory T cells can be found in lymph nodes, spleen and blood [6, 7]. However, limitations in cell numbers and availability often hamper the simultaneous analysis of regulatory T cells in the various compartments of the immune system. Although their presence in blood indicates that regulatory T cells migrate, it is not known whether these cells, just like naïve $CD4^+$ and $CD8^+$ T cells, continuously re-circulate through the different lymphoid compartments. Furthermore, questions about where regulatory T cells are being activated, and where they exert their regulatory function in naïve and antigen-stimulated immune systems, largely remain to be answered. In this chapter, we will discuss the activation and distribution of regulatory T cells by using recent data on regulatory T cells in naïve and arthritic rats as a guideline.

Origin of regulatory T cells

$CD4^+CD25^+$ naturally occurring regulatory T cells that suppress aberrant or surplus T-cell responses [7, 8] develop and acquire their suppressive function in the thymus [9], after which they migrate to the periphery. Regarding their role in suppression of organ-specific autoimmune diseases, regulatory T cells were thought to be biased to recognition of self-antigens [10, 11]. Nevertheless, the T-cell receptor (TCR) repertoire of regulatory T cells was demonstrated to be broad [11, 12], suggesting that they can recognize a wide variety of antigens. Besides the existence of regulatory T

Regulatory T Cells in Inflammation, edited by Leonie S. Taams, Arne N. Akbar and Marca H.M. Wauben

cells with thymic origin, several lines of evidence suggest that T cells can also acquire regulatory activity in the periphery. Peripherally induced regulatory T cells are able to suppress T-cell responses to foreign antigens like tetanus toxoid and microbes [13, 14], and play a role in preventing graft rejection by suppressing responses to alloantigens [15, 16]. Regulatory T cells can be induced *in vivo via* a "tolerogenic" route (e.g. *via* mucosal antigen administration) (reviewed in [17, 18]). *In vitro* antigen presentation in a "tolerogenic" form, e.g. by antigen-presenting cells incubated with vitamin D_3 and dexamethasone [19], in the presence of transforming growth factor (TGF)-β or interleukin (IL)-10 [20, 21], or by immature or plasmacytoid dendritic cells [22], was shown to be successful in the generation of regulatory T cells. So far it is not known whether these so-called "antigen-induced" or "adaptive" regulatory T cells form a separate lineage of regulatory T cells from the thymus-derived regulatory T cells. "Adaptive" regulatory T cells could originate from a population of non-regulatory $CD4^+$ T cells, or expand from thymus-derived, naturally occurring regulatory T cells.

Activation of regulatory T cells in naïve immune systems

Evidence is accumulating that $CD4^+CD25^+$ regulatory T cells isolated from naïve lymphoid organs resemble a heterogeneous population containing several phenotypically and functionally different regulatory T-cell subsets [23–25]. Recurring observations of $CD4^+CD25^+$ regulatory T-cell subsets with differential expression of activation markers, such as CD62L, CD38, CD45RB and CD134 (OX-40) [25–27] indicate the presence of regulatory T cells with different levels of activation.

Although naturally occurring regulatory T cells that leave the thymus have a partially activated phenotype, *in vitro* stimulation *via* TCR triggering is assumed to be a prerequisite for freshly isolated $CD4^+CD25^+$ T cells to suppress T-cell responses [3, 4, 7, 9]. The requirement of TCR stimulation *in vitro* suggests that the same requirement applies to regulatory T cells *in vivo*. However, several studies report the capacity of activated regulatory T cells to suppress T cells in an antigen-independent fashion [4, 25, 26]. Analysis of the co-existence of resting and activated regulatory T cells that differ in their suppressive activity within freshly isolated $CD4^+CD25^+$ T-cell populations is in many studies not evaluated because of the type of suppression assays used. In most studies, suppression assays in which regulatory and responder T cells are simultaneously activated by polyclonal stimuli, like anti-CD3 monoclonal antibody (mAb), are used. Therefore differences in suppressive activity of regulatory T-cell subpopulations selected according to expression of activation markers, such as CD62L, CD38 or CD45RB, were not revealed [26–28]. Nevertheless, several studies recently confirmed the presence of phenotypically different populations of naïve/resting and activated/effector regulatory T cells in mice [23, 24]. The $CD4^+CD25^+CD134^{int}CD62L^{hi}CD45RB^{int}$ resting regulatory-T-cell population described by Fisson et al. [23] was found to differentiate towards the $CD4^{+}CD25^{+}CD134^{hi}CD62L^{int}CD45RB^{low}$ activated regulatory-T-cell phenotype upon antigen recognition in peripheral tissue. Huehn et al. [24] recently discriminated naïve and effector/memory regulatory T cells based on expression of the integrin $\alpha_E\beta_7$, and also found that the activated subset of regulatory T cells displayed lower levels of CD62L and CD45RB compared to the naïve cells. Other *in vitro* and *in vivo* studies confirm the down-regulation of CD62L and CD45RB, and up-regulation of CD134 on regulatory T cells upon activation of these cells [29–33]. However, some studies report the re-expression of CD62L on regulatory T cells several days after activation [30, 31]. The most potent regulatory T cells within human blood have been described to reside in the $CD25^{hi}CD62L^{hi}$ fraction of $CD4^{+}$ T cells [34–36]. Interestingly, Baecher-Allan et al. [37] found that freshly isolated, human peripheral-blood-derived $CD25^{hi}CD62L^{hi}$ regulatory T cells can suppress weakly stimulated responder T cells without being activated *in vitro*. These results suggest that the human $CD25^{hi}CD62L^{hi}$ regulatory T cells in blood may be regarded as effector/memory rather than naïve regulatory T cells.

We recently found that a subpopulation of $CD4^{+}CD25^{+}$ regulatory T cells in naïve rats exhibited an overall activated phenotype, characterized by decreased expression of CD62L and CD45RC, and increased levels of CD134 (OX-40 antigen) and MHC class II expression [25]. Importantly, we found that freshly isolated subpopulations of regulatory T cells that had been dissected based on the expression of CD134 differed in their capacity to suppress T-cell responses. In a co-culture assay with responder rat clonal T cells and antigen-presenting cells expressing the cognate antigen for the responder T cells, the $CD134^{+}$ regulatory T cells suppressed the antigen-specific proliferation of the responder T cells *in vitro* without additional TCR stimulation. In contrast, $CD4^{+}CD25^{+}CD134^{-}$ regulatory T cells had suppressive activity only during simultaneous activation of responder and regulatory T cells, i.e. in a co-culture of these cells stimulated with anti-TCR mAb and antigen-presenting cells, or after *in vitro* pre-activation [25]. The freshly isolated $CD4^{+}CD25^{+}CD134^{+}$ T cells showed a similar suppressive capacity as *in vitro* pre-activated total $CD4^{+}CD25^{+}$ T cells. Based on these results, we propose that the $CD134^{+}$ subset of $CD4^{+}CD25^{+}$ regulatory T cells represents an *in vivo* activated and highly suppressive regulatory-T-cell subset.

Altogether these results indicate that in naïve immune systems the thymus-derived $CD4^{+}CD25^{+}$ regulatory-T-cell pool contains the whole spectrum, from quiescent cells to recently activated cells.

Activation of regulatory T cells in antigen-stimulated immune systems

In the experimental autoimmune model of adjuvant arthritis, a monophasic disease model for rheumatoid arthritis in rats induced by immunization with *Mycobacteri-*

um tuberculosis [38], we found an increased number of the highly suppressive CD4$^+$CD25$^+$CD134$^+$ T-cell subset, previously identified in naïve rats [25], even before clinical disease onset. Their number remained elevated compared to naïve rats also after clinical signs of arthritis subsided. Interestingly it appeared that inflammatory and regulatory CD4$^+$ T cells activated during the course of adjuvant arthritis could be identified on the basis of differential expression of the activation markers CD25 and CD134, since activated autoaggressive CD4$^+$ T cells expressed either CD25 or CD134, while activated CD4$^+$ T cells co-expressing both molecules were strongly suppressive and could interfere in disease induction in contrast to the single CD25$^+$ subset (M.H.M. Wauben, personal communication). Comparable observations have been made in an arthritis model in mice in which only an activated subset of CD25$^+$ regulatory T cells (integrin $\alpha_E\beta_7^+$) could interfere in disease development [24]. An important question to be answered is: what is the origin and antigen-specificity of these regulatory T cells? In the adjuvant arthritis model the CD4$^+$CD25$^+$CD134$^+$ regulatory T cells could have been induced in the periphery by mycobacterial or self-antigens presented during the inflammatory process or, as suggested before [34–36], developed from antigen-activated non-regulatory T cells. Alternatively, activation of the naturally occurring, thymus-derived regulatory T cells in the periphery could have augmented their suppressive capacity resulting in the down-regulation of the strong autoaggressive T-cell response induced by immunization with *M. tuberculosis*.

These data indicate, as could be expected from the observations made in naïve immune systems, that the CD4$^+$CD25$^+$ regulatory-T-cell population in activated immune systems is heterogeneous. Besides the presence of the whole spectrum of thymus-derived CD4$^+$CD25$^+$ regulatory T cells, antigen-induced CD4$^+$CD25$^+$ regulatory T cells, and recently activated CD4$^+$CD25$^+$ effector T cells, further complicate the analysis of immunoregulation, and increase the need for the definition of discriminating markers.

Migration and site-specific activation of regulatory T cells

Besides the lack of knowledge with respect to the antigen specificity of regulatory T cells, the migratory behavior and the site of activation of these cells is largely unknown. Just like non-regulatory CD4$^+$ T cells, CD4$^+$CD25$^+$ regulatory T cells may be re-circulating through peripheral lymphoid organs and blood, retained in lymph nodes upon antigen recognition, and leaving the lymph nodes in an activated state. Similar to non-regulatory T cells, expression of CD62L [39] and the chemokine receptor CCR7 [40] is a likely requirement for regulatory T cells to be able to enter lymph nodes *via* high endothelial venules (HEV).

We analyzed the distribution of CD4$^+$CD25$^+$ regulatory T cells in naive rats in different compartments of the immune system, and observed that CD134$^-$ non-activated regulatory T cells were almost equally distributed over the different lymphoid organs. In contrast, the CD134-expressing activated subset of $CD4^+CD25^+$ regulatory T cells was mainly present in lymph nodes and spleen, whereas only a very small number of these cells could be found in peripheral blood (Fig. 1) [25]. These data indicate that resting regulatory T cells display a genuine migratory phenotype. In contrast, activated $CD134^+$ regulatory T cells appeared to be retained in the lymph nodes and spleen. In accordance with our data, transferred TCR transgenic regulatory T cells have been described to proliferate or accumulate in draining lymph nodes after administration of cognate antigen in adjuvant [29, 41, 42]. Similar results were obtained upon transfer of TCR transgenic regulatory T cells into animals expressing cognate antigen as self-protein in a peripheral tissue [23, 41]. These data suggest that, upon exposure of cognate antigen, either being expressed as a self-protein or presented as a "foreign" antigen, regulatory T cells are activated and gather in lymphoid organs. This accumulation could be the result of recruitment, expansion and/or retention of regulatory T cells at these sites. Intriguingly, in a paper by Battaglia et al. [43] describing functional and phenotypical characteristics of regulatory T cells in human lymph nodes, it was shown that the total population of $CD4^+CD25^+$ regulatory T cells in lymph nodes had reduced CD62L and CD45RO expression and higher MHC class II expression compared to peripheral-blood regulatory T cells. These data indeed point towards accumulation of activated regulatory T cells in lymph nodes.

Although we found that $CD4^+CD25^+CD134^+$ T cells were present in many compartments of the immune system, mucosa-draining lymph nodes, such as the mesenteric and submandibular lymph nodes, contained the highest proportion of these cells as part of the $CD4^+$ population (Fig. 2) [25]. The percentage of $CD134^+$ regulatory T cells within the $CD4^+$ T-cell population of non-mucosa-draining lymph nodes and spleen was similar. Furthermore, $CD4^+CD25^+CD134^+$ T cells in mucosa-draining lymph nodes expressed the highest level of CD134, whereas the expression was low on cells detected in blood (Fig. 2) [25]. The mucosa-draining lymph nodes are known to be active sites of tolerance induction and function to prevent responses of T cells upon encounter of harmless antigens, such as food proteins and commensal bacteria [44, 45]. Unique factors that dominate within these lymph nodes supporting tolerance induction [46] may include the presence of specific subsets of mucosal DC that favor Th2 responses or the induction of regulatory T cells [47–49]. The population of activated regulatory T cells in the mucosa-draining lymph nodes could be diverse. It has been speculated that naturally occurring regulatory T cells may for example be activated at this location by high levels of bacterial heat-shock proteins (HSPs) that show high homology to self-HSP [50]. Furthermore, presentation of antigen in the tolerizing environment of these mucosal lymph nodes can lead to high numbers of effector T cells that acquire a regulatory phenotype.

We and others [23, 25, 41] found that, within the naïve immune system, peripheral lymph nodes that do not drain mucosal sites and the spleen also contain acti-

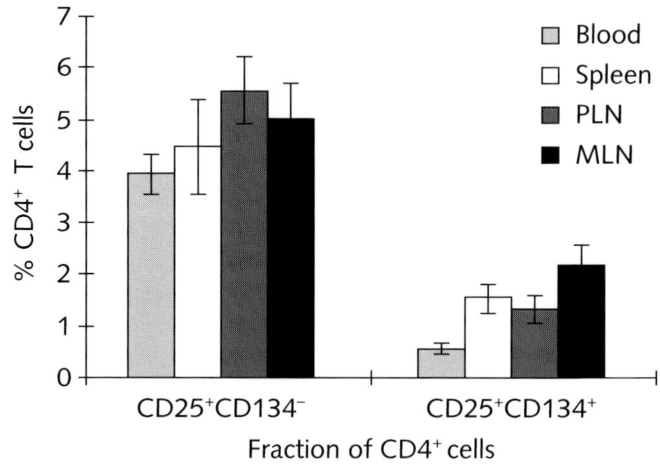

Figure 1
Distribution of CD134⁻ and CD134⁺CD4⁺CD25⁺ regulatory T cells over various compartments of the naïve immune system.
The percentages of CD25⁺CD134⁺ and CD25⁺CD134⁻ T cells within the CD4⁺ fractions of peripheral blood, spleen, popliteal and brachial lymph nodes (PLN) and mesenteric lymph nodes (MLN) of naïve rats were determined by flow-cytometric analysis. Shown are the mean percentages ± S.D. from five individual rats.

vated regulatory T cells. It has been proposed that self-antigen-specific regulatory T cells can be activated by dendritic cells that migrate to lymph nodes and present tissue antigen acquired from dead cells or captured from live cells [23]. Since activated regulatory T cells have reduced levels of CD62L (L-selectin), which is needed to enter lymph nodes *via* HEV (reviewed in [39], it is unlikely that the activated regulatory T cells in non-mucosa-draining lymph nodes are the result of migration of cells previously activated in mucosa-draining lymph nodes.

In naïve rats we found that activated regulatory T cells in spleen, lymph nodes and blood shared the $CD4^+CD25^+CD134^+CD45RB^{lo}$ phenotype. The activated regulatory T cells present in the spleen, however, displayed lower CD62L-expression levels compared to their counterparts in other immune compartments (E.P.J. Boot, E.N.M. Nolte'-t Hoen, J.P.A. Wagenaar-Hilbers, J.H.M. van Bilsen, G.J.A. Arkestejin, G. Storm, L.A. Everse, W. van Eden, M.H.M. Wauben; unpublished observations). The role of the spleen, in regulatory-T-cell activation and expansion is still unclear. Accumulation of regulatory T cells in the spleen has been observed after

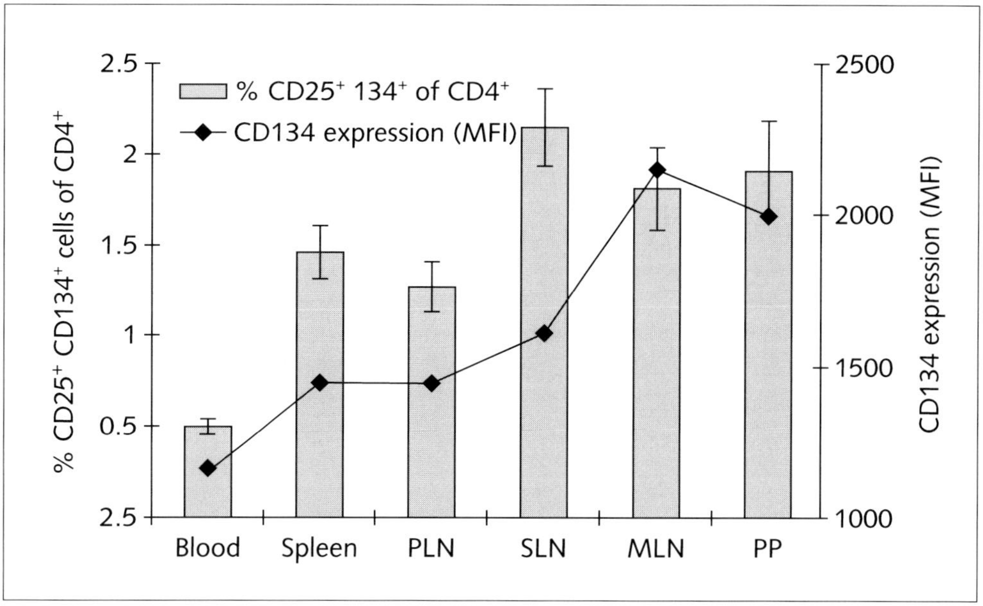

Figure 2
Amount and CD134 expression level of $CD4^+CD25^+CD134^+$ regulatory T cells in various compartments of the naïve immune system.
The percentages of $CD25^+CD134^+$ T cells within the $CD4^+$ fractions were determined in blood, spleen, mucosa-draining submandibular lymph nodes (SLN), mesenteric lymph nodes (MLN), Peyer's patches (PP) and non-mucosa-draining, popliteal and brachial lymph nodes (PLN) of naïve rats (mean percentage ± S.D. from five rats). The CD134 expression levels of these six different lymphoid organs were compared. Expression-level data are presented as the mean fluorescence intensity (MFI) of CD134 staining of the $CD4^+CD25^+CD134^+$ cells from five individual rats.

oral- or nasal-tolerance induction [51] or activation in the steady state [23]. In contrast to lymph nodes, which drain (mucosal) organs and skin, the spleen is known to drain blood-borne antigens. The presentation of these antigens could lead to activation of regulatory T cells. Additionally, since CD62L expression is not required for lymphocytes to enter the spleen [52], this organ could play a role in retention, proliferation and perpetuation of activated regulatory T cells. Furthermore as shown for non-regulatory $CD4^+$ T cells [53], the spleen may also function as a "memory organ" for regulatory T cells. To dissect the role of the individual lymphoid compartments it is challenging to investigate the suppressive activities of regulatory T cells that reside in these different compartments of the immune system.

The continuous presence of activated regulatory T cells, which suppress T cells in an antigen-independent fashion, may indicate that T-cell suppression in lymphoid organs occurs by default. This hypothesis is supported by the observation that removal of regulatory $CD25^{+}CD4^{+}$ T cells can result in autoimmunity [1, 2] and can abrogate immunological unresponsiveness to syngeneic tumors *in vivo* and *in vitro* [54]. The presence of activated regulatory T cells could therefore provide a threshold for T-cell activation, which can be overcome in case of strong TCR stimulation [37, 55]. The higher amount of activated regulatory T cells in mucosa-draining lymph nodes may indicate a higher threshold of T-cell activation, necessary to prevent activation of T cells by relatively high doses of harmless antigen draining to these lymph nodes.

Migration of regulatory T cells to effector sites

Not only the activation site, but also the effector site of regulatory T cells is important to reveal. The increased presence of activated regulatory T cells in peripheral lymphoid organs of naïve animals suggests that these organs represent major effector sites of activated regulatory T cells. However, under inflammatory conditions such as *Leishmania* infection [14], viral infection [56] or autoimmune disease [57–60] accumulation of regulatory T cells has been observed in inflamed peripheral tissue. The effector sites for regulatory T cells involved in the maintenance of homeostasis or the control of inflammatory conditions may very well differ. During steady-state "housekeeping" the major effector site of regulatory T cells would be the lymphoid organs, where aberrant activation of autoreactive T cells is likely to take place. However, during inflammatory conditions, migration of regulatory T cells to inflamed tissues maybe important to reduce effector functions of (auto-aggressive) T cells and to modulate antigen-presenting cells. The ability of activated regulatory T cells to exit lymph nodes and migrate to (inflamed) peripheral tissues largely depends on the expression of chemokine receptors (reviewed in [61]). Human peripheral-blood $CD4^{+}CD25^{+}$ regulatory T cells express CCR4 and CCR8 to a higher extent than non-regulatory T cells, and respond to the inflammatory cytokines CCL1, CCL22 and CCL17, which could guide the regulatory T cells to mature dendritic cells and the skin [62–64]. Mouse splenic regulatory T cells were shown to express CCR5, used for trafficking towards activated antigen-presenting cells producing CCL4 [65]. Two murine studies describe chemokine receptor expression differences between subpopulations of regulatory T cells [24, 66]. In both studies, the regulatory T cell population with the most-activated phenotype (defined as $CD62L^{-}$ in [66] and as α_E $integrin^{+}CD62L^{int}CD45RB^{lo}CD134^{+}$ in [24]) shows expression of CCR2, CCR4 and CXCR3 and migration towards the inflammatory cytokines such as CCL17, CCL20 and CXCL9. The resting regulatory T cells, however, expressed CCR7 and migrated towards secondary lymphoid tissue chemokines, CCL21 and CCL19.

These data indicate that, although isolated from lymphoid organs, activated regulatory T cells are often well equipped to enter (inflamed) peripheral tissues, whereas resting regulatory T cells are prone to subside in the lymphoid organs. Furthermore, it is tempting to speculate that there is a division of labor between thymus-derived regulatory T cells acting at a systemic level by re-circulating through lymphoid tissues poised to sustain tolerance, and adaptive regulatory T cells acting locally by migrating into inflamed peripheral tissues and suppressing on-going inflammatory responses.

References

1 Sakaguchi S, Sakaguchi N, Asano M, Itoh M, Toda M (1995) Immunologic self-tolerance maintained by activated T cells expressing IL-2 receptor alpha-chains (CD25). Breakdown of a single mechanism of self-tolerance causes various autoimmune diseases. *J Immunol* 155: 1151–1164

2 Asano M, Toda M, Sakaguchi N, Sakaguchi S (1996) Autoimmune disease as a consequence of developmental abnormality of a T cell subpopulation. *J Exp Med* 184: 387–396

3 Thornton AM, Shevach EM (1998) $CD4^+CD25^+$ immunoregulatory T cells suppress polyclonal T cell activation *in vitro* by inhibiting interleukin 2 production. *J Exp Med* 188: 287–296

4 Takahashi T, Kuniyasu Y, Toda M, Sakaguchi N, Itoh M, Iwata M, Shimizu J, Sakaguchi S (1998) Immunologic self-tolerance maintained by $CD25^+CD4^+$ naturally anergic and suppressive T cells: induction of autoimmune disease by breaking their anergic/suppressive state. *Int Immunol* 10: 1969–1980

5 Read S, Mauze S, Asseman C, Bean A, Coffman R, Powrie F (1998) $CD38^+$ CD45RB(low) $CD4^+$ T cells: a population of T cells with immune regulatory activities *in vitro*. *Eur J Immunol* 28: 3435–3447

6 Mason D (2001) T-cell-mediated control of autoimmunity. Arthritis Res 3: 133–135

7 Shevach EM (2002) $CD4^+$ $CD25^+$ suppressor T cells: more questions than answers. *Nat Rev Immunol* 2: 389–400

8 Sakaguchi S (2000) Regulatory T cells: key controllers of immunologic self-tolerance. *Cell* 101: 455–458

9 Sakaguchi S, Sakaguchi N, Shimizu J, Yamazaki S, Sakihama T, Itoh M, Kuniyasu Y, Nomura T, Toda M, Takahashi T (2001) Immunologic tolerance maintained by $CD25^+$ $CD4^+$ regulatory T cells: their common role in controlling autoimmunity, tumor immunity, and transplantation tolerance. *Immunol Rev* 182: 18–32

10 Jordan MS, Boesteanu A, Reed AJ, Petrone AL, Holenbeck AE, Lerman, MA, Naji A,

Caton AJ (2001) Thymic selection of CD4+CD25+ regulatory T cells induced by an agonist self-peptide. *Nat Immunol* 2: 301–306
11 Hsieh CS, Liang Y, Tyznik AJ, Self SG, Liggitt D, Rudensky AY (2004) Recognition of the peripheral self by naturally arising CD25+ CD4+ T cell receptors. *Immunity* 21: 267–277
12 Pacholczyk R, Kraj P, Ignatowicz L (2002) Peptide specificity of thymic selection of CD4+CD25+ T cells. *J Immunol* 168: 613–620
13 Taams LS, Vukmanovic-Stejic M, Smith J, Dunne PJ, Fletcher JM, Plunkett FJ, Ebeling SB, Lombardi G, Rustin MH, Bijlsma JW et al (2002) Antigen-specific T cell suppression by human CD4+CD25+ regulatory T cells. *Eur J Immunol* 32: 1621–1630
14 Belkaid Y, Piccirillo CA, Mendez S, Shevach EM, Sacks DL (2002) CD4+CD25+ regulatory T cells control Leishmania major persistence and immunity. *Nature* 420: 502–507
15 Taylor PA, Noelle RJ, Blazar BR (2001) CD4(+)CD25(+) immune regulatory cells are required for induction of tolerance to alloantigen *via* costimulatory blockade. *J Exp Med* 193: 1311–1318
16 Hara M, Kingsley CI, Niimi M, Read S, Turvey SE, Bushell AR, Morris PJ, Powrie F, Wood KJ (2001) IL-10 is required for regulatory T cells to mediate tolerance to alloantigens *in vivo*. *J Immunol* 166: 3789–3796
17 Bluestone JA, Abbas AK (2003) Natural *versus* adaptive regulatory T cells. *Nat Rev Immunol* 3: 253–257
18 von Herrath MG, Harrison LC (2003) Antigen-induced regulatory T cells in autoimmunity. *Nat Rev Immunol* 3: 223–232
19 Barrat FJ, Cua DJ, Boonstra A, Richards DF, Crain C, Savelkoul HF, de Waal-Malefyt R, Coffman RL, Hawrylowicz CM, O'Garra A (2002) *In vitro* generation of interleukin 10-producing regulatory CD4(+) T cells is induced by immunosuppressive drugs and inhibited by T helper type 1 (Th1)- and Th2-inducing cytokines. *J Exp Med* 195: 603–616
20 Hoffmann P, Eder R, Kunz-Schughart LA, Andreesen R, Edinger M (2004) Large-scale *in vitro* expansion of polyclonal human CD4(+)CD25^{high} regulatory T cells. *Blood* 104: 895–903
21 Levings MK, Sangregorio R, Roncarolo MG (2001) Human CD25(+)CD4(+) T regulatory cells suppress naive and memory T cell proliferation and can be expanded *in vitro* without loss of function. *J Exp Med* 193: 1295–1302
22 Wakkach A, Fournier N, Brun V, Breittmayer JP, Cottrez F, Groux H (2003) Characterization of dendritic cells that induce tolerance and T regulatory 1 cell differentiation *in vivo*. *Immunity* 18: 605–617
23 Fisson S, Darrasse-Jeze G, Litvinova E, Septier F, Klatzmann D, Liblau R, Salomon BL (2003) Continuous activation of autoreactive CD4+ CD25+ regulatory T cells in the steady state. *J Exp Med* 198: 737–746
24 Huehn J, Siegmund K, Lehmann JC, Siewert C, Haubold U, Feuerer M, Debes GF, Lauber J, Frey O, Przybylski GK et al (2004) Developmental stage, phenotype, and

migration distinguish naive- and effector/memory-like CD4+ regulatory T cells. *J Exp Med* 199: 303–313

25 Nolte-'t Hoen EN, Wagenaar-Hilbers JP, Boot EP, Lin CH, Arkesteijn GJ, van Eden W, Taams LS, Wauben MHM (2004) Identification of a CD4(+)CD25(+) T cell subset committed *in vivo* to suppress antigen-specific T cell responses without additional stimulation. *Eur J Immunol* 34: 3016–3027

26 Thornton AM, Shevach EM (2000) Suppressor effector function of CD4+CD25+ immunoregulatory T cells is antigen nonspecific. *J Immunol* 164: 183–190

27 Kuniyasu Y, Takahashi T, Itoh M, Shimizu J, Toda G, Sakaguchi S (2000) Naturally anergic and suppressive CD25(+)CD4(+) T cells as a functionally and phenotypically distinct immunoregulatory T cell subpopulation. *Int Immunol* 12: 1145–1155

28 Lehmann J, Huehn J, de la Rosa M, Maszyna F, Kretschmer U, Krenn V, Brunner M, Scheffold A, Hamann A (2002) Expression of the integrin alpha Ebeta 7 identifies unique subsets of CD25+ as well as CD25− regulatory T cells. *Proc Natl Acad Sci USA* 99: 13031–13036

29 Gavin MA, Clarke SR, Negrou E, Gallegos A, Rudensky A (2002) Homeostasis and anergy of CD4(+)CD25(+) suppressor T cells *in vivo*. *Nat Immunol* 3: 33–41

30 Yamazaki S, Iyoda T, Tarbell K, Olson K, Velinzon K, Inaba K, Steinman RM (2003) Direct expansion of functional CD25+ CD4+ regulatory T cells by antigen-processing dendritic cells. *J Exp Med* 198: 235–247

31 Unger WW, Hauet-Broere F, Jansen W, van Berkel LA, Kraal G, Samsom JN (2003) Early events in peripheral regulatory T cell induction *via* the nasal mucosa. *J Immunol* 171: 4592–4603

32 Alyanakian MA, You S, Damotte D, Gouarin C, Esling A, Garcia C, Havouis S, Chatenoud L, Bach JF (2003) Diversity of regulatory CD4+T cells controlling distinct organ-specific autoimmune diseases. *Proc Natl Acad Sci USA* 100: 15806–15811

33 McHugh RS, Whitters MJ, Piccirillo CA, Young DA, Shevach EM, Collins M, Byrne M C (2002) CD4(+)CD25(+) immunoregulatory T cells: gene expression analysis reveals a functional role for the glucocorticoid-induced TNF receptor. *Immunity* 16: 311–323

34 Baecher-Allan C, Brown JA, Freeman GJ, Hafler DA (2001) CD4+CD25high regulatory cells in human peripheral blood. *J Immunol* 167: 1245–1253

35 Ng WF, Duggan PJ, Ponchel F, Matarese G, Lombardi G, Edwards AD, Isaacs JD, Lechler RI (2001) Human CD4(+)CD25(+) cells: a naturally occurring population of regulatory T cells. *Blood* 98: 2736–2744

36 Wing K, Ekmark A, Karlsson H, Rudin A, Suri-Payer E (2002) Characterization of human CD25+ CD4+ T cells in thymus, cord and adult blood. *Immunology* 106: 190–199

37 Baecher-Allan C, Viglietta V, Hafler DA (2002) Inhibition of human CD4(+)CD25(+high) regulatory T cell function. *J Immunol* 169: 6210–6217

38 Pearson CM (1956) Development of arthritis, periarthritis and periostitis in rats given adjuvant. *Proc Soc Exp Biol Med* 91: 95

39 Kraal G, Mebius RE (1997) High endothelial venules: lymphocyte traffic control and controlled traffic. *Adv Immunol* 65: 347–395
40 Sallusto F, Lanzavecchia A (2000) Understanding dendritic cell and T-lymphocyte traffic through the analysis of chemokine receptor expression. *Immunol Rev* 177: 134–140
41 Walker LS, Chodos A, Eggena M, Dooms H, Abbas AK (2003) Antigen-dependent proliferation of $CD4^+$ $CD25^+$ regulatory T cells *in vivo*. *J Exp Med* 198: 249–258
42 Klein L, Khazaie K, von Boehmer H (2003) *In vivo* dynamics of antigen-specific regulatory T cells not predicted from behavior *in vitro*. *Proc Natl Acad Sci USA* 100: 8886–8891
43 Battaglia A, Ferrandina G, Buzzonetti A, Malinconico P, Legge F, Salutari V, Scambia G, Fattorossi A (2003) Lymphocyte populations in human lymph nodes. Alterations in $CD4^+$ $CD25^+$ T regulatory cell phenotype and T-cell receptor Vbeta repertoire. *Immunology* 110: 304–312
44 Weiner HL (1997) Oral tolerance: immune mechanisms and treatment of autoimmune diseases. *Immunol Today* 18: 335–343
45 Kraal G, Wolvers DA (2001) Regulation of immunological mucosal tolerance. *Arch Immunol Ther Exp (Warsz)* 49 (Suppl 1): S1–S6
46 Wolvers DA, Coenen-de Roo CJ, Mebius RE, van der Cammen MJ, Tirion F, Miltenburg AM, Kraal G (1999) Intranasally induced immunological tolerance is determined by characteristics of the draining lymph nodes: studies with OVA and human cartilage gp-39. *J Immunol* 162: 1994–1998
47 Weiner HL (2001) The mucosal milieu creates tolerogenic dendritic cells and T(R)1 and T(H)3 regulatory cells. *Nat Immunol* 2: 671–672
48 Iwasaki A, Kelsall BL (2001) Unique functions of $CD11b^+$, CD8 $alpha^+$, and double-negative Peyer's patch dendritic cells. *J Immunol* 166: 4884–4890
49 Akbari O, DeKruyff RH, Umetsu DT (2001) Pulmonary dendritic cells producing IL-10 mediate tolerance induced by respiratory exposure to antigen. *Nat Immunol* 2: 725–731
50 van Eden W, van der Zee R, Paul AG, Prakken BJ, Wendling U, Anderton SM, Wauben MH (1998) Do heat shock proteins control the balance of T-cell regulation in inflammatory diseases? *Immunol Today* 19: 303–307
51 Unger WW, Jansen W, Wolvers DA, van Halteren AG, Kraal G, Samsom JN (2003) Nasal tolerance induces antigen-specific $CD4^+CD25^-$ regulatory T cells that can transfer their regulatory capacity to naive $CD4^+$ T cells. *Int Immunol* 15: 731–739
52 Nolte MA, Hamann A, Kraal G, Mebius RE (2002) The strict regulation of lymphocyte migration to splenic white pulp does not involve common homing receptors. *Immunology* 106: 299–307
53 Nolte MA, Hoen EN, van Stijn A, Kraal G, Mebius RE (2000) Isolation of the intact white pulp. Quantitative and qualitative analysis of the cellular composition of the splenic compartments. *Eur J Immunol* 30: 626–634
54 Shimizu J, Yamazaki S, Sakaguchi S (1999) Induction of tumor immunity by removing $CD25^+CD4^+$ T cells: a common basis between tumor immunity and autoimmunity. *J Immunol* 163: 5211–5218

55 George TC, Bilsborough J, Viney JL, Norment AM (2003) High antigen dose and activated dendritic cells enable Th cells to escape regulatory T cell-mediated suppression *in vitro*. *Eur J Immunol* 33: 502–511
56 Suvas S, Azkur AK, Kim BS, Kumaraguru U, Rouse BT (2004) $CD4^+CD25^+$ regulatory T cells control the severity of viral immunoinflammatory lesions. *J Immunol* 172: 4123–4132
57 Lepault F, Gagnerault MC (2000) Characterization of peripheral regulatory $CD4^+$ T cells that prevent diabetes onset in nonobese diabetic mice. *J Immunol* 164: 240–247
58 Green EA, Gorelik L, McGregor CM, Tran EH, Flavell RA (2003) $CD4^+CD25^+$ T regulatory cells control anti-islet $CD8^+$ T cells through TGF-beta-TGF-beta receptor interactions in type 1 diabetes. *Proc Natl Acad Sci USA* 100: 10878–10883
59 Cao D, Malmstrom V, Baecher-Allan C, Hafler D, Klareskog L, Trollmo C (2003) Isolation and functional characterization of regulatory $CD25brightCD4^+$ T cells from the target organ of patients with rheumatoid arthritis. *Eur J Immunol* 33: 215–223
60 De Kleer IM, Wedderburn LR, Taams LS, Patel A, Varsani H, Klein M, De Jager W, Pugayung G, Giannoni F, Rijkers G et al (2004) CD4(+)CD25(bright) Regulatory T cells actively regulate inflammation in the joints of patients with the remitting form of juvenile idiopathic arthritis. *J Immunol* 172: 6435–6443
61 D'Ambrosio D, Sinigaglia F, Adorini L (2003) Special attractions for suppressor T cells. *Trends Immunol* 24: 122–126
62 Iellem A, Mariani M, Lang R, Recalde H, Panina-Bordignon P, Sinigaglia F, D'Ambrosio D (2001) Unique chemotactic response profile and specific expression of chemokine receptors CCR4 and CCR8 by CD4(+)CD25(+) regulatory T cells. *J Exp Med* 194: 847–853
63 Iellem A, Colantonio L, D'Ambrosio D (2003) Skin-versus gut-skewed homing receptor expression and intrinsic CCR4 expression on human peripheral blood $CD4^+CD25^+$ suppressor T cells. *Eur J Immunol* 33: 1488–1496
64 Colantonio L, Iellem A, Sinigaglia F, D'Ambrosio D (2002) Skin-homing CLA^+ T cells and regulatory $CD25^+$ T cells represent major subsets of human peripheral blood memory T cells migrating in response to CCL1/I-309. *Eur J Immunol* 32: 3506–3514
65 Bystry RS, Aluvihare V, Welch KA, Kallikourdis M, Betz AG (2001) B cells and professional APCs recruit regulatory T cells *via* CCL4. *Nat Immunol* 2: 1126–1132
66 Szanya V, Ermann J, Taylor C, Holness C, Fathman CG (2002) The subpopulation of $CD4^+CD25^+$ splenocytes that delays adoptive transfer of diabetes expresses L-selectin and high levels of CCR7. *J Immunol* 169: 2461–2465

Regulatory T cells and the innate immune system

Kevin J. Maloy and Fiona Powrie

Sir William Dunn School of Pathology, University of Oxford, South Parks Road, Oxford, OX1 3RE, UK

Why regulate innate immune responses?

The traditional components of the innate immune system are leukocytes of the myeloid and granulocytic lineages; namely macrophages, monocytes, dendritic cells (DCs), natural killer (NK) cells, neutrophils, eosinophils, basophils and mast cells [1]. Allied to these are the complement system and tissue cells that constitutively produce factors that inhibit microbial colonization, such as the mucus-secreting goblet cells and defensin-secreting paneth cells of the gastrointestinal tract [2]. As our understanding of how immune responses are co-ordinated against pathogenic organisms has improved, it has become clear that many peripheral tissue cells, such as epithelial and endothelial cells, can also contribute to innate immune responses, primarily through the production of cytokines and chemokines that attract and activate inflammatory cells. Thus, most cell types in the body are able to contribute to innate immune defences, particularly those at mucosal surfaces where pathogens are most frequently encountered and this must be borne in mind when thinking about how innate immune activation may be regulated.

The purpose of the innate immune system is to rapidly respond to foreign organisms, the majority of which are recognized and destroyed by innate immunity without the need for inducing adaptive responses [1, 3]. However, pathogenic organisms have evolved ways to avoid elimination by innate effector mechanisms, necessitating an appropriate adaptive immune response. Under these circumstances, the innate immune system plays two vital roles: firstly to restrict replication and spread of pathogens during the early phase of infection and secondly to provide a range of molecular signals that direct appropriate adaptive immune responses against the pathogen. Immune responses, especially those involving innate immune effector mechanisms, are intrinsically linked to inflammation, meaning that prolonged innate immune responses are likely to lead to destruction of host tissue. This is not usually a problem, because as the pathogen is eradicated the stimuli for the immune response decrease and the inflammation resolves. By contrast, in situations where

Regulatory T Cells in Inflammation, edited by Leonie S. Taams, Arne N. Akbar and Marca H.M. Wauben

the pathogen avoids elimination and establishes chronic infection, prolonged activation of innate immunity is likely. Similarly, whenever a self-antigen becomes the target of the adaptive immune response, the production of cytokines or autoantibodies can lead to persistent activation of innate immune effector mechanisms. In both cases the end result is the same – sustained inflammation and destruction of host tissues. In order to avoid this, it is advantageous for the host to have regulatory mechanisms in place that maintain a balance between limiting pathogen spread while preserving host tissue integrity.

The renaissance of regulatory T (T_R) cells was driven by observations that $CD25^+CD4^+$ T_R cells were able to inhibit a range of autoimmune and inflammatory diseases in a variety of experimental models [4–6]. Most studies on $CD25^+CD4^+$ T_R cells have concentrated on their ability to inhibit pathological adaptive immune responses, particularly those mediated by other T cells. However, as noted above, suppression of other T cells is unlikely to be able to completely resolve all types of immune pathology *in vivo*. As we learn more about $CD25^+CD4^+$ T_R cells, it is becoming clear that they use a variety of mechanisms to prevent harmful inflammatory responses, including pathways that can clearly impede the activation of innate immunity.

Innate immune activation

The innate immune system uses a range of invariant receptors, termed pathogen-recognition receptors (PRRs), to recognize a limited set of conserved molecules present on foreign microbes, known as pathogen-associated molecular patterns (PAMPs) [1]. There are several different types of PRR; some are expressed on the cell surface or in intracellular compartments while others are secreted into the bloodstream and tissue fluids. PRRs mediate distinct functions, including opsonization, phagocytosis, production of inflammatory mediators and activation of the complement cascade.

Perhaps the most important family of PRRs are the Toll-like receptors (TLRs), a set of evolutionarily ancient receptors, of which 13 have so far been identified in mammals [7, 8]. Individual TLRs recognize distinct molecular patterns that are characteristic of different classes of pathogenic organisms. For example, TLR2 recognizes peptidoglycan found in Gram-positive bacteria [9, 10], while TLR4 recognizes lipopolysaccharide (LPS) [11] and TLR5 recognizes flagellin [12], both of which are components of Gram-negative bacteria. TLR9 recognizes unmethylated CpG motifs that are abundant in prokaryotic and viral DNA [13]. Viral infections can also be detected through the recognition of single- or double-stranded RNA, by TLR7 [14, 15] or TLR3 [16], respectively. All TLRs possess a conserved cytoplasmic domain (TIR) that is crucial for initiating the signalling cascades that elicit inflammatory responses [17]. Downstream signalling progresses through a common pathway that is dependent on the adaptor molecule MyD88, eventually resulting in activation of the transcription factor nuclear factor κB(NF-κB), which translocates into the nucleus and activates transcription of inflammatory cytokines. More recently, it has been shown that some TLRs, such as TLR3 and TLR4, can additionally signal *via* MyD88-independent pathways, by utilizing alternative adaptor molecules that lead to the activation of IRF-3, a transcription factor that induces interferon (IFN)-β secretion [18–21]. Thus, it appears that fundamental differences in the biological responses elicited by different TLRs are due to differential activation of the adaptor proteins that control downstream signalling pathways [17, 22].

Although TLRs share similar signal transduction pathways, they show different patterns of expression on different cell types and are also found in distinct cellular locations. Most leukocytes of the innate immune system express a range of TLRs, particularly phagocytic cells such as monocytes, macrophages and neutrophils [8]. In addition, several other cell types may express TLRs, especially those at mucosal surfaces, such as intestinal epithelial cells [23–25]. The immediate consequence of TLR ligation in most cells types is activating innate immune inflammatory responses. However, it is now well established that TLRs also play a fundamental role in shaping subsequent adaptive immune responses, primarily through their expression on the DCs that control T-cell activation.

DCs, TLRs and T_R cells

DCs are the sentinels of the immune system [26, 27]. They detect pathogens through expression of PRRs, especially TLRs. Activation of TLRs on DCs triggers a series of events that are crucial for efficient priming of T-cell responses [28]. Firstly, the DCs undergo a programme of maturation during which they simultaneously up-regulate the surface co-stimulatory molecules involved in antigen presentation, leave the tissue where they have encountered the pathogen and migrate into the T-cell areas of the local draining lymph node. TLR-activated DCs also produce inflammatory cytokines such as interleukin (IL)-6, IFN-α/β, tumour necrosis factor (TNF)-α and IL-12. Thus, by presenting high levels of antigenic peptides and co-stimulatory molecules, together with cytokine "cocktails", TLR-activated DCs direct the activation and differentiation of appropriate types of anti-pathogen effector T-cell response. Interestingly, the production of inflammatory cytokines by TLR-activated DCs is dependent on the MyD88 signalling pathway, while maturation of surface phenotype and migration into the draining lymph node is not.

Although DC migration is enhanced by TLR ligation, even in the absence of inflammation there is constitutive low-level migration of DCs from peripheral tissues into lymphoid tissues [29–32]. Since these constitutively migrating DCs maintain an immature phenotype and do not secrete pro-inflammatory cytokines, it has been suggested that their function might be to elicit tolerogenic T-cell responses towards self-antigens [33]. The preferential activation of naturally occurring $CD25^+CD4^+$ T_R cells by immature DCs *in vivo* remains to be demonstrated, although recent studies have shown that immunization with antigen-bearing DCs triggers activation and proliferation of antigen-specific $CD25^+CD4^+$ T_R cells [34, 35]. In addition, there is growing evidence that immature DCs can drive the development of T cells with a regulatory phenotype, often termed T_R1 cells [36–40]. Although T_R1 cells typically secrete IL-10 and have been shown to be able to inhibit experimental autoimmune diseases, recent evidence suggests that they represent a distinct population to the naturally occurring $CD25^+CD4^+$ T_R cells. For example, while development $CD25^+CD4^+$ T_R cells occurs in the thymus and is directed by the transcription factor Foxp3 [41], IL-10-secreting T_R1 cells may differentiate from peripheral $CD25^-CD4^+$ T cells through a Foxp3-independent process [42]. In addition, while the antigen specificity of $CD25^+CD4^+$ T_R cells remains elusive, induction of pathogen-specific T_R1 cells is frequently observed during chronic infections [43]. It is possible that T_R1 cells may develop under the influence of $CD25^+CD4^+$ T_R cells [44] and additional studies are necessary in order to fully understand their interactions *in vivo*.

If constitutively migrating DCs prime self-antigen specific T_R cells in the draining lymph nodes, then one must ask how this suppressor activity can be overcome to permit immune responses against pathogens. A recent study has provided strong evidence that this may be achieved through TLR signalling in DCs [45]. They observed that freshly isolated DCs, pulsed with anti-CD3 antibodies, stimulated the proliferation of co-cultured $CD25^-CD4^+$ T cells *in vitro*, but that this was suppressed by the addition of $CD25^+CD4^+$ T_R cells to the culture. Importantly, when the DCs were exposed to TLR ligands, such as LPS or CpG DNA, then the $CD25^+CD4^+$ T_R cells were no longer able to inhibit the proliferation of the co-cultured $CD25^-CD4^+$ T cells. By utilizing MyD88-deficient DC and cytokine-blocking antibodies, they could further demonstrate that TLR-activated DC overcame $CD25^+CD4^+$ T_R activity by secreting soluble factors including IL-6, which rendered the $CD25^-CD4^+$ T cells refractory to suppression. Further evidence that TLR signalling on DC can overcome $CD25^+CD4^+$ TR activity was obtained in a study examining $CD8^+$ T-cell responses to "neo-self"-antigens [46]. In these experiments, haemagglutinin (HA)-specific $CD8^+$ T cells were tolerized by adoptive transfer into transgenic mice expressing high levels of HA. Immunization of these mice with recombinant viral vectors expressing HA reversed the $CD8^+$ T cell tolerance and elicited potent HA-specific $CD8^+$ T cell effector responses. By contrast, immunization with DCs expressing HA peptides was unable to break $CD8^+$ T-cell tolerance, unless the recipients were concomitantly depleted of $CD25^+CD4^+$ T_R cells [46]. Furthermore, when administered together with TLR ligands, such as LPS or CpG DNA, DCs expressing HA peptides could also restore $CD8^+$ T cell responses, even in the presence of $CD25^+CD4^+$ T_R cells. Taken together, these studies show that a dynamic equilibrium exists between activation of effector T-cell and T_R-cell responses by DCs and that the TLR-triggered activation of DCs is a crucial factor in overcoming $CD25^+CD4^+$ T_R-cell regulation to shift the balance towards effector T-cell responses.

Given that TLRs play a crucial role in directing adaptive immune responses, then modulation of TLR signalling could provide a means for microbes to avoid eliciting harmful inflammatory responses. Indeed, evidence has recently emerged that a number of pathogenic organisms possess molecules that appear to trigger TLR signalling pathways that drive the development of pathogen-specific T_R cells [43]. For example, studies with *Bordetella pertussis* have found that TLR4 signalling in DCs exposed to *B. pertussis* results in the production not only of pro-inflammatory cytokines, but also in the secretion of large amounts of IL-10 [47]. This innate IL-10 in turn drives the development of B. pertussis-specific T_R1 cells and together they prevent excessive inflammatory pathology during the acute phase of infection. Similarly, a virulence antigen from *Yersinia* spp. that is known to be involved in immunosuppression, LcrV, was found to be able to trigger IL-10 production by macrophages *via* a TLR2/CD14-dependent signalling mechanism [48]. TLR2 signalling has also been recently implicated in suppressing immunity against *Candida albicans*. Here, it was found that TLR2-deficient mice were more resistant to disseminated *Candida* infection than wild-type mice and that this correlated with decreased production of IL-10 and lower numbers of $CD25^+CD4^+$ T_R cells [49]. Finally, a lysophosphatidylserine lipid moiety extracted from *Schistosoma mansoni* can also bind to TLR2 and modulate DCs to drive the development of IL-10-producing regulatory T cells [50], which play a critical role in protecting the host from excessive immune pathology [51]. Since non-pathogenic microbes also possess PAMPs that can trigger TLRs it is perhaps not surprising that some negative-feedback pathways exist to limit TLR-induced inflammatory responses [17]. In an experimental model of chemically induced inflammatory bowel disease (IBD), it was recently reported that administration of a cocktail of non-pathogenic, commensal "probiotic" bacteria had a protective effect on the disease [52]. Intriguingly, the beneficial effect of these probiotics was dependent on TLR9 signalling and could be reproduced using bacterial DNA that possessed CpG motifs. Thus, it appears that TLR signalling may sometimes contribute to the avoidance of severe immune pathology, having mutual benefits for the host and microbe.

The final twist in the complex world of T_R cells, TLRs and DCs is the recent observation that $CD25^+CD4^+$ T_R cells may themselves express a variety of TLRs, including TLR4, TLR5 and TLR7 [53]. Furthermore, exposure of $CD25^+CD4^+$ T_R cells to bacterial LPS led to enhanced proliferation and survival *in vitro*, especially when given in conjugation with T-cell receptor (TCR) ligation. These results suggest that $CD25^+CD4^+$ T_R cells may respond directly to bacterial infection, although whether this is involved in the control of inflammatory responses *in vivo* remains to be established.

The studies described above underscore the complex interactions that can occur between TLR signalling, DC activation and T_R cell function. Ligation of TLRs on DCs elicits not only acute pro-inflammatory responses and drives effector T-cell responses, but also sets in progress natural feedback pathways, including the production of IL-10 and induction of regulatory T cells. The delicate equilibrium between responsiveness and regulation may be essential for the effective control of chronic infections and the avoidance of excessive immune pathology.

Control of innate immune pathology by T_R cells

The aetiology and pathogenesis of most autoimmune and inflammatory disorders are complex, usually involving sustained activation of the both innate and adaptive immune effector mechanisms. This makes it difficult to dissect out any direct effects that T_R cells may have on innate immune activation, from those that may occur indirectly, through inhibition of pathogenic T- and B-cell responses. However, it is clear that innate immune effector mechanisms are associated with pathology in many diseases. For example, infiltrating neutrophils and monocytes play a key role in the gastritis observed in some *Helicobacter pylori*-infected individuals [54–56]. Similarly, it is well established that activated monocytes and macrophages, by secreting large amounts of TNF-α, play a central role in immune pathology in the joints of rheumatoid arthritis (RA) patients [57, 58]. Interestingly, $CD25^+CD4^+$ T_R cells capable of suppressing the proliferation of autologous $CD25^-CD4^+$ T cells are enriched within the inflamed joints of RA patients [59]. However, a recent study suggests that $CD25^+CD4^+$ T_R cells isolated from patients with active RA are functionally defective – although they inhibited T cell proliferation *in vitro*, they were unable to suppress the production of pro-inflammatory cytokines by $CD25^-CD4^+$ effector T cells or monocytes [60]. Furthermore, treatment of RA patients with anti-TNF-α therapy both restored the ability of $CD25^+CD4^+$ T_R cells to suppress pro-inflammatory cytokine production and led to an increase in the number $CD25^+CD4^+$ T_R cells. Additional evidence that $CD25^+CD4^+$ T_R cells may play an important protective role in rheumatic disease comes from a study on patients with juvenile idiopathic arthritis (JIA). Patients with the relatively benign form of disease, oligoarticular JIA, harboured higher numbers of $CD25^+CD4^+$ T_R cells in blood and synovial fluid than patients with the more destructive extended oligoarticular JIA [61]. Together, these results suggest that restoration or enhancement of $CD25^+CD4^+$ T_R cell function could be a useful therapeutic tool for the treatment of arthritic disease.

Some of the most useful models for studying induction and regulation of immune pathology *in vivo* are the rodent models of IBD that share features of the pathology found in human patients with Crohn's disease (CD) or ulcerative colitis (UC) [62]. Although the aetiology of IBD is largely unknown, the current prevailing hypothesis is that they are caused by dysregulated immune responses towards components of the intestinal bacterial flora and innate immune activation is a strong feature of such pathology [63]. An important role for innate immunity in human IBD is further supported by the recent findings that humans with mutations in *NOD2*, a gene encoding a cytosolic PRR that recognizes a muramyl dipeptide (MDP) motif present in bacterial peptidoglycan, show increased susceptibility to CD [64–66]. This finding was somewhat surprising, because recognition of peptidoglycan by NOD2 induces NF-κB activation and production of pro-inflammatory cytokines [66], and therefore mutations in *NOD2* might be expected to result in decreased inflammatory responses. However, a recent study found that peripheral blood mononuclear cells from CD patients harbouring *NOD2* mutations exhibited decreased production of IL-10 and TGF-β after peptidoglycan stimulation [67]. Complimentary findings were made using cells from NOD2-deficient mice, which exhibited greatly enhanced TLR2-mediated NF-κB activation and IL-12 production after stimulation with peptidoglycan [68]. Together, these studies suggest that NOD2 signalling normally down-modulates TLR2-triggered inflammatory responses and offer an explanation of how mutations in *NOD2* could lead to excessive intestinal inflammation in CD patients.

Direct evidence that $CD25^+CD4^+$ T_R cells inhibit innate immune pathology has emerged from studies of murine models of IBD that are based on infection with the pathogenic bacterium *Helicobacter hepaticus*. *H. hepaticus* inhabits the cecum and colon [69] and although it does not usually cause disease in immune-competent mice, several reports have shown that it is capable of triggering typhocolitis in a range of immune-deficient strains [70–74]. By utilizing lymphocyte-deficient 129SvEv $RAG2^{-/-}$ mice, we demonstrated that *H. hepaticus* infection triggered the development of intestinal inflammation mediated by cells of the innate immune system [73]. This innate immune pathology was driven by pro-inflammatory cytokines and was associated with the accumulation of neutrophils and monocytic cells, both locally in the intestinal lamina propria and systemically in the spleen. Importantly, adoptively transferred $CD25^+CD4^+$ T_R cells completely suppressed this innate immune pathology and reduced the accumulation and activation of innate immune cells in the spleen [73]. These results were confirmed by Erdman et al. [74], who additionally demonstrated that $CD25^+CD4^+$ T_R cells also inhibited the development of colon cancer in *H. hepaticus*-infected 129SvEv $RAG2^{-/-}$ mice. Parallel studies by the same investigators identified a further association between *H. hepaticus* infection, innate immune activation and intestinal inflammation, as mice lacking the p50 subunit of NF-κB and heterozygous for the p65 subunit, NF-κBp$50^{-/-}$p$65^{+/-}$, developed typhocolitis following infection with *H. hepaticus* [72]. Interestingly, although $CD4^+$ T cells isolated from both wild-type and NF-κBp$50^{-/-}$p$65^{+/-}$ mice suppressed *H. hepaticus*-triggered inflammation in $RAG^{-/-}$ recipients, neither population of $CD4^+$ T cells was able to prevent the development of colitis in *H. hepaticus*-infected $RAG^{-/-}$ NF-κBp$50^{-/-}$p$65^{+/-}$ recipients [75]. Thus, NF-κBp50/p65 expression by cells of the innate immune system is required in order for T_R cells to be able to regulate the intestinal inflammatory responses triggered by *H. hepaticus*. The molecular basis behind this surprising anti-inflammatory activity of NF-κBp50/p65 remains to be elucidated; however, when compared to wild-type cells, NF-κB

$p50^{-/-}p65^{+/-}$ macrophages produced significantly elevated levels of IL-12p40 antibodies *in vitro* and treatment with anti-IL-12p40 antibodies abrogated colitis in *H. hepaticus*-infected $RAG^{-/-}$ NF-κB$p50^{-/-}p65^{+/-}$ mice [75].

Immune-suppressive cytokines played a key role in controlling innate pathology, as transforming growth factor (TGF)-β was required and the production of IL-10 by $CD25^{+}CD4^{+}$ T_R cells was crucial for the protection from *H. hepaticus*-induced innate intestinal inflammation and colon cancer [73, 76]. These results support previous evidence of the central role of IL-10 in preventing bacterially-driven intestinal inflammation, as $IL\text{-}10^{-/-}$ mice developed severe colitis after infection with *H. hepaticus* [71, 77]. Furthermore, the spontaneous development of colitis in $RAG^{-/-}$ mice colonized with an intestinal flora that included *H. hepaticus* was ameliorated by treatment with recombinant IL-10 [78].

IL-10 is a pleiotropic cytokine and the molecular pathways through which it suppresses innate immune responses remain to be clearly defined. However there is ample evidence that it suppresses macrophage activation, enhances neutrophil apoptosis and can inhibit the production of inflammatory chemokines and cytokines by monocytes, macrophages and neutrophils [79, 80]. Like most cytokine receptors, IL-10 receptor signalling proceeds through the activation of Janus-activated kinase (JAK) tyrosine kinases that in turn activate members of the signal transducer and activator of transcription (STAT) family [79, 80]. The transcription factor STAT3 plays a critical role in the anti-inflammatory actions of IL-10, as transgenic mice selectively lacking STAT3 activity in myeloid cells (monocytes, macrophages and neutrophils) also develop severe enterocolitis [81, 82]. Interestingly, further analysis of these STAT3 mutant mice indicated that colitis development was dependent on the production of IL-12p40 and was also attenuated in TLR4/STAT3 double mutants [83]. These results suggest that, in the absence of IL-10-driven STAT3 signalling, myeloid cells become hyperactivated through TLRs and produce high amounts of the pro-inflammatory cytokines that drive colitis. Although the pathways by which STAT3 mediates anti-inflammatory responses are still poorly understood, recent studies suggest that the suppressors of cytokine signalling (SOCS) family of cytoplasmic proteins are central components of negative-feedback regulatory loops that prevent excessive cytokine-driven inflammatory responses. SOCS proteins are induced following cytokine or TLR signalling and function to attenuate JAK/STAT signal transduction [84, 85]. For example, in macrophages, IL-10 activates STAT3 which in turn leads to the expression of SOCS3, which has been shown to inhibit signalling by IL-6 and related pro-inflammatory cytokines [84, 85].

Outlook

The data discussed in this chapter highlights the importance of regulation of innate immune responses in preventing excessive immune pathology and illustrates the plasticity of immune regulatory networks *in vivo*. Pathogenic microbes that trigger TLRs on innate immune cells can transiently overcome T_R cell activity, allowing potent effector immune responses to occur. Conversely, $CD25^+CD4^+$ T_R cells constitute one mechanism by which innate inflammatory responses may be controlled, but there are other complimentary regulatory pathways that are not yet fully understood. There is a growing realization that the signalling cascades elicited through TLRs and cytokines do not only induce the production of pro-inflammatory molecules, but also set in progress negative-feedback mechanisms that limit the magnitude of inflammatory responses. These regulatory pathways often include the production of IL-10, which activates anti-inflammatory responses and can further enhance the development of T_R cells. Some microbes have evolved mechanisms that hijack these regulatory pathways, often leading to induction of pathogen-specific T_R cells and establishment of persistent infection. As we learn more about how the complex signalling pathways elicited by pro-inflammatory and anti-inflammatory stimuli interact, we should be able to develop therapeutic agents that target defined molecules to either enhance or suppress innate immune activation.

Acknowledgements
KJM is a Wellcome Trust Career Development Fellow and FP is a Wellcome Trust Senior Research Fellow.

References

1 Janeway CA Jr, Medzhitov R (2002) Innate immune recognition. *Annu Rev Immunol* 20: 197–216
2 Ganz T (2003) Defensins: antimicrobial peptides of innate immunity. *Nat Rev Immunol* 3: 710–720
3 Medzhitov R, Janeway C Jr (2000) Innate immunity. *N Engl J Med* 343: 338–344
4 Maloy KJ, Powrie F (2001) Regulatory T cells in the control of immune pathology. *Nat Immunol* 2: 816–822
5 Shevach EM (2002) $CD4^+$ $CD25^+$ suppressor T cells: more questions than answers. *Nat Rev Immunol* 2: 389–400
6 Sakaguchi S (2004) Naturally arising $CD4^+$ regulatory t cells for immunologic self-tolerance and negative control of immune responses. *Annu Rev Immunol* 22: 531–562
7 Medzhitov R (2001) Toll-like receptors and innate immunity. *Nat Rev Immunol* 1: 135–145
8 Takeda K, Kaisho T, Akira S (2003) Toll-like receptors. *Annu Rev Immunol* 21: 335–376
9 Schwandner R, Dziarski R, Wesche H, Rothe M, Kirschning CJ (1999) Peptidoglycan-

and lipoteichoic acid-induced cell activation is mediated by toll-like receptor 2. *J Biol Chem* 274: 17406–17409
10 Takeuchi O, Hoshino K, Kawai T, Sanjo H, Takada H, Ogawa T, Takeda K, Akira S (1999) Differential roles of TLR2 and TLR4 in recognition of gram-negative and gram-positive bacterial cell wall components. *Immunity* 11: 443–451
11 Poltorak A, He X, Smirnova I, Liu MY, Van Huffel C, Du X, Birdwell D, Alejos E, Silva M, Galanos C et al (1998) Defective LPS signaling in C3H/HeJ and C57BL/10ScCr mice: mutations in Tlr4 gene. *Science* 282: 2085–2088
12 Hayashi F, Smith KD, Ozinsky A, Hawn TR, Yi EC, Goodlett DR, Eng JK, Akira S, Underhill DM, Aderem A (2001) The innate immune response to bacterial flagellin is mediated by Toll-like receptor 5. *Nature* 410: 1099–1103
13 Hemmi H, Takeuchi O, Kawai T, Kaisho T, Sato S, Sanjo H, Matsumoto M, Hoshino K, Wagner H, Takeda K et al (2000) A Toll-like receptor recognizes bacterial DNA. *Nature* 408: 740–745
14 Diebold SS, Kaisho T, Hemmi H, Akira S, Reis e Sousa C (2004) Innate antiviral responses by means of TLR7-mediated recognition of single-stranded RNA. *Science* 303: 1529–1531
15 Heil F, Hemmi H, Hochrein H, Ampenberger F, Kirschning C, Akira S, Lipford G, Wagner H, Bauer S (2004) Species-specific recognition of single-stranded RNA *via* toll-like receptor 7 and 8. *Science* 303: 1526–1529
16 Alexopoulou L, Holt AC, Medzhitov R, Flavell RA (2001) Recognition of double-stranded RNA and activation of NF-kappaB by Toll-like receptor 3. *Nature* 413: 732–738
17 Akira S, Takeda K (2004) Toll-like receptor signalling. *Nat Rev Immunol* 4: 499–511
18 Yamamoto M, Sato S, Hemmi H, Hoshino K, Kaisho T, Sanjo H, Takeuchi O, Sugiyama M, Okabe M, Takeda K et al (2003) Role of adaptor TRIF in the MyD88-independent toll-like receptor signaling pathway. *Science* 301: 640–643
19 Hoebe K, Du X, Georgel P, Janssen E, Tabeta K, Kim SO, Goode J, Lin P, Mann N, Mudd S et al (2003) Identification of Lps2 as a key transducer of MyD88-independent TIR signalling. *Nature* 424: 743–748
20 Yamamoto M, Sato S, Hemmi H, Uematsu S, Hoshino K, Kaisho T, Takeuchi O, Takeda K, Akira S (2003) TRAM is specifically involved in the Toll-like receptor 4-mediated MyD88-independent signaling pathway. *Nat Immunol* 4: 1144–1150
21 Fitzgerald KA, Rowe DC, Barnes BJ, Caffrey DR, Visintin A, Latz E, Monks B, Pitha PM, Golenbock DT (2003) LPS-TLR4 signaling to IRF-3/7 and NF-kappaB involves the toll adapters TRAM and TRIF. *J Exp Med* 198: 1043–1055
22 Beutler B (2004) Inferences, questions and possibilities in Toll-like receptor signalling. *Nature* 430: 257–263
23 Cario E, Podolsky DK (2000) Differential alteration in intestinal epithelial cell expression of toll-like receptor 3 (TLR3) and TLR4 in inflammatory bowel disease. *Infect Immun* 68: 7010–7017
24 Hornef MW, Frisan T, Vandewalle A, Normark S, Richter-Dahlfors A (2002) Toll-like

receptor 4 resides in the Golgi apparatus and colocalizes with internalized lipopolysaccharide in intestinal epithelial cells. *J Exp Med* 195: 559–570

25 Gewirtz AT, Navas TA, Lyons S, Godowski PJ, Madara JL (2001) Cutting edge: bacterial flagellin activates basolaterally expressed TLR5 to induce epithelial proinflammatory gene expression. *J Immunol* 167: 1882–1885

26 Banchereau J, Steinman RM (1998) Dendritic cells and the control of immunity. *Nature* 392: 245–252

27 Banchereau J, Briere F, Caux C, Davoust J, Lebecque S, Liu YJ, Pulendran B, Palucka K (2000) Immunobiology of dendritic cells. *Annu Rev Immunol* 18: 767–811

28 Reis e Sousa C (2004) Toll-like receptors and dendritic cells: for whom the bug tolls. *Semin Immunol* 16: 27–34

29 Kurts C, Heath WR, Carbone FR, Allison J, Miller JF, Kosaka H (1996) Constitutive class I-restricted exogenous presentation of self antigens *in vivo*. *J Exp Med* 184: 923–930.

30 Adler AJ, Marsh DW, Yochum GS, Guzzo JL, Nigam A, Nelson WG, Pardoll DM (1998) $CD4^+$ T cell tolerance to parenchymal self-antigens requires presentation by bone marrow-derived antigen-presenting cells. *J Exp Med* 187: 1555–1564

31 Huang FP, Platt N, Wykes M, Major JR, Powell TJ, Jenkins CD, MacPherson GG (2000) A discrete subpopulation of dendritic cells transports apoptotic intestinal epithelial cells to T cell areas of mesenteric lymph nodes. *J Exp Med* 191: 435–444

32 Hemmi H, Yoshino M, Yamazaki H, Naito M, Iyoda T, Omatsu Y, Shimoyama S, Letterio JJ, Nakabayashi T, Tagaya H et al (2001) Skin antigens in the steady state are trafficked to regional lymph nodes by transforming growth factor-beta1-dependent cells. *Int Immunol* 13: 695–704

33 Steinman RM, Hawiger D, Nussenzweig MC (2003) Tolerogenic dendritic cells. *Annu Rev Immunol* 21: 685–711

34 Walker LS, Chodos A, Eggena M, Dooms H, Abbas AK (2003) Antigen-dependent proliferation of $CD4^+$ $CD25^+$ regulatory T cells *in vivo*. *J Exp Med* 198: 249–258

35 Yamazaki S, Iyoda T, Tarbell K, Olson K, Velinzon K, Inaba K, Steinman RM (2003) Direct expansion of functional $CD25^+$ $CD4^+$ regulatory T cells by antigen-processing dendritic cells. *J Exp Med* 198: 235–247

36 Roncarolo MG, Levings MK, Traversari C (2001) Differentiation of T regulatory cells by immature dendritic cells. *J Exp Med* 193: F5–F9

37 Jonuleit H, Schmitt E, Schuler G, Knop J, Enk AH (2000) Induction of interleukin 10-producing, nonproliferating CD4(+) T cells with regulatory properties by repetitive stimulation with allogeneic immature human dendritic cells. *J Exp Med* 192: 1213–1222

38 Dhodapkar MV, Steinman RM, Krasovsky J, Munz C, Bhardwaj N (2001) Antigen-specific inhibition of effector T cell function in humans after injection of immature dendritic cells. *J Exp Med* 193: 233–238

39 Menges M, Rossner S, Voigtlander C, Schindler H, Kukutsch NA, Bogdan C, Erb K, Schuler G, Lutz MB (2002) Repetitive injections of dendritic cells matured with tumor

necrosis factor alpha induce antigen-specific protection of mice from autoimmunity. *J Exp Med* 195: 15–21

40 Wakkach A, Fournier N, Brun V, Breittmayer JP, Cottrez F, Groux H (2003) Characterization of dendritic cells that induce tolerance and T regulatory 1 cell differentiation *in vivo*. *Immunity* 18: 605–617

41 Hori S, Sakaguchi S (2004) Foxp3: a critical regulator of the development and function of regulatory T cells. *Microbes Infect* 6: 745–751

42 Vieira PL, Christensen JR, Minaee S, O'Neill EJ, Barrat FJ, Boonstra A, Barthlott T, Stockinger B, Wraith DC, O'Garra A (2004) IL-10-secreting regulatory T cells do not express Foxp3 but have comparable regulatory function to naturally occurring $CD4^+CD25^+$ regulatory T cells. *J Immunol* 172: 5986–5993

43 Mills KH, McGuirk P (2004) Antigen-specific regulatory T cells – their induction and role in infection. *Semin Immunol* 16: 107–117

44 Jonuleit H, Schmitt E (2003) The regulatory T cell family: distinct subsets and their interrelations. *J Immunol* 171: 6323–6327

45 Pasare C, Medzhitov R (2003) Toll pathway-dependent blockade of $CD4^+CD25^+$ T cell-mediated suppression by dendritic cells. *Science* 299: 1033–1036

46 Yang Y, Huang CT, Huang X, Pardoll DM (2004) Persistent Toll-like receptor signals are required for reversal of regulatory T cell-mediated CD8 tolerance. *Nat Immunol* 5: 508–515

47 Higgins SC, Lavelle EC, McCann C, Keogh B, McNeela E, Byrne P, O'Gorman B, Jarnicki A, McGuirk P, Mills KH (2003) Toll-like receptor 4-mediated innate IL-10 activates antigen-specific regulatory T cells and confers resistance to *Bordetella pertussis* by inhibiting inflammatory pathology. *J Immunol* 171: 3119–3127

48 Sing A, Rost D, Tvardovskaia N, Roggenkamp A, Wiedemann A, Kirschning CJ, Aepfelbacher M, Heesemann J (2002) *Yersinia* V-antigen exploits toll-like receptor 2 and CD14 for interleukin 10-mediated immunosuppression. *J Exp Med* 196: 1017–1024

49 Netea MG, Sutmuller R, Hermann C, Van der Graaf CA, Van der Meer JW, van Krieken JH, Hartung T, Adema G, Kullberg BJ (2004) Toll-like receptor 2 suppresses immunity against *Candida albicans* through induction of IL-10 and regulatory T cells. *J Immunol* 172: 3712–3718

50 van der Kleij D, Latz E, Brouwers JF, Kruize YC, Schmitz M, Kurt-Jones EA, Espevik T, de Jong EC, Kapsenberg ML, Golenbock DT et al (2002) A novel host-parasite lipid cross-talk. Schistosomal lyso-phosphatidylserine activates toll-like receptor 2 and affects immune polarization. *J Biol Chem* 277: 48122–48129

51 Hesse M, Piccirillo CA, Belkaid Y, Prufer J, Mentink-Kane M, Leusink M, Cheever AW, Shevach EM, Wynn TA (2004) The pathogenesis of schistosomiasis is controlled by cooperating IL-10-producing innate effector and regulatory T cells. *J Immunol* 172: 3157–3166

52 Rachmilewitz D, Katakura K, Karmeli F, Hayashi T, Reinus C, Rudensky B, Akira S, Takeda K, Lee J, Takabayashi K et al (2004) Toll-like receptor 9 signaling mediates the

anti-inflammatory effects of probiotics in murine experimental colitis. *Gastroenterology* 126: 520–528
53 Caramalho I, Lopes-Carvalho T, Ostler D, Zelenay S, Haury M, Demengeot J (2003) Regulatory T cells selectively express toll-like receptors and are activated by lipopolysaccharide. *J Exp Med* 197: 403–411
54 Crabtree JE (1996) Immune and inflammatory responses to Helicobacter pylori infection. *Scand J Gastroenterol* (Suppl) 215: 3–10
55 Ernst PB, Gold BD (2000) The disease spectrum of Helicobacter pylori: the immunopathogenesis of gastroduodenal ulcer and gastric cancer. *Annu Rev Microbiol* 54: 615–640
56 Dixon MF (2001) Pathology of gastritis and peptic ulceration. In: HL Mobley, GL Mendz, SL Hazell (eds): *Helicobacter pylori: physiology and genetics*. ASM Press, Washington, DC, 459–469
57 Lee DM, Weinblatt ME (2001) Rheumatoid arthritis. *Lancet* 358: 903–911
58 Firestein GS (2003) Evolving concepts of rheumatoid arthritis. *Nature* 423: 356–361
59 Cao D, Malmstrom V, Baecher-Allan C, Hafler D, Klareskog L, Trollmo C (2003) Isolation and functional characterization of regulatory CD25brightCD4$^+$ T cells from the target organ of patients with rheumatoid arthritis. *Eur J Immunol* 33: 215–223
60 Ehrenstein MR, Evans JG, Singh A, Moore S, Warnes G, Isenberg DA, Mauri C (2004) Compromised function of regulatory T cells in rheumatoid arthritis and reversal by anti-TNFα therapy. *J Exp Med* 200: 277–285
61 de Kleer IM, Wedderburn LR, Taams LS, Patel A, Varsani H, Klein M, de Jager W, Pugayung G, Giannoni F, Rijkers G et al (2004) CD4$^+$CD25(bright) regulatory T cells actively regulate inflammation in the joints of patients with the remitting form of juvenile idiopathic arthritis. *J Immunol* 172: 6435–6443
62 Strober W, Fuss IJ, Blumberg RS (2002) The immunology of mucosal models of inflammation. *Annu Rev Immunol* 20: 495–549
63 Podolsky DK (2002) Inflammatory bowel disease. *N Engl J Med* 347: 417–429
64 Hugot JP, Chamaillard M, Zouali H, Lesage S, Cezard JP, Belaiche J, Almer S, Tysk C, O'Morain CA, Gassull M et al (2001) Association of NOD2 leucine-rich repeat variants with susceptibility to Crohn's disease. *Nature* 411: 599–603
65 Ogura Y, Bonen DK, Inohara N, Nicolae DL, Chen FF, Ramos R, Britton H, Moran T, Karaliuskas R, Duerr RH et al (2001) A frameshift mutation in NOD2 associated with susceptibility to Crohn's disease. *Nature* 411: 603–606
66 Girardin SE, Hugot JP, Sansonetti PJ (2003) Lessons from Nod2 studies: towards a link between Crohn's disease and bacterial sensing. *Trends Immunol* 24: 652–658
67 Netea MG, Kullberg BJ, De Jong DJ, Franke B, Sprong T, Naber TH, Drenth JP, Van Der Meer JW (2004) NOD2 mediates anti-inflammatory signals induced by TLR2 ligands: implications for Crohn's disease. *Eur J Immunol* 34: 2052–2059
68 Watanabe T, Kitani A, Murray PJ, Strober W (2004) NOD2 is a negative regulator of Toll-like receptor 2-mediated T helper type 1 responses. *Nat Immunol* 5: 800–808
69 Fox JG, Dewhirst FE, Tully JG, Paster BJ, Yan L, Taylor NS, Collins MJ, Jr., Gorelick

PL, Ward JM (1994) *Helicobacter hepaticus* sp. nov., a microaerophilic bacterium isolated from livers and intestinal mucosal scrapings from mice. *J Clin Microbiol* 32: 1238–1245

70 Cahill RJ, Foltz CJ, Fox JG, Dangler CA, Powrie F, Schauer DB (1997) Inflammatory bowel disease: an immunity-mediated condition triggered by bacterial infection with *Helicobacter hepaticus*. *Infect Immun* 65: 3126–3131

71 Kullberg MC, Ward JM, Gorelick PL, Caspar P, Hieny S, Cheever A, Jankovic D, Sher A (1998) *Helicobacter hepaticus* triggers colitis in specific-pathogen-free interleukin-10 (IL-10)-deficient mice through an IL-12- and gamma interferon-dependent mechanism. *Infect Immun* 66: 5157–5166

72 Erdman S, Fox JG, Dangler CA, Feldman D, Horwitz BH (2001) Typhlocolitis in NF-kappa B-deficient mice. *J Immunol* 166: 1443–1447

73 Maloy KJ, Salaun L, Cahill R, Dougan G, Saunders NJ, Powrie F (2003) $CD4^{+}CD25^{+}$ T(R) cells suppress innate immune pathology through cytokine-dependent mechanisms. *J Exp Med* 197: 111–119

74 Erdman SE, Poutahidis T, Tomczak M, Rogers AB, Cormier K, Plank B, Horwitz BH, Fox JG (2003) $CD4^{+}$ $CD25^{+}$ regulatory T lymphocytes inhibit microbially induced colon cancer in Rag2-deficient mice. *Am J Pathol* 162: 691–702

75 Tomczak MF, Erdman SE, Poutahidis T, Rogers AB, Holcombe H, Plank B, Fox JG, Horwitz BH (2003) NF-kappa B is required within the innate immune system to inhibit microflora-induced colitis and expression of IL-12 p40. *J Immunol* 171: 1484–1492

76 Erdman SE, Rao VP, Poutahidis T, Ihrig MM, Ge Z, Feng Y, Tomczak M, Rogers AB, Horwitz BH, Fox JG (2003) CD4(+)CD25(+) regulatory lymphocytes require interleukin 10 to interrupt colon carcinogenesis in mice. *Cancer Res* 63: 6042–6050

77 Kullberg MC, Rothfuchs AG, Jankovic D, Caspar P, Wynn TA, Gorelick PL, Cheever AW, Sher A (2001) *Helicobacter hepaticus*-induced colitis in interleukin-10-deficient mice: cytokine requirements for the induction and maintenance of intestinal inflammation. *Infect Immun* 69: 4232–4241

78 von Freeden-Jeffry U, Davidson N, Wiler R, Fort M, Burdach S, Murray R (1998) IL-7 deficiency prevents development of a non-T cell non-B cell-mediated colitis. *J Immunol* 161: 5673–5680

79 Moore KW, de Waal Malefyt R, Coffman RL, O'Garra A (2001) Interleukin-10 and the interleukin-10 receptor. *Annu Rev Immunol* 19: 683–765

80 Pestka S, Krause CD, Sarkar D, Walter MR, Shi Y, Fisher PB (2004) Interleukin-10 and related cytokines and receptors. *Annu Rev Immunol* 22: 929–979

81 Takeda K, Clausen BE, Kaisho T, Tsujimura T, Terada N, Forster I, Akira S (1999) Enhanced Th1 activity and development of chronic enterocolitis in mice devoid of Stat3 in macrophages and neutrophils. *Immunity* 10: 39–49

82 Welte T, Zhang SS, Wang T, Zhang Z, Hesslein DG, Yin Z, Kano A, Iwamoto Y, Li E, Craft JE et al (2003) STAT3 deletion during hematopoiesis causes Crohn's disease-like pathogenesis and lethality: a critical role of STAT3 in innate immunity. *Proc Natl Acad Sci USA* 100: 1879–1884

83 Kobayashi M, Kweon MN, Kuwata H, Schreiber RD, Kiyono H, Takeda K, Akira S (2003) Toll-like receptor-dependent production of IL-12p40 causes chronic enterocolitis in myeloid cell-specific Stat3-deficient mice. *J Clin Invest* 111: 1297–1308
84 Kubo M, Hanada T, Yoshimura A (2003) Suppressors of cytokine signaling and immunity. *Nat Immunol* 4: 1169–1176
85 Alexander WS, Hilton DJ (2004) The role of suppressors of cytokine signaling (SOCS) proteins in regulation of the immune response. *Annu Rev Immunol* 22: 503–529

Part II
The potential use of regulatory T cells in immunotherapy

Exploiting the potential of regulatory T cells in the control of type 1 diabetes

Nadia Giarratana, Giuseppe Penna, Silvia Gregori, Kenn C. Daniel and Luciano Adorini

BioXell, Via Olgettina 58, I-20132 Milano, Italy

Introduction

The non-obese diabetic (NOD) mouse spontaneously develops autoimmune diabetes, and represents a reliable model for human type 1 diabetes (T1D) [1–3]. Several cellular mechanisms have been implicated in T1D development in the NOD mouse, including interleukin (IL)-12-dependent Th1 cells [4], cytotoxic $CD8^+$ lymphocytes and macrophages [5], and defective immunoregulation [6]. The activation of pathogenic Th1 cells specific for pancreatic autoantigens could reflect defective thymic deletion of autoreactive T-cell clones [7], inefficient mechanisms of peripheral tolerance [6], alteration in IL-12 expression [8], or impaired $CD4^+CD25^+$ suppressor T cells [9].

Among the various populations of regulatory/suppressor T cells described, naturally occurring thymic and peripheral $CD4^+$ T cells that co-express CD25 ($CD25^+$Ts) are currently the most actively investigated [10]. $CD25^+$Ts prevent the activation and proliferation of potentially autoreactive T cells that have escaped thymic deletion [11]. They fail to proliferate and secrete cytokines in response to polyclonal or antigen-specific stimulation, and are not only anergic but also inhibit the activation of responsive T cells [10]. Although CD25, CD152, and glucocorticoid-induced tumor necrosis factor (TNF)-receptor family-related gene (GITR) are markers of $CD25^+$Ts cells, they are also expressed by activated T cells [10]. A more faithful marker distinguishing $CD25^+$Ts from recently activated $CD4^+$ T cells is Foxp3, a member of the forkhead family of transcription factors that is required for $CD25^+$Ts development and is sufficient for their suppressive function [12–14]. $Foxp3^+CD25^+$Ts cells play an important role in preventing the induction of several autoimmune diseases, such as the autoimmune syndrome induced by day-3 thymectomy in genetically susceptible mice [11], inflammatory bowel disease [15], and T1D in thymectomized rats [16] and in NOD mice [9, 17]. $CD25^+$Ts are reduced in NOD compared to other mouse strains, and this reduc-

Regulatory T Cells in Inflammation, edited by Leonie S. Taams, Arne N. Akbar and Marca H.M. Wauben

tion could be a factor in their susceptibility to T1D [9, 18]. $CD25^+$Ts and effector T cells coexist within the pancreatic lesion before the onset of T1D, and several factors, such as blockade of inducible co-stimulator (ICOS) [19], can perturb this balance, precipitating autoimmunity. A defect in peripheral regulatory cells affecting both $CD25^+$Ts and NK cells has been described also in T1D patients [20], and autoreactive T cells in diabetics are skewed to a proinflammatory Th1 phenotype lacking the IL-10-secreting T cells found in non-diabetic, HLA-matched controls [21].

Suppression of T1D by adoptive transfer of regulatory T cells

Adoptive transfer of $CD25^+$Ts inhibits T1D induced in immunodeficient NOD mice by diabetogenic T cells [9, 18, 22], indicating the potential of a cell-based therapy in controlling, and possibly reversing, disease progression. Although these data have established a preclinical proof of concept for this approach, the heterogeneity within the $CD25^+$Ts population and the low precursor frequency of cells specific for islet autoantigens makes it quite inefficient, usually requiring a relatively high ratio of $CD25^+$Ts to T effector cells to effectively inhibit T1D development.

To overcome these limitations, two different strategies for *in vitro* expansion of antigen-specific $CD25^+$Ts have been pursued: one based on direct stimulation of $CD25^+$Ts with a cocktail of inducing agents and growth factors, and an alternative one based on dendritic-cell (DC)-induced expansion of $CD25^+$Ts. The first approach utilizes a combination of anti-CD3, anti-CD28, and IL-2 to expand *in vitro*-purified $CD25^+$Ts from BDC2.5 T-cell receptor (TCR) transgenic mice [23]. The expanded $CD25^+$Ts are very efficient suppressor cells, and they cannot only prevent T1D transfer, but also stably reverse it after disease onset in the majority of recipients, a key point for the clinical applicability of this approach. Alternatively, IL-2 and transforming growth factor (TGF)-β can also be used to generate large numbers of $CD25^+$Ts *ex vivo* from naïve T cells [24]. Interestingly, adoptively transferred $CD25^+$Ts induced *ex vivo* by IL-2/TGF-β could generate new $CD25^+$Ts *in vivo*, thus sustaining long-term suppressive effects [25].

The second approach utilizes DCs from NOD mice to directly expand $CD25^+$Ts from BDC2.5 TCR transgenic mice [26, 27]. In this case, $CD25^+$Ts expanded by DCs pulsed with an autoantigenic peptide mimotope could suppress T1D development very efficiently, and very low numbers of $CD25^+$Ts, as low as 5×10^3/recipient, were sufficient to inhibit disease in about 50% of the recipients [27]. Reversal of overt T1D has not yet been demonstrated in this model, although DC-expanded $CD25^+$Ts could still block diabetes development when transferred 15 days after diabetogenic cells [27].

Both cell-based approaches may have potential clinical applications in the treatment of T1D and other autoimmune diseases, although the second one could be applied only to diseases in which the inciting autoantigens can be identified. Since autoantigens involved in the pathogenesis of T1D are fairly well known [21], induction of functional $CD25^{+}$Ts with the correct antigen specificity should not be an issue. However, several other problems, common to any cell-based treatment, will need to be solved before *ex vivo* expansion of $CD25^{+}$Ts can be translated to the clinic. In addition, $CD25^{+}$Ts will be likely obtained from the patient's peripheral blood, requiring efficient protocols able to selectively expand *ex vivo* the islet antigen-specific $CD25^{+}$Ts to be reinfused.

Treatment with immunomodulatory agents can enhance regulatory T cells and inhibit T1D development

Although adoptive transfer of *in vitro*-generated $CD25^{+}$Ts could represent a viable option for the treatment of T1D, the development of pharmacologic agents able to induce and/or enhance $CD25^{+}$Ts in affected individuals would have distinct advantages. Along these lines, a superagonistic antibody specific for CD28 has been shown to preferentially expand $CD25^{+}$Ts *in vivo*, leading to a 20-fold increase of this T-cell subset within 3 days in response to a single antibody dose [28]. An interesting example of this class of agents is represented by anti-CD3 monoclonal antibodies. CD3-specific monoclonal antibodies, which have been in clinical use for over two decades in the treatment of acute allograft rejection, have shown the capacity to induce operational tolerance to islet cell autoantigens in mouse models of T1D [29]. CD3-specific antibodies induce long-term disease remission in spontaneous [30, 31] and virus-induced [32] T1D models by rapidly clearing pathogenic T cells from the pancreatic islets, and subsequently promoting induction of TGF-β-dependent $CD25^{+}$Ts cells [33]. These results are consistent with the observation that a transient pulse of TGF-β in the islets during the priming phase of T1D is sufficient to inhibit disease onset by promoting the expansion of intra-islet $Foxp3^{+}CD25^{+}$Ts cells [34].

Anti-CD3 monoclonal antibodies have also been tested clinically in T1D patients. Administration of humanized, Fc-mutated antibodies to the human CD3ε chain could induce preservation of β-cell function in recently diagnosed T1D patients for up to 24 months following a short course of treatment [35], associated with the generation of IL-10-producing $CD4^{+}$ T cells that could be involved in the beneficial effect on islet β-cell destruction [36]. Although the effects of a short course of anti-CD3 treatment are relatively long-lasting, it remains to be seen for how long they can be sustained, and if they can induce insulin independence. Mitogenicity and immunogenicity, two major adverse effects of anti-CD3 monoclonal antibodies, are considerably decreased in the genetically engineered antibody versions used in clinical trials [29]. This has been achieved by eliminating the CH_2 domain glycosylation site, as in the ChAgly CD3 monoclonal antibody (mAb) [37], or by mutating the Fc region, as in the huOKT3γ1 Ala-Ala mAb [38]. However, adverse events associated with inhibition of immune responses could still occur, and this has caused the suspension of the second phase II trial in new-onset T1D patients treated with the huOKT3γ1 Ala-Ala antibody. This antibody has also been used, in combination with sirolimus, to induce immunosuppression in islet-transplantation recipients [39]. The four insulin-independent patients, out of the six transplanted, showed prolonged $CD4^+$ T-cell lymphocytopenia, inverted CD4/CD8 ratios, and increases in the percentage of $CD25^+$Ts cells. These cells suppressed the *in vitro* proliferative response to donor cells and, to a lesser extent, to third-party cells. Severe adverse events were limited to a transient rash in one recipient and to temporary neutropenia in three, suggesting that a combination of appropriate immunosuppressive agents can enhance $CD25^+$Ts cells able to control immune responses without inducing major adverse events.

Treatment with vitamin D receptor ligands leads to enhancement of $CD25^+$Ts inhibiting T1D development

In addition to biological agents, small organic molecules can also induce and/or enhance $CD25^+$Ts cells [40]. In this respect, the activated form of vitamin D_3, 1,25-dihydroxyvitamin D_3 [1,25$(OH)_2D_3$], has received considerable attention. 1,25$(OH)_2D_3$ is a secosteroid hormone that has, in addition to its central function in calcium and bone metabolism, important effects on the growth and differentiation of many cell types, and pronounced immunoregulatory properties [41–45]. The biological effects of 1,25$(OH)_2D_3$ are mediated by the vitamin D receptor (VDR), a member of the superfamily of nuclear hormone receptors functioning as a ligand-activated transcription factor that binds to specific DNA sequence elements, vitamin D-responsive elements, in vitamin D-responsive genes and ultimately influences their rate of RNA polymerase II-mediated transcription [46].

Antigen-presenting cells (APCs), and notably DCs, express the VDR and are key targets of VDR ligands, both *in vitro* and *in vivo*. A number of studies has clearly demonstrated that 1,25$(OH)_2D_3$ and its analogues inhibit the differentiation and maturation of DCs [47–52]. These studies, performed on either monocyte-derived DCs from human peripheral blood or bone marrow-derived mouse DCs, have consistenly shown that *in vitro* treatment of DCs with 1,25$(OH)_2D_3$ and its analogues leads to down-regulated expression of the co-stimulatory molecules CD40, CD80, and CD86, and to decreased IL-12 and enhanced IL-10 production, resulting in decreased T-cell activation. The block of maturation, coupled with abrogation of IL-12 and strongly enhanced production of IL-10, highlight the important functional effects of 1,25$(OH)_2D_3$ and its analogues on DCs and are, at least in part, responsible for the induction of DCs with tolerogenic properties. The combination of these effects can explain the capacity of VDR ligands to promote DCs with tolerogenic properties that favor $CD25^{+}$Ts induction. DCs are able to synthesize $1,25(OH)_2D_3$ *in vitro* as a consequence of increased 1α-hydroxylase expression [53], and this could also contribute to promote $CD25^{+}$Ts induction. It is also possible that $1,25(OH)_2D_3$ may contribute to the physiological control of immune responses, and perhaps also be involved in maintaining tolerance to self-antigens, as suggested by the enlarged lymph nodes containing a higher frequency of mature DCs in VDR-deficient mice [54].

The prevention of DC differentiation and maturation as well as the modulation of their activation and survival leading to DCs with tolerogenic phenotype and function appears thus to play an important role in the immunoregulatory activity of VDR ligands, although they can also exert, in combination with dexamethasone, direct regulatory effects on T cells [55, 56]. Interestingly, in this case IL-10-secreting regulatory T cells are induced but they fail to express Foxp3, although they show suppressive activity comparable to $CD25^{+}$Ts cells [57], suggesting that these two cell types may have distinct origins.

$1,25(OH)_2D_3$ and its analogues can also induce DCs with tolerogenic properties *in vivo*, as demonstrated in models of allograft rejection by oral administration directly to the recipient [58] or by adoptive transfer of *in vitro*-treated DCs [54]. We have found that a short treatment with $1,25(OH)_2D_3$ induces tolerance to fully mismatched mouse islet allografts that is stable to challenge with donor-type spleen cells and allows acceptance of donor-type vascularized heart grafts [58]. This effect is enhanced by co-administration of mycophenolate mofetil, a selective inhibitor of T- and B-cell proliferation, that also has effects similar to $1,25(OH)_2D_3$ on DCs [59]. Graft acceptance is associated with an increased percentage of $CD25^{+}$Ts cells in the spleen and in the draining lymph node, and they can protect 100% of syngeneic recipients from islet allograft rejection [58].

$CD25^{+}$Ts cells able to inhibit the T cell response to the pancreatic autoantigen IA-2, and to significantly delay disease transfer by pathogenic $CD4^{+}CD25^{-}$ T cells, are also induced by treatment of adult NOD mice with the VDR agonist BXL-219 [60]. Importantly, treatment of adult NOD mice with BXL-219 arrests insulitis, blocks the progression of Th1-cell infiltration into the pancreatic islets, and inhibits T1D development at non-hypercalcemic doses [60]. Although the T1D and islet-transplantation models are quite different, in both cases administration of VDR ligands doubles the number of $CD25^{+}$Ts cells, both in the spleen and in pancreatic lymph nodes [58, 60]. Treatment of pre-diabetic adult NOD mice with BXL-219 appears also to enhance intra-islet $CD25^{+}$Ts cells, as indicated by the over 2-fold increase of Foxp3 transcripts observed in islets from BXL-219-treated NOD mice compared to vehicle-treated controls (Fig. 1). The strong response of islet cells to systemic treatment with BXL-219 is shown by the marked up-regulation of transcripts encoding 25-hydroxyvitamin D-24-hydroxylase (CYP24, Fig. 1), a primary vitamin D-responsive gene [46].

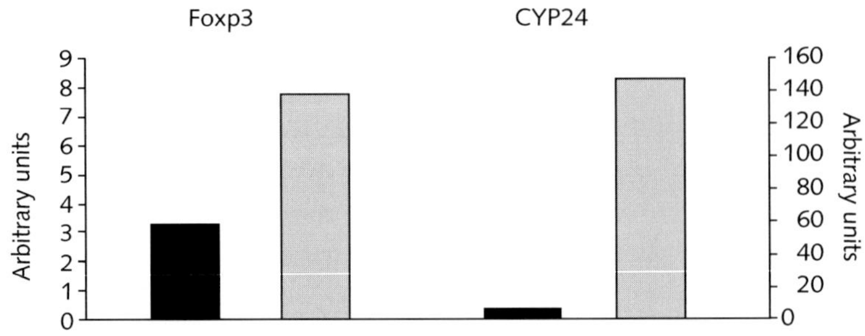

Figure 1
Foxp3 up-regulation in NOD islets is induced by BXL-219 administration.
Eight-week-old NOD female mice were treated orally daily for 1 week with vehicle (black bars) or with 0.1 μg/kg BXL-219 (grey bars). Foxp3 and CYP24 mRNA levels, shown as arbitrary units normalized to GAPDH expression, were detected by real-time reverse transcriptase (RT)-PCR in purified islets tested immediately after isolation.

Modulation of chemokine production by VDR ligands can affect $CD25^+$Ts and effector T cell recruitment in T1D pathogenesis

In both islet-transplantation and T1D models, treatment with VDR ligands has a profound effect on the migration of effector T cells, preventing their entry into the pancreatic islets [58, 60]. Thus, VDR ligand-induced down-regulation of chemokine production by islet cells could represent an important mechanism of action leading to inhibition of T1D development. We have found that transcripts encoding all Toll-like receptors (TLRs) are expressed by mouse and human islet cells and they are functional, as demonstrated by the marked up-regulation of chemokine production following TLR engagement by specific ligands [61], suggesting that TLR-mediated up-regulation of proinflammatory chemokine production like CXCL10, CCL2 and CCL5 by islet cells plays an important role in the early events leading to leukocyte infiltration into the pancreatic islets. The constitutive and inducible production by mouse and human islet cells of CXCL10, a ligand for CXCR3 expressed by Th1 cells [62], was most prominent. CXCL10 has been implicated in human T1D, as elevated serum levels have been observed in diabetes patients and in autoantibody-positive subjects at risk of developing the disease [63, 64]. In addition, our results show that CCL5, the ligand for CCR5, another chemokine receptor expressed by Th1 cells [62], is also constitutively produced by islet cells and is markedly up-regulated following TLR ligation. Mouse islet cells produce, besides CCL5, the CCL2 and

CCL3 ligands able to recruit $CCR1^+$ and $CCR2^+$ macrophages [61]. CCL2 has been shown to be produced also by human islet cells, and it appears to play an important role in the clinical outcome of islet transplantation in T1D patients [65]. Islet-produced CXCL10, CCL5, CCL2 and CCL3 could also recruit immature dendritic cells [62]. Thus, pancreatic β cells, as well as other islet cell types, produce chemokines potentially able to attract the pathogenic cells ultimately responsible for β-cell death.

BXL-219 significantly down-regulates *in vitro* and *in vivo* proinflammatory chemokine production by islet cells, inhibiting T-cell recruitment into the pancreatic islets and T1D development [61]. The inhibition of CXCL10 may be particularly relevant, consistent with the decreased recruitment of Th1 cells into sites of inflammation by treatment with an anti-CXCR3 antibody [66], and with the substantial delay of T1D development observed in CXCR3-deficient mice [67]. The inhibition of islet chemokine production by BXL-219 treatment *in vivo* persists after restimulation with TLR ligands and is associated with up-regulation of IκBα transcription, an inhibitor of nuclear factor κB (NF-κB), and with arrest of NF-κBp65 nuclear translocation [61], highlighting a novel mechanism of action exerted by VDR ligands potentially relevant for the treatment of T1D and other autoimmune diseases. These observations expand the known mechanisms of action exerted by vitamin D analogs in the treatment of T1D and other autoimmune diseases, which include arrest of DC maturation, inhibition of Th1-cell responsiveness, and enhancement of regulatory T cells [42, 43, 45]. In addition to modulating chemokine production in target tissues such as pancreatic islets, it is also possible that VDR ligands can affect the migration of $CD25^+$Ts cells by regulating their chemokine receptor expression, a hypothesis that we are currently testing.

Human $CD25^+$Ts cells express CCR4, CCR8, and CCR5, displaying a rather unique chemokine receptor profile [68]. A similar situation exists also in the mouse: $CD4^+CD25^+$Ts cells from either BALB/c or NOD mice selectively migrate in response to CCR4 and CCR8 ligands, whereas $CD4^+CD25^-$ T cells migrate preferentially to CCL19 and CCL21 (Fig. 2A). Purified splenic $CD4^+CD25^+$ T cells from both BALB/c and NOD mice show a comparable suppressive activity, possibly with a slightly higher efficacy of NOD $CD4^+CD25^+$ T cells (Fig. 2B). To address the functional role of $CD4^+$ cells migrated in response to CCL1, we have first transferred these cells into NOD.SCID recipients and shown that they fail to transfer disease (Fig. 3A). Next, we could show their suppressive activity *in vivo* by the significantly reduced T1D development following co-transfer into NOD.SCID recipients of CD25-depleted spleen cells from recently diabetic NOD mice together with a 10-fold-lower number of $CD4^+$ cells migrated in response to CCL1 (Fig. 3B).

We have recently shown that, in contrast to the high production by circulating human myeloid DCs (M-DCs), the CCR4 ligands CCL17 and CCL22 are poorly produced by plasmacytoid DCs (P-DCs) [69]. It is noteworthy that blood-borne M-DCs, in contrast to P-DCs, constitutively produce CCL17 and CCL22 *ex vivo* [69].

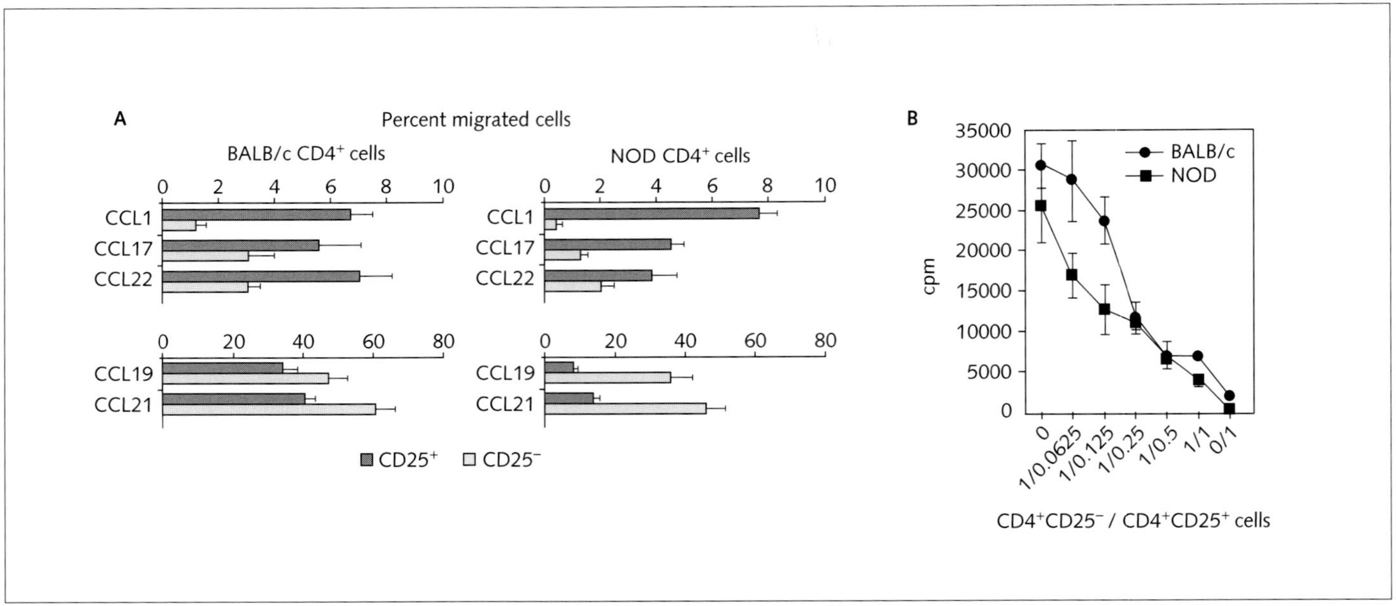

Figure 2
Mouse $CD4^+CD25^+$ Ts cells selectively migrate to CCR8 and CCR4 ligands.
(A) Chemotactic responses of splenic $CD4^+CD25^+$ (black bars) and $CD4^+CD25^-$ (grey bars) T cells from 8-week-old NOD female mice to CCL1, CCL17, CCL22, CCL19, and CCL21 (500 ng/ml) were calculated as the percentage of cells migrated in response to chemokines versus total input cells, with the percentage of spontaneously migrated cells subtracted. The results show means ± S.E. from a representative experiment performed in triplicate.
(B) $CD4^+CD25^+$ T cells from NOD and BALB/c mice suppress the alloantigen-induced proliferation of $CD4^+CD25^-$ T cells. $CD4^+CD25^-$ T cells from 8-week-old NOD (■) or BALB/c (●) mice (2×10^5 cells/well) were incubated with T-cell-depleted C57BL/6 cells (10^5 cells/ well) in the presence of increasing numbers of $CD4^+CD25^+$ T cells (up to 2×10^5 cells/well) isolated from 8-week-old NOD (■) or BALB/c (●) mice. Proliferation of alloreactive T cells was measured by [^{3}H]-thymidine incorporation. Results represent mean c.p.m. ± S.E. from triplicate cultures from a representative experiment.

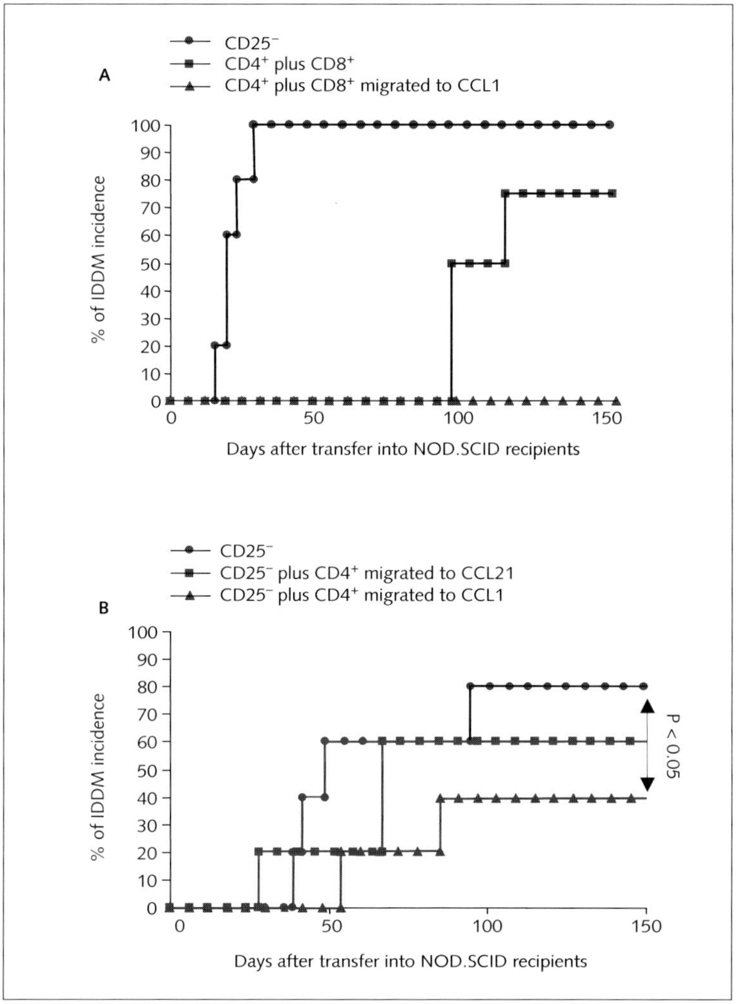

Figure 3
T cells migrated in response to CCL1 inhibit diabetes transfer into NOD.SCID mice.
(A) NOD.SCID mice were injected with 4×10^5 positively selected CD4⁺ and CD8⁺ T splenocytes from 8-week-old prediabetic NOD mice (■; n = 4), with 4×10^5 CD4⁺ and CD8⁺ T splenocytes selectively migrated in response to CCL1 (▲; n = 4), or with 5×10^6 CD25-depleted splenocytes from 16-week-old diabetic NOD mice (●; n = 5). (B) NOD.SCID mice were injected with 10^6 CD25-depleted splenocytes from 16-week-old diabetic NOD mice (●, n = 5), or co-injected with 10^6 CD25-depleted splenocytes from 16-week-old diabetic NOD mice and 10^5 CD4⁺ splenocytes from 8-week-old NOD mice, selectively migrated in response to CCL1 (▲; n = 5) or CCL21 (■; n = 4). Mice were considered diabetic after two consecutive glycemia measurements above 200 mg/dl.

The high production of CCR4 ligands by M-DCs suggests their capacity to selectively recruit at sites of inflammation T cells with regulatory properties or with a Th2 phenotype, whereas P-DCs, by preferentially secreting CCR1/CCR5 ligands, would mostly recruit effector T cells and, in particular, Th1-type cells. Thus, the selective constitutive production of CCR4 ligands by immature M-DCs could lead to the preferential attraction of $CD25^+$Ts cells, a mechanism expected to favor tolerance induction. Intriguingly, the production of CCL22 is markedly enhanced by $1,25(OH)_2D_3$ in blood M-DCs but not P-DCs (G. Penna et al., manuscript in preparation), suggesting a positive effect of vitamin D analogs on the production of chemokines able to recruit $CD25^+$Ts cells. VDR-ligand-regulated chemokine production by distinct DC subsets may therefore selectively attract $CD25^+$Ts cells expressing the appropriate receptors. Experiments to address this issue *in vivo* are currently ongoing.

Besides maintaining peripheral immunological tolerance in homeostatic conditions, $CD25^+$Ts cells could limit ongoing inflammatory responses. Inflammatory signals strongly induce maturation and influx of both M-DCs and P-DCs into secondary lymphoid tissues [70], and maturation of M-DCs and P-DCs enhances their production of several proinflammatory chemokines that can potentially attract different T-cell subsets. Interestingly, maturing P-DCs, similarly to activated B cells [71], produce large quantities of the CCR5 ligand CCL4 [69]. Thus, in analogy with the proposed role for CCL4 in $CD25^+$Ts-cell attraction by activated B cells [71], mature P-DCs could secrete chemokines recruiting these cells to limit ongoing inflammatory responses.

Conclusions

Manipulation of $CD25^+$Ts cells by adoptive transfer of cells generated *ex vivo*, or by their induction directly in patients, offers potentially rewarding immunotherapeutic opportunities in the treatment of T1D, as well as other autoimmune diseases. In this respect, VDR ligands are interesting immunomodulatory agents able to directly target DCs and T cells, leading to the inhibition of pathogenic effector T cells and enhancing the frequency of T cells with suppressive properties, effects that appear to be largely mediated *via* induction of tolerogenic DCs. Multiple mechanisms probably contribute to induction of DC tolerogenicity by VDR ligands, and analysis of inhibitory receptor expression, such as ILT3, by tolerogenic DCs may represent a fruitful area for further research [40]. VDR ligands can also modulate chemokine secretion by the target organ in inflammatory conditions, inhibiting the recruitment of pathogenic effector cells and enhancing the production of chemokines able to recruit regulatory/suppressor T cells. The immunoregulatory activities of VDR ligands, coupled with the absence of major side effects once calcemia is under control, have been translated into effective immunointervention in a variety of autoimmune disease models [43], highlighting their potential applicability in chronic inflammatory conditions sustained by autoreactive immune responses, such as T1D. In addition to the topical treatment of psoriasis, a Th1-mediated autoimmune disease of the skin where topically applied VDR ligands are the most commonly used drugs, these agents may find a broader application in the treatment of inflammatory conditions, where their immunomodulatory effects leading to enhanced $CD25^{+}$Ts cells could turn out to be highly beneficial.

References

1 Kikutani H, Makino S (1992) The murine autoimmune diabetes model: NOD and related strains. *Adv Immunol* 51: 285–322

2 Atkinson MA, Leiter EH (1999) The NOD mouse model of type 1 diabetes: as good as it gets? *Nat Med* 5: 601–604

3 Adorini L, Gregori S, Harrison LC (2002) Understanding autoimmune diabetes: insights from mouse models. *Trends Mol Med* 8: 31–38

4 Trembleau S, Penna G, Bosi E, Mortara A, Gately MK, Adorini L (1995) IL-12 administration induces Th1 cells and accelerates autoimmune diabetes in NOD mice. *J Exp Med* 181: 817–821

5 Benoist C, Mathis D (1997) Cell death mediators in autoimmune diabetes – no shortage of suspects. *Cell* 89: 1–3

6 Delovitch TL, Singh B (1997) The non-obese diabetic mouse as a model of autoimmune diabetes: immune dysregulation gets the NOD. *Immunity* 7: 727–738

7 Ridgway WM, Fasso M, Lanctot A, Garvey C, Fathman CG (1996) Breaking self-tolerance in nonobese diabetic mice. *J Exp Med* 183: 1657–1662

8 Ymer SI, Huang D, Penna G, Gregori S, Branson K, Adorini L, Morahan G (2002) Polymorphisms in the Il12b gene affect structure and expression of IL-12 in NOD and other autoimmune-prone mouse strains. *Genes Immun* 3: 151–157

9 Salomon B, Lenschow DJ, Rhee L, Ashourian N, Singh B, Sharpe A, Bluestone JA (2000) B7/CD28 costimulation is essential for the homeostasis of the $CD4^{+}CD25^{+}$ immunoregulatory T cells that control autoimmune diabetes. *Immunity* 12: 431–440

10 Shevach EM (2002) $CD4^{+}$ $CD25^{+}$ suppressor T cells: more questions than answers. Nat *Rev Immunol* 2: 389–400

11 Sakaguchi S (2000) Regulatory T cells: key controllers of immunologic self-tolerance. *Cell* 101: 455–458

12 Hori S, Nomura T, Sakaguchi S (2003) Control of regulatory T cell development by the transcription factor Foxp3. *Science* 299: 1057–1061

13 Fontenot JD, Gavin MA, Rudensky AY (2003) Foxp3 programs the development and function of $CD4^{+}CD25^{+}$ regulatory T cells. *Nat Immunol* 4: 330–336

14 Khattri R, Cox T, Yasayko SA, Ramsdell F (2003) An essential role for Scurfin in $CD4^{+}CD25^{+}$ T regulatory cells. *Nat Immunol* 4: 337–342

15 Read S, Malmstrom V, Powrie F (2000) Cytotoxic T lymphocyte-associated antigen 4 plays an essential role in the function of CD25(+)CD4(+) regulatory cells that control intestinal inflammation. *J Exp Med* 192: 295–302

16 Stephens LA, Mason D (2000) CD25 is a marker for $CD4^+$ thymocytes that prevent autoimmune diabetes in rats, but peripheral T cells with this function are found in both $CD25^+$ and $CD25^-$ subpopulations. *J Immunol* 165: 3105–3110

17 Wu AJ, Hua H, Munson SH, McDevitt HO (2002) Tumor necrosis factor-alpha regulation of $CD4^+CD25^+$ T cell levels in NOD mice. *Proc Natl Acad Sci USA* 99: 12287–12292

18 Gregori S, Giarratana N, Smiroldo S, Adorini L (2003) Dynamics of pathogenic and suppressor T cells in autoimmune diabetes development. *J Immunol* 171: 4040–4047

19 Herman AE, Freeman GJ, Mathis D, Benoist C (2004) $CD4^+CD25^+$ T regulatory cells dependent on ICOS promote regulation of effector cells in the prediabetic lesion. *J Exp Med* 199: 1479–1489

20 Kukreja A, Cost G, Marker J, Zhang C, Sun Z, Lin-Su K, Ten S, Sanz M, Exley M, Wilson B et al. (2002) Multiple immuno-regulatory defects in type-1 diabetes. *J Clin Invest* 109: 131–140

21 Arif S, Tree TI, Astill TP, Tremble JM, Bishop AJ, Dayan CM, Roep BO, Peakman M (2004) Autoreactive T cell responses show proinflammatory polarization in diabetes but a regulatory phenotype in health. *J Clin Invest* 113: 451–463

22 Szanya V, Ermann J, Taylor C, Holness C, Fathman CG (2002) The subpopulation of $CD4^+CD25^+$ splenocytes that delays adoptive transfer of diabetes expresses L-selectin and high levels of CCR7. *J Immunol* 169: 2461–2465

23 Tang Q, Henriksen KJ, Bi M, Finger EB, Szot G, Ye J, Masteller EL, McDevitt H, Bonyhadi M, Bluestone JA (2004) *In vitro*-expanded antigen-specific regulatory T cells suppress autoimmune diabetes. *J Exp Med* 199: 1455–1465

24 Horwitz DA, Zheng SG, Gray JD, Wang JH, Ohtsuka K, Yamagiwa S (2004) Regulatory T cells generated *ex vivo* as an approach for the therapy of autoimmune disease. *Semin Immunol* 16: 135–143

25 Zheng SG, Wang JH, Gray JD, Soucier H, Horwitz DA (2004) Natural and induced $CD4^+CD25^+$ cells educate $CD4^+CD25^-$ cells to develop suppressive activity: the role of IL-2, TGF-beta, and IL-10. *J Immunol* 172: 5213–5221

26 Yamazaki S, Iyoda T, Tarbell K, Olson K, Velinzon K, Inaba K, Steinman RM (2003) Direct expansion of functional $CD25^+$ $CD4^+$ regulatory T cells by antigen-processing dendritic cells. *J Exp Med* 198: 235–247

27 Tarbell KV, Yamazaki S, Olson K, Toy P, Steinman RM (2004) $CD25^+$ $CD4^+$ T cells, expanded with dendritic cells presenting a single autoantigenic peptide, suppress autoimmune diabetes. *J Exp Med* 199: 1467–1477

28 Lin CH, Hunig T (2003) Efficient expansion of regulatory T cells *in vitro* and *in vivo* with a CD28 superagonist. *Eur J Immunol* 33: 626–638

29 Chatenoud L (2003) CD3-specific antibody-induced active tolerance: from bench to bedside. *Nat Rev Immunol* 3: 123–132

30 Chatenoud L, Thervet E, Primo J, Bach JF (1994) Anti-CD3 antibody induces long-term remission of overt autoimmunity in nonobese diabetic mice. *Proc Natl Acad Sci USA* 91: 123–127

31 Chatenoud L, Primo J, Bach JF (1997) CD3 antibody-induced dominant self tolerance in overtly diabetic NOD mice. *J Immunol* 158: 2947–2954

32 von Herrath MG, Coon B, Wolfe T, Chatenoud L (2002) Nonmitogenic CD3 antibody reverses virally induced (rat insulin promoter-lymphocytic choriomeningitis virus) autoimmune diabetes without impeding viral clearance. *J Immunol* 168: 933–941

33 Belghith M, Bluestone JA, Barriot S, Megret J, Bach JF, Chatenoud L (2003) TGF-beta-dependent mechanisms mediate restoration of self-tolerance induced by antibodies to CD3 in overt autoimmune diabetes. *Nat Med* 9: 1202–1208

34 Peng Y, Laouar Y, Li MO, Green EA, Flavell RA (2004) TGF-beta regulates *in vivo* expansion of Foxp3-expressing $CD4^+CD25^+$ regulatory T cells responsible for protection against diabetes. *Proc Natl Acad Sci USA* 101: 4572–4577

35 Herold KC, Hagopian W, Auger JA, Poumian-Ruiz E, Taylor L, Donaldson D, Gitelman SE, Harlan DM, Xu D, Zivin RA, Bluestone JA (2002) Anti-CD3 monoclonal antibody in new-onset type 1 diabetes mellitus. *N Engl J Med* 346: 1692–1698

36 Herold KC, Burton JB, Francois F, Poumian-Ruiz E, Glandt M, Bluestone JA (2003) Activation of human T cells by FcR nonbinding anti-CD3 mAb, hOKT3gamma1(Ala-Ala). *J Clin Invest* 111: 409–418

37 Bolt S, Routledge E, Lloyd I, Chatenoud L, Pope H, Gorman SD, Clark M, Waldmann H (1993) The generation of a humanized, non-mitogenic CD3 monoclonal antibody which retains *in vitro* immunosuppressive properties. *Eur J Immunol* 23: 403–411

38 Alegre ML, Peterson LJ, Xu D, Sattar HA, Jeyarajah DR, Kowalkowski K, Thistlethwaite JR, Zivin RA, Jolliffe L, Bluestone JA (1994) A non-activating "humanized" anti-CD3 monoclonal antibody retains immunosuppressive properties *in vivo*. *Transplantation* 57: 1537–1543

39 Hering BJ, Kandaswamy R, Harmon JV, Ansite JD, Clemmings SM, Sakai T, Paraskevas S, Eckman PM, Sageshima J, Nakano M et al (2004) Transplantation of cultured islets from two-layer preserved pancreases in type 1 diabetes with anti-CD3 antibody. *Am J Transplant* 4: 390–401

40 Adorini L, Giarratana N, Penna G (2004) Pharmacological induction of tolerogenic dendritic cells and regulatory T cells. *Semin Immunol* 16: 127–134

41 Deluca HF, Cantorna MT (2001) Vitamin D: its role and uses in immunology. *FASEB J* 15: 2579–2585

42 Mathieu C, Adorini L (2002) The coming of age of 1,25-dihydroxyvitamin D(3) analogs as immunomodulatory agents. *Trends Mol Med* 8: 174–179

43 Adorini L (2002) Immunomodulatory effects of vitamin D receptor ligands in autoimmune diseases. *Int Immunopharmacol* 2: 1017–1028

44 Adorini L (2002) 1,25-Dihydroxyvitamin D3 analogs as potential therapies in transplantation. *Curr Opin Invest Drugs* 3: 1458–1463

45 Griffin MD, Xing N, Kumar R (2003) Vitamin D and its analogs as regulators of immune activation and antigen presentation. *Annu Rev Nutr* 23: 117–145

46 Carlberg C, Polly P (1998) Gene regulation by vitamin D3. *Crit Rev Eukaryot Gene Expr* 8: 19–42

47 Penna G, Adorini L (2000) 1,25-dihydroxyvitamin D3 inhibits differentiation, maturation, activation and survival of dendritic cells leading to impaired alloreactive T cell activation. *J Immunol* 164: 2405–2411

48 Piemonti L, Monti P, Sironi M, Fraticelli P, Leone BE, Dal Cin E, Allavena P, Di Carlo V (2000) Vitamin D3 affects differentiation, maturation, and function of human monocyte-derived dendritic cells. *J Immunol* 164: 4443–4451

49 Griffin MD, Lutz WH, Phan VA, Bachman LA, McKean DJ, Kumar R (2000) Potent inhibition of dendritic cell differentiation and maturation by vitamin D analogs. *Biochem Biophys Res Commun* 270: 701–708

50 Berer A, Stockl J, Majdic O, Wagner T, Kollars M, Lechner K, Geissler K, Oehler L (2000) 1,25-Dihydroxyvitamin D(3) inhibits dendritic cell differentiation and maturation *in vitro*. *Exp Hematol* 28: 575–583

51 Canning MO, Grotenhuis K, de Wit H, Ruwhof C, Drexhage HA (2001) 1-alpha,25-Dihydroxyvitamin D3 (1,25(OH)(2)D(3)) hampers the maturation of fully active immature dendritic cells from monocytes. *Eur J Endocrinol* 145: 351–357

52 van Halteren AG, van Etten E, de Jong EC, Bouillon R, Roep BO, Mathieu C (2002) Redirection of human autoreactive T-cells Upon interaction with dendritic cells modulated by TX527, an analog of 1,25 dihydroxyvitamin D(3). *Diabetes* 51: 2119–2125

53 Hewison M, Freeman L, Hughes SV, Evans KN, Bland R, Eliopoulos AG, Kilby MD, Moss PA, Chakraverty R (2003) Differential regulation of vitamin D receptor and its ligand in human monocyte-derived dendritic cells. *J Immunol* 170: 5382–5390

54 Griffin MD, Lutz W, Phan VA, Bachman LA, McKean DJ, Kumar R (2001) Dendritic cell modulation by 1alpha,25 dihydroxyvitamin D3 and its analogs: A vitamin D receptor-dependent pathway that promotes a persistent state of immaturity *in vitro* and *in vivo*. *Proc Natl Acad Sci USA* 98: 6800–6805

55 Boonstra A, Barrat FJ, Crain C, Heath VL, Savelkoul HF, O'Garra A (2001) 1alpha,25-Dihydroxyvitamin D3 has a direct effect on naive $CD4^+$ T Cells to Enhance the Development of Th2 Cells. *J Immunol* 167: 4974–4980

56 Barrat FJ, Cua DJ, Boonstra A, Richards DF, Crain C, Savelkoul HF, de Waal-Malefyt R, Coffman RL, Hawrylowicz CM, O'Garra A (2002) *In vitro* generation of interleukin 10-producing regulatory CD4(+) T cells is induced by immunosuppressive drugs and inhibited by T helper type 1 (Th1)- and Th2-inducing cytokines. *J Exp Med* 195: 603–616

57 Vieira PL, Christensen JR, Minaee S, O'Neill EJ, Barrat FJ, Boonstra A, Barthlott T, Stockinger B, Wraith DC, O'Garra A (2004) IL-10-secreting regulatory T cells do not express Foxp3 but have comparable regulatory function to naturally occurring $CD4^+CD25^+$ regulatory T cells. *J Immunol* 172: 5986–5993

58 Gregori S, Casorati M, Amuchastegui S, Smiroldo S, Davalli AM, Adorini L (2001) Reg-

ulatory T cells induced by 1α,25-Dihydroxyvitamin D_3 and mycophenolate mofetil treatment mediate transplantation tolerance. *J Immunol* 167: 1945–1953

59 Mehling A, Grabbe S, Voskort M, Schwarz T, Luger TA, Beissert S (2000) Mycophenolate mofetil impairs the maturation and function of murine dendritic cells. *J Immunol* 165: 2374–2381

60 Gregori G, Giarratana N, Smiroldo S, Uskokovic M, Adorini L (2002) A 1α,25-Dihydroxyvitamin D_3 analog enhances regulatory T cells and arrests autoimmune diabetes in NOD mice. *Diabetes* 51: 1367–1374

61 Giarratana N, Penna G, Amuchastegui S, Mariani R, Daniel KC, Adorini L (2004) A vitamin D analog downregulates proinflammatory chemokine production by pancreatic islets inhibiting T cell recruitment and type 1 diabetes development. *J Immunol* 173: 2280–2287

62 Rossi D, Zlotnik A (2000) The biology of chemokines and their receptors. *Annu Rev Immunol* 18: 217–242

63 Shimada A, Morimoto J, Kodama K, Suzuki R, Oikawa Y, Funae O, Kasuga A, Saruta T, Narumi S (2001) Elevated serum IP-10 levels observed in type 1 diabetes. *Diabetes* Care 24: 510–515

64 Nicoletti F, Conget I, Di Mauro M, Di Marco R, Mazzarino MC, Bendtzen K, Messina A, Gomis R (2002) Serum concentrations of the interferon-gamma-inducible chemokine IP-10/CXCL10 are augmented in both newly diagnosed Type I diabetes mellitus patients and subjects at risk of developing the disease. *Diabetologia* 45: 1107–1110

65 Piemonti L, Leone BE, Nano R, Saccani A, Monti P, Maffi P, Bianchi G, Sica A, Peri G, Melzi R et al. (2002) Human pancreatic islets produce and secrete MCP-1/CCL2: relevance in human islet transplantation. *Diabetes* 51: 55–65

66 Xie JH, Nomura N, Lu M, Chen SL, Koch GE, Weng Y, Rosa R, Di Salvo J, Mudgett J, Peterson LB et al. JA (2003) Antibody-mediated blockade of the CXCR3 chemokine receptor results in diminished recruitment of T helper 1 cells into sites of inflammation. *J Leukoc Biol* 73: 771–780

67 Frigerio S, Junt T, Lu B, Gerard C, Zumsteg U, Hollander GA, Piali L (2002) Beta cells are responsible for CXCR3-mediated T-cell infiltration in insulitis. *Nat Med* 8: 1414–1420

68 D'Ambrosio D, Sinigaglia F, Adorini L (2003) Special attractions for suppressor T cells. *Trends Immunol* 24: 122–126

69 Penna G, Vulcano M, Roncari A, Facchetti F, Sozzani S, Adorini L (2002) Differential chemokine production by myeloid and plasmacytoid dendritic cells. *J Immunol* 169: 6673–6676

70 Shortman K, Liu YJ (2002) Mouse and human dendritic cell subtypes. *Nat Rev* Immunol 2: 151–161

71 Bystry RS, Aluvihare V, Welch KA, Kallikourdis M, Betz AG (2001) B cells and professional APCs recruit regulatory T cells *via* CCL4. *Nat Immunol* 2: 1126–1132

Regulatory T cells in type 1 autoimmune diabetes

Sophie Candon and Lucienne Chatenoud

INSERM U580, Hôpital Necker, 161 Rue de Sèvres, 75743 Paris Cedex 15, France

Introduction

Insulin-dependent diabetes mellitus (IDDM) is caused by the progressive and selective destruction of insulin-producing pancreatic β cells by autoreactive T lymphocytes. Although the pathogenesis of autoimmune IDDM has been studied extensively, the precise mechanisms involved in the initiation and progression of β cell destruction remain unclear. Animal models such as the non-obese diabetic (NOD) mouse have greatly enhanced our understanding of the pathogenic mechanisms involved in this disease. The NOD mouse model of autoimmune diabetes is characterized by an inappropriate response against self proteins resulting in T-cell-mediated destruction of the pancreatic β islets as well as other endocrine tissues. The disease results from the activation of both β-cell-specific $CD4^+$ and $CD8^+$ T lymphocytes, as demonstrated by their ability to induce disease upon adoptive transfer into syngeneic immunodeficient hosts (i.e. NOD neonates, adult irradiated NOD mice, NOD.SCID mice). T cells reactive to islet antigens including insulin [1–4], glutamic acid decarboxylase (GAD) [5–8], a peptide (p277) of the heat-shock protein HSP60 [9], the tyrosine phosphatase IA-2 [10, 11] and, more recently, the islet-specific glucose-6-phosphatase-catalytic-subunit-related protein (IGRP) [12] are detected in lymphoid organs and/or islet-infiltrating cells of NOD mice. T-cell clones specific for these autoantigens have been derived which transfer diabetes as efficiently as polyclonal diabetogenic $CD4^+$ and $CD8^+$ cells. The development of the disease is also B-cell-dependent [13–15], although autoantibodies against various islet autoantigens found in both NOD mice and patients with IDDM are not pathogenic [16, 17]. There is convincing evidence to suggest that B cells could play a particular role in the presentation of candidate autoantigens.

Regulatory T Cells in Inflammation, edited by Leonie S. Taams, Arne N. Akbar and Marca H.M. Wauben

Interestingly, there is a significant delay (4–8 weeks) between the first signs of autoreactive T-cell activation at 3–4 weeks of age when intra-islet inflammatory infiltration begins and the onset of significant islet-cell destruction resulting in overt diabetes at 3–5 months of age. Data have been accumulated to suggest that this "pre-diabetic" phase is due to T-cell-mediated regulatory mechanisms countering the anti-islet immune response [18]. There have been many attempts to better define the nature of this immune regulation and compelling evidence has been collected over the last two decades showing that unique subsets of regulatory $CD4^+$ T cells are involved in the control of pathogenic effectors leading, at least in part, to this delayed occurrence of disease.

Various subsets of regulatory T cells were initially described and defined essentially according to their phenotype and ability to produce immunoregulatory/suppressive cytokines, i.e. interleukin (IL)-10-producing $CD4^+$ T cells (T_R1) [19], transforming growth factor (TGF)-β-secreting $CD4^+$ T cells (Th3) [20], IL4-producing natural killer T (NKT) cells [21], and IL-10- and/or IL-4-producing $CD8^+$ γδ T cells [22]. Some of these cells are induced upon antigen exposure and at present their respective roles in the regulation of normal and autoimmune responses are still ill-defined. A major breakthrough was the demonstration by Nishizuka and Sakakura [23] that removal of the thymus at 3 days of age induced a polyautoimmune syndrome. Later on, Sakaguchi et al. [24] identified in the thymus a $CD4^+CD25^+$ regulatory-T-cell population that is exported to the periphery after 3 days of age and is capable of suppressing T-cell-mediated organ-specific autoimmunity. These results were confirmed and extended by other groups and $CD4^+CD25^+$ regulatory T cells are now widely viewed as a major component responsible for the regulation of T-cell responses not only to self-antigens, but also to various non-self-antigens including tumor antigens. Two general categories of $CD4^+CD25^+$ regulatory T cells have recently been proposed [25–27]. The first one is represented by natural $CD4^+CD25^+$ regulatory T cells. They are generated from the normal process of T-cell maturation in the thymus and exert their function of prevention and/or repression of responses to self-antigens in the periphery, in an antigen-non-specific, contact-dependent, cytokine-independent manner. The second category of $CD4^+CD25^+$ regulatory T cells is also derived from thymic precursors, but is generated in the periphery through activation of mature naïve $CD4^+CD25^-$ T cells. These antigen-induced suppressors function *in vitro* and *in vivo* primarily by secreting immunoregulatory cytokines, such as IL-10 and TGF-β. It is however important to mention that other subsets of $CD4^+CD25^-$ regulatory T cells involved in the control of autoimmune or alloimmune responses have also been identified. These $CD25^-$ T-cell subsets, like induced $CD25^+$ regulatory T cells, Th3 cells and T_R1 cells (which may or may not express CD25), suppress T-cell responses in an IL-10- and/or TGF-β-dependent manner [28–30]. These various regulatory T cells, $CD25^+$ *versus* $CD25^-$, natural *versus* induced, have all been identified over the past few years in the NOD mouse model of type 1 diabetes.

Evidence for defective regulation of immune responses to self-antigens during diabetes

Presence of types of regulatory T cells in the NOD mouse

$CD4^+CD25^+$ regulatory T cells

Initial evidence for the existence of specialized T-cell suppressor subsets in the NOD model stems from co-transfer experiments showing that $CD4^+$ T splenocytes from young pre-diabetic mice fully prevent disease induced by transfer of diabetogenic cells into immunodeficient recipients [31, 32]. More recent data have showed that removal of $CD4^+$ T cells expressing the L-selectin receptor (CD62L) from the spleen of pre-diabetic mice causes an enrichment of diabetogenic lymphocytes whereas the protective or suppressor capacity concentrates within the $CD4^+CD62L^+$ compartment [33, 34]. Similar observations were made using CD25 as a marker of regulatory T cells [35, 36]. In fact, regulatory cells are found within both the $CD62L^+$ and $CD25^+$ subsets of $CD4^+$ T lymphocytes and these two subsets of regulatory T cells only partially overlap in young NOD mice [37]. The presence of these cells in immunocompetent animals explains that diabetogenic cells can transfer diabetes only when adoptively transferred into syngeneic immunodeficient recipients devoid of functional regulatory T cells, i.e. NOD.SCID mice [38], NOD neonates, sublethally irradiated adult NOD mice or thymectomized adults treated with depleting anti-CD4 antibodies [31, 39–42]. Cyclophosphamide, an alkylating agent that has been shown to selectively affect T-cell dependent regulation, is able to trigger acute diabetes within 2 weeks when injected into young pre-diabetic NOD mice [18, 41, 43]. These regulatory T cells are present not only in peripheral lymphoid organs but also in the thymus [31, 44]. Thymectomy at weaning (3 weeks of age) of NOD females accelerates disease onset [44] and mature $HSA^+\alpha\beta TCR^+CD4^+CD62L^+$ thymocytes (where HSA is heat shock antigen and TCR is T-cell receptor) from pre-diabetic NOD fully prevent diabetes transfer [45]. Importantly, these regulatory thymocytes belong to the single positive $CD4^+$ subset and do not include NKT cells which express an activated phenotype and lack the CD62L receptor [46–51].

The functional absence of B7.1, B7.2, CD40, CD40L or CD28 results in a significant decrease in the frequency/function of $CD4^+CD25^+$ regulatory T cells in peripheral lymphoid compartments, and the induction of autoimmunity [52], suggesting that co-stimulation is required for the normal development of $CD4^+CD25^+$ regulatory T cells. NOD.$CD28^{-/-}$ and NOD.B7.1/$B7.2^{-/-}$ mice develop diabetes more rapidly than their NOD control littermates, while infusion of wild-type syngeneic $CD25^+$ T cells back into $CD28^{-/-}$ NOD recipients prevents disease, suggesting that $CD4^+CD25^+$ regulatory T cells control β-islet-cell autoimmunity [36]. Another evidence for the role of regulatory T cells in the NOD model comes from the TCR transgenic NOD.BDC2.5 mice [53]. These animals express a transgenic TCR from an islet antigen-specific CD4 T-cell clone, BDC2.5. These NOD.BDC2.5

mice do not develop diabetes but $CD4^+$ T cells expressing high levels of the transgenic TCR are able to transfer diabetes to NOD.SCID recipients, as opposed to CD4 T cells expressing low levels, due to the expression of both transgenic and endogenous TCR α-chains. The $BDC2.5^{low}$ $CD4^+$ T-cell subset, composed of $CD25^+$ and $CD25^-$ T cells, have functional features of regulatory T cells and appear late in the ontogeny in the thymus and peripheral lymphoid organs, coinciding with resistance to cyclophosphamide-induced diabetes. This suggests that a balance between diabetogenic T cells and regulatory T cells controls the onset of diabetes and that BDC2.5 mice remain diabetes-free because of a dominant regulatory-T-cell function.

The $CD4^+CD25^+$ T cells present in NOD mice share some similarities with those described in other models, such as the polyautoimmune syndrome induced by day 3 thymectomy in BALB/c mice [24, 54–56] or the diabetes observed after thymectomy and sublethal irradiation of PVG rats [57–59]. In particular, they express high levels of the transcription factor Foxp3 [60] (and L. Chatenoud et al., unpublished observations) considered at present as the most specific "candidate" marker for regulatory T cells [61, 62] as well as other cell-surface markers such as glucocorticoid-induced tumor necrosis factor (TNF) glucocorticoid-induced tumor necrosis factor receptor (GITR) [63, 64], membrane TGF-β [65] and CD103 [66, 67]. Moreover, *in vitro*, they effectively suppress the anti-CD3-induced proliferation of autologous $CD4^+CD25^-$ T cells [37].

Numerous questions about the mode of action of $CD4^+CD25^+$ regulatory T cells in normal and in autoimmune conditions still remain controversial. Notably, the antigen specificity driving thymic selection, differentiation and peripheral activation of natural regulatory T cells remains elusive. $CD4^+CD25^+$ natural regulatory T cells have a polyclonal repertoire and could, in principle, recognize a broad spectrum of self-antigens [68]. *In vitro*, they need to be activated through their TCR in order to acquire their suppressive ability but, *in vivo*, their requirements for activation as well as the nature of the activating antigen (self or non-self) are unclear [69]. In the NOD model, the *in vivo* suppressive ability of naturally occurring $CD4^+CD25^+$ regulatory T cells (isolated from the spleen or the thymus of wild-type NOD mice) upon co-transfer into immunocompromized hosts is rather inefficient, as compared to other models, requiring high numbers of regulatory T cells [34, 36, 45]. A low precursor frequency of regulatory T cells specific for pancreatic self-antigens is a likely explanation, since regulatory T cells specific for a pancreatic self-antigen isolated from NOD.BDC2.5 TCR transgenic mice inhibit more efficiently diabetes after activation and co-transfer with diabetogenic T cells in immunodeficient NOD, than activated polyclonal regulatory T cells from NOD mice [53, 70]. This is in keeping with the observation that adoptive transfer of freshly isolated or activated NOD $CD4^+CD25^+$ T cells into immunocompetent hosts, such as NOD.CD28$^{-/-}$, can delay diabetes only when high numbers of cells are infused ($>5 \times 10^6$ cells) [36], whereas low numbers (5×10^5) of activated BDC2.5

$CD25^+$ regulatory T cells completely prevent diabetes [70]. These results suggest that in the NOD model regulatory T cells need to be activated in an antigen-specific manner at the site of the autoimmune process in order to exert their suppressive function. Furthermore, $CD4^+CD25^+$ regulatory T cells responsive to a single autoantigen, such as those of BDC2.5 mice, can inhibit diabetes mediated by pathogenic T-cell responses to multiple autoantigens, illustrating an important functional feature that is the antigen-non-specific nature of T-cell-mediated suppression [71].

The presence in humans of $CD4^+CD25^+$ T cells exhibiting features similar to those of their mouse counterparts has recently been proposed [72–74]. Their suppressive function, tested in *in vitro* proliferation assays, seems to segregate with a minor subset of cells exhibiting the $CD25^{high}$ phenotype [75]. A quantitative defect of peripheral blood $CD4^+CD25^+$ T cells has been identified in patients with newly diagnosed type 1 diabetes but the *in vitro* suppressive ability of this particular T-cell subset was not tested [76]. Indeed, there are still too few reports on the functional analysis of $CD4^+CD25^+$ cells isolated from patients with autoimmune disease to draw any definitive conclusion. Defective suppressive activity was recently reported in patients with multiple sclerosis and autoimmune polyglandular syndrome type II [77, 78], suggesting that the lack of efficient regulatory activity might be one of the conditions allowing the development of autoimmunity. Functional and phenotypical properties of natural regulatory T cells still need to be addressed in IDDM patients as well as the potential role of other subsets of regulatory T cells.

$CD4^+CD25^-$ regulatory T cells

In our laboratory, we have observed that $CD4^+CD25^+$ regulatory T cells are not the only regulatory T-cell population in NOD mice. In fact, elimination of regulatory T cells in NOD mice requires depletion not only of $CD25^+$ T cells but also of $CD62L^+$ T cells [37]. We identified a $CD4^+CD25^-CD62L^+$ T-cell subset capable of inhibiting disease transfer when coinjected in NOD.SCID mice with diabetogenic T cells (L. Chatenoud et al., unpublished observations). $CD4^+CD25^-CD62L^+$ regulatory T cells have also been described in a diabetes model observed after thymectomy and sublethal irradiation of PVG rats [57–59]. Moreover, in the NOD.BDC2.5 TCR transgenic model, T cells expressing low levels of BDC2.5 TCR and displaying functional features of regulatory T cells were found in both the $CD25^+$ and $CD25^-$ subsets [53]. Recent data from our group have demonstrated that diabetes onset in BDC2.5-NOD.*rag*$^{-/-}$ mice, devoid of regulatory T cells, is completely prevented by infusion of polyclonal NOD $CD4^+CD62L^+$ T cells, but only marginally so by infusion of NOD $CD4^+CD25^+$ T cells [79]. At variance with $CD4^+CD25^+$ T cells, $CD4^+CD25^-CD62L^+$ T cells found in NOD mice do not express any suppressor activity *in vitro* [37] and they do not express GITR, membrane TGF-β or CD103 (L. Chatenoud et al., unpublished observations). Very low levels of Foxp3 expression were observed but one cannot draw any definitive conclusion since functionally relevant regulatory T cells may represent only a minority of the $CD4^+CD25^-CD62L^+$ population. Overall, it clearly appears that, in the NOD model, regulation can be afforded separately but also probably synergistically by $CD4^+CD25^+$ T cells and $CD25^-CD62L^+$ T cells.

NKT cells

A small population of T cells has been identified in NOD mice that express the cell-surface marker, NK1.1, and an invariant TCR repertoire (Vα14-Jα281) [49]. These cells, which recognize MHC-like molecules encoded by the CD1 family of molecules, produce large amounts of IL-4 during immune responses [49]. The numerical and functional deficiency of NKT cells in NOD mice [51] and the partial prevention of diabetes afforded by NKT-cell overexpression in Vα14-Jα281 transgenic NOD mice [80] argue for the possible implication of NKT cells in the control of diabetes in NOD mice. However, the evidence that these cells exert a direct immunoregulatory function in the natural disease is still controversial. Thus, only the Vα14-Jα28 transgenic lines with the largest numbers of NKT cells and the most vigorous production of IL-4 were protected from diabetes (and, even in this situation, the protection is only partial) mice [80]. Additionally, the effect of CD1d genetic deficiency on diabetes onset in NOD mice varies among laboratories: an acceleration of disease onset has been reported by some groups [81, 82] whereas no effect has been reported by others [83] (and A. Herbelin, personal communication). These conflicting results may well be reconciled by the recent demonstration in a model of oral tolerance to nickel, that tolerogenic antigen-presenting cells require signals from IL-4- and IL-10-producing NKT cells in order to generate regulatory T cells [84]. Thus, NKT cells in NOD mice may not act directly on diabetogenic T cells, but rather may transmit signals to antigen-presenting cells to promote the differentiation or the function of regulatory cells.

In humans, soon after the identification of the NKT-cell subset, several studies reported defects within the NKT compartment of human type 1 diabetes patients, suggesting that NKT-cell numbers might be a valuable predictor of susceptibility [76, 85]. Two other studies, however, found no association between type 1 diabetes and blood NKT-cell levels in humans and consequently rejected a link between the onset of diabetes and NKT-cell deficiency [86, 87]. However, it was recently suggested that NKT-cell levels in blood might poorly reflect those in other organs [88]. This raises important questions regarding the methods used to assay NKT-cell numbers in humans. Overall, although the question of the importance of NKT cells in human type 1 diabetes is still unsettled, they should not be discarded as they may be a potentially important clinical target.

Defects of $CD4^+CD25^+$ regulatory T cells

Two major checkpoints have been defined in the pathogenesis of diabetes in mouse models. The first checkpoint, between 3 and 4 weeks of age, concomitant with the appearance of measurable reactivity to pancreatic islet antigens, regulates the onset of insulitis and is controlled by antigen-presenting-cell subsets and the expression of adhesion molecules and integrins. The second checkpoint, active between 8 and 12 weeks, regulates the progression from "controlled", non-destructive insulitis to destructive insulitis and the development of glucose dysregulation and overt diabetes [89]. The demonstration that a reduction of $CD4^+CD25^+$ T cells in NOD mice as a consequence of genetic defects (NOD.$CD28^{-/-}$ or NOD.B7.1/$B7.2^{-/-}$) results in a more rapid transition from insulitis to overt diabetes [36], suggesting that $CD4^+CD25^+$ T-cell subset might be important for controlling checkpoint II. Autoimmunity in the NOD model may result either from a defect in either the initial number of $CD4^+CD25^+$ T cells generated from the thymus or the homeostasis and function of these cells in the periphery, or from the overriding of regulation by an uncontrollable activation and expansion of autoreactive T cells. A decrease in the numbers of $CD4^+CD25^+$ natural regulatory T cells in NOD mice as compared to other strains has been reported by some groups [36, 90], whereas a recent study found normal numbers of these cells in the thymus and lymphoid organs including pancreatic lymph nodes, of prediabetic 6-week-old NOD female, as compared to age-matched females from three non-autoimmune strains (CBA, BALB/c and B6) [91]. Thus, it is still not clear whether quantitative deficiencies in the $CD4^+CD25^+$ compartment may partly explain the loss of tolerance in the NOD mouse or not. However, recent data from Adorini's group and our own suggest that a decline in regulatory-T-cell function during the progression from insulitis to diabetes may contribute to disease establishment ([92] and L. Chatenoud et al., unpublished observations). In fact, $CD4^+CD25^+$ regulatory T cells from overtly diabetic NOD mice (16–20 weeks old) seem less potent suppressors than those from prediabetic 6–8-week-old mice. This was demonstrated both *in vivo* in co-transfer experiments of diabetogenic T cells (CD25-depleted splenocytes, $CD4^+CD25^-$ or $CD25^-CD62L^-$ purified T cells) with $CD4^+CD25^+$ subsets, and *in vitro* in alloantigen- or CD3-induced proliferation assays [92] (and L. Chatenoud et al., unpublished observations).

In the absence of a more in-depth knowledge of the molecular mechanisms involved in the suppression mediated by regulatory T cells, it is difficult to address why and how $CD4^+CD25^+$ T cells lose their supressive ability with aging. Interestingly, defects in the dendritic cells (DCs) of NOD mice have been reported; an imbalance towards CD8α-negative DCs is characteristic of the NOD strain [93] and the yield of generation of mature DCs from the bone marrow of NOD mice is reduced when compared to B10-H mice [94] or C57BL/6 mice [95]. Mature DCs have been shown to be critical for the expansion of $CD4^+CD25^+$ T cells *in vitro* [96].

It is thus possible that inefficient maturation of DCs in NOD mice leads to a reduction in the number and/or function of regulatory T cells.

Role of cytokines in T-cell-mediated suppression

The relative contribution of immunosuppressive cytokines such as IL-4, IL-10 and TGF-β1 in the suppressor function of $CD4^+CD25^+$ regulatory T cells has been vigorously debated in recent years. Immunosuppressive cytokines have been implicated in the *in vivo* function of antigen-induced regulatory T cells [25, 97]. By contrast, natural regulatory T cells, as described by Sakaguchi [71] in the day-3 thymectomy model, were shown to function *in vivo* in a cytokine-independent fashion. *In vitro*, the consensus view is that IL-4 and IL-10 are not required for natural regulatory-T-cell-mediated suppression since neutralization of these cytokines failed to abrogate *in vitro* suppression [25, 98] and that $CD4^+CD25^+$ T cells isolated from IL-4- or IL-10-deficient mice are as potent suppressors *in vitro* as their wild-type counterparts [98, 99]. However, it has been reported that naturally occuring $CD4^+CD25^+$ T cells express a membrane-bound form of TGF-β1 which could mediate suppression [65]. This has been challenged by the observation that neither neutralization of TGF-β1 nor TGF-β1 genetic deficiency has an effect on natural regulatory-T-cell-mediated suppression *in vitro* [25, 98].

In the NOD model, the protective effect of $CD4^+CD62L^+$ T cells is not abrogated by treatment with antibodies to IL-4 and IL-10 [34], even if these cytokines can inhibit diabetes onset on their own [100, 101]. Furthermore, IL-$4^{-/-}$ mice do not show accelerated diabetes onset [102]. However, we have observed that the protection from diabetes observed after transfer of $CD4^+CD25^+CD62L^+$ regulatory T cells from pre-diabetic mice is inhibited by an anti-TGF-β antibody (L. Chatenoud et al., unpublished observations). *In vitro*, we have found that the inhibition of $CD4^+CD25^-$ T cells mediated by $CD4^+CD25^+$ regulatory T cells isolated from young NOD mice was inhited by an anti-TGF-β antibody in a dose-dependent fashion (Chatenoud et al., unpublished observations). Overall, our results suggest that in the NOD model regulatory T cells displaying the phenotypic ($CD62L^+$, $GITR^+$, CTLA-4^+, $Foxp3^+$) and functional (cell-contact-dependency) features of naturally occuring regulatory T cells mediate suppression through mechanisms involving TGF-β *in vitro* and *in vivo*. This is in keeping with data showing that engagement of CTLA-4, one of the phenotypic and functional markers of regulatory T cells, induces TGF-β expression in $CD4^+$ T [103]. It is still not clear however whether TGF-β acts as a mediator of regulation or as a growth/differentiation factor for regulatory T cells.

As mentioned earlier, we found that the $CD4^+CD25^-CD62L^+$ T-cell subset was a potent *in vivo* suppressor of diabetes onset, although not anergic and not capable of inhibiting the proliferation of diabetogenic T cells *in vitro* [37]. These cells were shown to produce TGF-β (L. Chatenoud et al., unpublished observations), but no IL-4 or IL-10 [37]. The mechanisms by which this T-cell subset exerts its regulatory function *in vivo* still have to be characterized.

Restoring immunoregulation of autoimmune responses in diabetes: potential strategies

One major focus of research has been to develop approaches restoring tolerance to self-antigens in patients with autoimmune diseases without the need for general immunosuppression. The demonstration of a role for regulatory T cells in the development of diabetes in NOD mice has recently prompted strategies aimed at restoring and/or improving immunoregulatory capacities of the immune system.

In vivo induction of regulatory T cells

Administration of soluble autoantigens

The advantage of antigen-specific induction of regulatory T cells through delivery of autoantigens is that the protective effect of such induced regulatory T cells is associated with a bystander effect so that regulatory T cells specific for a particular autoantigen can suppress immune responses to other autoantigens presented locally in the target tissue. Regulatory T cells can be experimentally induced by exogenous administration of autoantigens (such as GAD, HSP60 or insulin) in a tolerogenic (soluble) form or by a tolerogenic (mucosal) route. Such induced regulatory T cells are usually capable of protecting syngenic immunocompetent NOD mice from diabetes and do so through cytokine-dependent mechanisms [97, 104, 105]. Several of the reported tolerizing protocols have led to Th2 immune deviation [106–109] and were shown to mediate transferable protection to naive recipient [110]. Mucosal (oral or intranasal) administration of autoantigens favors antigenic presentation through mucosal DCs which seem to preferentially induce regulatory T cells resembling T_R1 and Th3 cells [111]. The success of mucosal immunization in experimental models of autoimmune diseases and in the NOD model has led to two clinical trials of oral administration of insulin in diabetic patients [112, 113]. None of these trials has shown significant clinical benefit. Several explanations have been proposed, including clinical intervention at a too-advanced stage of disease (as compared to experimental settings), insufficient dose or bioavailability of the antigen, insufficient numbers of induced regulatory T cells, concomitant activation of pathogenic T cells and inefficient homing of the regulatory T cells into the target organ, although it is not clear whether regulatory T cells were induced at all. The lack of any clinical benefit in these trials highlights the limitations in the interpretation of data from animal models of autoimmunity and the necessity for a better understanding of the cellular and molecular mechanisms underlying immunoregulation *in vivo*.

Anti-CD3 monoclonal antibody

Anti-T-cell monoclonal antibodies have a dual capacity to promote both immunosuppression and immune tolerance. While their immunosuppressive effect is transient, these antibodies can reverse established autoimmunity or prevent an autoimmune relapse by the establishment of a long-lasting antigen-specific unresponsiveness that persists in the absence of immunosuppression [114]. Antibodies to CD3 have unique tolerogenic properties in NOD mice [115, 116]. Treatment of NOD mice presenting with full-blown diabetes with low doses of the CD3 antibody 145-2C11 induces diabetes remission in the absence of exogenous insulin supply in 60–80% of mice [115, 116]. The remission is complete and definitive (follow-up of > 6 months) and relies on an antigen-specific effect since CD3-antibody-treated NOD mice, unlike untreated diabetic animals, do not destroy syngeneic islet grafts and are fully responsive to other unrelated tissue antigens (e.g. alloantigens) [115]. The CD3-antibody-induced tolerance is most effective in mice with recent-onset disease (14–20 weeks old) before β cells are completely destroyed. The tolerogenic effect is observed with intact anti-CD3 antibodies that are mitogenic [115] and with F(ab')2 forms that are non-mitogenic and well tolerated [116]. The mechanism of protection is not the result of a massive deletion of autoreactive cells in protected animals. Instead, it seems that immunoregulatory mechanisms resembling those present in young pre-diabetic NOD mice are induced and/or restored in anti-CD3-treated animals [117–119]. Anti-CD3-protected mice exhibit a peripheral, non-destructive insulitis and a single injection of cyclophosphamide rapidly and reproducibly reverses the anti-CD3-induced tolerance, indicating the active nature of the tolerance restored in the animals. Co-transfer experiments have demonstrated the presence in the spleen of CD3-treated mice of $CD4^+CD62L^+$ cells (L. Chatenoud et al., unpublished observations) and $CD4^+CD25^+$ cells [119] that inhibit very effectively the transfer of disease by diabetogenic T lymphocytes to NOD.SCID mice. These anti-CD3-induced protective $CD4^+CD25^+$ cells are also found in the lymph nodes and especially in the draining lymph nodes of the target organ. Their suppressive potential is also functional *in vitro* when tested in classical co-cultures with splenic diabetogenic $CD3^+$ T cells or $CD4^+CD25^-$ T cells from overtly diabetic NOD mice. Suppression is TGF-β-dependent, both *in vitro* and *in vivo* [119]. Overall, these results indicate that upon treatment with an anti-CD3 antibody, a subset of regulatory $CD4^+CD25^+$ T cells is induced that is capable of restoring long-lasting tolerance to β islet antigens in recent-onset diabetic mice.

These results in the NOD model have led to two clinical trials of anti-CD3 treatment in patients with recent-onset diabetes; one phase I trial has shown some encouraging results [120] while the other, a phase II placebo-controlled multicentric trial, is still on-going.

Cytokines

As discussed previously, $CD4^+$ regulatory T cells that control spontaneous autoimmune diabetes in NOD mice exert their function independently of Th2 cytokines. However, recent studies have suggested that IL-4 and IL-10 may promote their differentiation and/or increase their functional abilities [121, 122]. We have observed that retroviral transduction of $CD62L^+$ splenocytes from overtly diabetic or pre-diabetic NOD with murine IL-4 potentiate their capacity to inhibit disease transfer [121]. Similarly, and despite paradoxical effects of the administration of IL-10 to NOD mice on the course of diabetes [101, 123, 124], it has been shown that systemic treatment with a non-pathogenic recombinant adeno-associated virus expressing IL-10 prevents diabetes in NOD mice [125] and that this effect is associated with a diminished capacity of splenocytes to transfer disease into immunodeficient animals and, interestingly, with an increase of the percentage of $CD4^+CD25^+$ T cells in the spleen [122]. Some effects on antigen-presenting cells and particularly on $CD11c^+$ DCs were also suggested.

A role of TGF-β in the development of regulatory T cells *in vivo* have been suggested by the observation that transgenic expression of TGF-β restricted to pancreatic islets of NOD mice inhibits the development of diabetes and promote the expansion of $CD4^+CD25^+$ T cells [126]. These data open new perspectives for cytokine-based gene therapy in autoimmune diabetes.

Pharmacological agents

It is now clear that DCs can be tolerogenic, depending essentially on their maturation state [127]. Immature DCs are tolerogenic and have been found to induce T cells with suppressive activity [128]. Several immunosuppressive agents have been shown to induce DCs with tolerogenic phenotype and function [129]. Among them, an analog of vitamin D_3, 1α,25-dihydroxycholecalciferol or calcitriol, can induce tolerogenic DCs both *in vitro* and *in vivo* and it has been suggested that the tolerogenic DCs induced by a short treatment with $1,25(OH)_2D_3$ are responsible for the induction of $CD4^+CD25^+$ regulatory T cells capable of mediating transplantation tolerance [130]. Treatment of NOD mice with calcitriol arrests the progression of insulitis and prevents diabetes development [131]. This effect is associated with an enhanced frequency of $CD4^+CD25^+$ regulatory T cells in the pancreatic lymph nodes that are able to inhibit the T-cell response to the pancreatic autoantigen IA-2 and to significantly delay diabetes transfer by pathogenic $CD4^+CD25^-$ cells. It remains to be determined if such properties of $1,25(OH)_2D_3$ could be exploited in the treatment of IDDM.

Other agents acting on DC differentiation and maturation, such as Flt3 ligand, may also prove good candidates for the treatment of IDDM. It was recently shown that treatment of NOD mice with Flt3 ligand significantly decreased insulitis and progression to diabetes and was associated with a significant increase in myeloid DCs and *in vivo* induction of $CD4^+CD25^+$ cells in the pancreatic lymph node [132].

Ex vivo expansion of regulatory T cells

Although $CD4^+CD25^+$ regulatory T cells were shown to be anergic *in vitro* following TCR stimulation, multiple methods of manipulating *ex vivo* both their numbers and their activation status are now available [27] and some of them have recently been applied to the therapy of spontaneous type 1 diabetes in the NOD model. $CD4^+CD25^+$ regulatory T cells isolated from NOD or NOD.BDC2.5 TCR transgenic mice and expanded *in vitro* with a combination of anti-CD3 and anti-CD28 antibodies and high doses of IL-2 [70] were shown to maintain the phenotypical features of conventional regulatory T cells ($CD25^+$, $CD62L^+$, CTLA-4^+, $GITR^+$, $Foxp3^+$), to produce IL-10 and TGF-β, unlike their fresh, non-expanded counterparts, and to display suppressive ability both *in vitro* and *in vivo*. More strikingly, regulatory T cells expanded from transgenic BDC2.5 islet-antigen-specific $CD4^+$ lymphocytes could prevent diabetes when transferred into non-lymphopenic NOD.$CD28^{-/-}$ mice and were able to reverse diabetes in new-onset diabetic NOD mice, whereas expanded polyclonal NOD regulatory T cells in similar numbers could not. This strongly suggests, as discussed previously, that to exert their suppressive function, expanded regulatory T cells, despite TCR stimulation during their expansion need to be reactivated in an antigen-specific manner at the inflammatory site. Functional autoantigen-specific regulatory T cells have also been produced through *in vitro* culture of $CD4^+CD25^+$ T cells from NOD.BDC2.5 mice with mature, bone marrow NOD DCs pulsed with a mimetope peptide capable of stimulating BDC2.5 T cells, in the presence of IL-2 [95], or through immunization of NOD mice with insulin B-chain peptide B:9-23 followed by *in vitro* culture of $CD4^+$ T cells with B:9-23 peptide [133]. Overall, these results indicate that the efficacy of regulatory T-cell-based immune therapy for IDDM will be critically dependent on the antigen specificity of regulatory T cells.

Conclusion

Over the last two decades the NOD model has provided a better comprehension of the mechanisms involved in the pathogenesis of type 1 diabetes and organ-specific autoimmune diseases in general. Recently it proved also to be a major tool for the identification and characterization of the unique role of regulatory T cells during the early phases of the disease. Type 1 diabetes in NOD mice is the result of a complex quantitative and/or qualitative balance between different subsets of regulatory T cells and pathogenic autoreactive T cells. $CD4^+CD25^+$ T cells, displaying phenotypic and functional features of regulatory T cells, as well as $CD4^+CD25^-CD62L^+$ T cells, less-well-characterized, are both capable of immunoregulation *in vivo*. The $CD25^-CD62L^+$ subset, whose relationship to natural regulatory T cells or antigen-induced regulatory T cells is unclear, seem the most potent regulators. Both subsets appear to exert their suppressive function in a TGF-β-dependent manner. There is compelling evidence to show that a progressive and age-dependent remodelling of the balance between regulatory and pathogenic T cells underlies the disease progression. Strategies aimed at up-regulating or restoring the functional abilities of regulatory T cells have been applied successfully to NOD mice and have been proved to prevent and even in some cases to reverse diabetes and induce long-term remission. Clinical trials based on such strategies are underway. A better understanding of the mechanisms responsible for the loss of sensitivity to T-cell-mediated suppression will provide the basis for optimization of immunotherapeutic approaches.

References

1 Daniel D, Gill RG, Schloot N, Wegmann D (1995) Epitope specificity, cytokine production profile and diabetogenic activity of insulin-specific T cell clones isolated from NOD mice. *Eur J Immunol* 25: 1056–1062

2 Wegmann DR, Gill RG, Norbury-Glaser M, Schloot N, Daniel D (1994) Analysis of the spontaneous T cell response to insulin in NOD mice. *J Autoimmun* 7: 833–843

3 Wegmann DR, Norbury-Glaser M, Daniel D (1994) Insulin-specific T cells are a predominant component of islet infiltrates in pre-diabetic NOD mice. *Eur J Immunol* 24: 1853–1857

4 Daniel D, Wegmann DR (1996) Protection of nonobese diabetic mice from diabetes by intranasal or subcutaneous administration of insulin peptide B-(9-23). *Proc Natl Acad Sci USA* 93: 956–960

5 Baekkeskov S, Aanstoot HJ, Christgau S, Reetz A, Solimena M, Cascalho M, Folli F, Richter-Olesen H, De Camilli P, Camilli PD (1990) Identification of the 64K autoantigen in insulin-dependent diabetes as the GABA-synthesizing enzyme glutamic acid decarboxylase. *Nature* 347: 151–156

6 Honeyman MC, Cram DS, Harrison LC (1993) Glutamic acid decarboxylase 67-reactive T cells: a marker of insulin-dependent diabetes. *J Exp Med* 177: 535–540

7 Panina-Bordignon P, Lang R, van Endert PM, Benazzi E, Felix AM, Pastore RM, Spinas GA, Sinigaglia F (1995) Cytotoxic T cells specific for glutamic acid decarboxylase in autoimmune diabetes. *J Exp Med* 181: 1923–1927

8 Tisch R, Yang XD, Singer SM, Liblau RS, Fugger L, McDevitt HO (1993) Immune

response to glutamic acid decarboxylase correlates with insulitis in non-obese diabetic mice. *Nature* 366: 72–75

9 Elias D, Prigozin H, Polak N, Rapoport M, Lohse AW, Cohen IR (1994) Autoimmune diabetes induced by the beta-cell toxin STZ. Immunity to the 60-kDa heat shock protein and to insulin. *Diabetes* 43: 992–998

10 Dotta F, Dionisi S, Viglietta V, Tiberti C, Matteoli MC, Cervoni M, Bizzarri C, Marietti G, Testi M, Multari G et al. (1999) T-cell mediated autoimmunity to the insulinoma-associated protein 2 islet tyrosine phosphatase in type 1 diabetes mellitus. *Eur J Endocrinol* 141: 272–278

11 Trembleau S, Penna G, Gregori S, Magistrelli G, Isacchi A, Adorini L (2000) Early Th1 response in unprimed nonobese diabetic mice to the tyrosine phosphatase-like insulinoma-associated protein 2, an autoantigen in type 1 diabetes. *J Immunol* 165: 6748–6755

12 Lieberman SM, Evans AM, Han B, Takaki T, Vinnitskaya Y, Caldwell JA, Serreze DV, Shabanowitz J, Hunt DF, Nathenson SG et al. (2003) Identification of the beta cell antigen targeted by a prevalent population of pathogenic $CD8^+$ T cells in autoimmune diabetes. *Proc Natl Acad Sci USA* 100: 8384–8388

13 Serreze DV, Chapman HD, Varnum DS, Hanson MS, Reifsnyder PC, Richard SD, Fleming SA, Leiter EH, Shultz LD (1996) B lymphocytes are essential for the initiation of T cell-mediated autoimmune diabetes: analysis of a new "speed congenic" stock of NOD.Ig mu null mice. *J Exp Med* 184: 2049–2053

14 Akashi T, Nagafuchi S, Anzai K, Kondo S, Kitamura D, Wakana S, Ono J, Kikuchi M, Niho Y, Watanabe T (1997) Direct evidence for the contribution of B cells to the progression of insulitis and the development of diabetes in non-obese diabetic mice. *Int Immunol* 9: 1159–1164

15 Noorchashm H, Noorchashm N, Kern J, Rostami SY, Barker CF, Naji A (1997) B-cells are required for the initiation of insulitis and sialitis in nonobese diabetic mice. *Diabetes* 46: 941–946

16 Yu J, Yu L, Bugawan TL, Erlich HA, Barriga K, Hoffman M, Rewers M, Eisenbarth GS (2000) Transient antiislet autoantibodies: infrequent occurrence and lack of association with "genetic" risk factors. *J Clin Endocrinol Metab* 85: 2421–2428

17 Myers MA, Laks MR, Feeney SJ, Mandel TE, Koulmanda M, Bone A, Barley J, Rowley MJ, Mackay IR (1998) Antibodies to ICA512/IA-2 in rodent models of IDDM. *J Autoimmun* 11: 265–272

18 Delovitch TL, Singh B (1997) The nonobese diabetic mouse as a model of autoimmune diabetes: immune dysregulation gets the NOD. *Immunity* 7:727–738.

19 Roncarolo MG, Bacchetta R, Bordignon C, Narula S, Levings MK (2001) Type 1 T regulatory cells. *Immunol Rev* 182: 68–79

20 Weiner HL (2001) Induction and mechanism of action of transforming growth factor-beta-secreting Th3 regulatory cells. *Immunol Rev* 182: 207–214

21 Kronenberg M, Gapin L (2002) The unconventional lifestyle of NKT cells. *Nat Rev Immunol* 2: 557–568

22 Harrison LC, Dempsey-Collier M, Kramer DR, Takahashi K (1996) Aerosol insulin

induces regulatory CD8 gamma delta T cells that prevent murine insulin-dependent diabetes. *J Exp Med* 184: 2167–2174
23 Nishizuka Y, Sakakura T (1969) Thymus and reproduction: sex-linked dysgenesia of the gonad after neonatal thymectomy in mice. *Science* 166: 753–755
24 Sakaguchi S, Sakaguchi N, Asano M, Itoh M, Toda M (1995) Immunologic self-tolerance maintained by activated T cells expressing IL-2 receptor alpha-chains (CD25). Breakdown of a single mechanism of self-tolerance causes various autoimmune diseases. *J Immunol* 155: 1151–1164
25 Piccirillo CA, Shevach EM (2004) Naturally-occurring $CD4^+CD25^+$ immunoregulatory T cells: central players in the arena of peripheral tolerance. Semin Immunol 16: 81–88
26 Bluestone JA, Abbas AK (2003) Natural *versus* adaptive regulatory T cells. *Nat Rev Immunol* 3: 253-257
27 Horwitz DA, Zheng SG, Gray JD, Wang JH, Ohtsuka K, Yamagiwa S (2004) Regulatory T cells generated *ex vivo* as an approach for the therapy of autoimmune disease. *Semin Immunol* 16: 135–143
28 Groux H, O'Garra A, Bigler M, Rouleau M, Antonenko S, de Vries JE, Roncarolo MG (1997) A $CD4^+$ T-cell subset inhibits antigen-specific T-cell responses and prevents colitis. *Nature* 389: 737–742
29 Levings MK, Roncarolo MG (2000) T-regulatory 1 cells: a novel subset of CD4 T cells with immunoregulatory properties. *J Allergy Clin Immunol* 106: S109–S112
30 Weiner HL (2001) Oral tolerance: immune mechanisms and the generation of Th3-type TGF-beta-secreting regulatory cells. *Microbes Infect* 3: 947–954
31 Boitard C, Yasunami R, Dardenne M, Bach JF (1989) T cell-mediated inhibition of the transfer of autoimmune diabetes in NOD mice. *J Exp Med* 169: 1669–1680
32 Hutchings PR, Cooke A (1990) The transfer of autoimmune diabetes in NOD mice can be inhibited or accelerated by distinct cell populations present in normal splenocytes taken from young males. *J Autoimmun* 3: 175–185
33 Lepault F, Gagnerault MC, Faveeuw C, Bazin H, Boitard C (1995) Lack of L-selectin expression by cells transferring diabetes in NOD mice: insights into the mechanisms involved in diabetes prevention by Mel-14 antibody treatment. *Eur J Immunol* 25: 1502–1507
34 Lepault F, Gagnerault MC (2000) Characterization of peripheral regulatory $CD4^+$ T cells that prevent diabetes onset in nonobese diabetic mice. *J Immunol* 164: 240–247
35 Sakaguchi S (2000) Regulatory T cells: key controllers of immunologic self-tolerance. *Cell* 101: 455–458
36 Salomon B, Lenschow DJ, Rhee L, Ashourian N, Singh B, Sharpe A, Bluestone JA (2000) B7/CD28 costimulation is essential for the homeostasis of the $CD4^+CD25^+$ immunoregulatory T cells that control autoimmune diabetes. *Immunity* 12: 431–440
37 Alyanakian MA, You S, Damotte D, Gouarin C, Esling A, Garcia C, Havouis S, Chatenoud L, Bach JF (2003) Diversity of regulatory $CD4^+$ T cells controlling distinct organ-specific autoimmune diseases. *Proc Natl Acad Sci USA* 100: 15806–15811

38 Shimada A, Charlton B, Rohane P, Taylor-Edwards C, Fathman CG (1996) Immune regulation in type 1 diabetes. *J Autoimmun* 9: 263–269
39 Bendelac A, Carnaud C, Boitard C, Bach JF (1987) Syngeneic transfer of autoimmune diabetes from diabetic NOD mice to healthy neonates. Requirement for both L3T4$^+$ and Lyt-2$^+$ T cells. *J Exp Med* 166: 823–832
40 Wicker LS, Miller BJ, Mullen Y (1986) Transfer of autoimmune diabetes mellitus with splenocytes from nonobese diabetic (NOD) mice. *Diabetes* 35: 855–860
41 Bach JF (1994) Insulin-dependent diabetes mellitus as an autoimmune disease. *Endocr Rev* 15: 516–542
42 Sempe P, Richard MF, Bach JF, Boitard C (1994) Evidence of CD4$^+$ regulatory T cells in the non-obese diabetic male mouse. *Diabetologia* 37: 337–343
43 Charlton B, Bacelj A, Slattery RM, Mandel TE (1989) Cyclophosphamide-induced diabetes in NOD/WEHI mice. Evidence for suppression in spontaneous autoimmune diabetes mellitus. *Diabetes* 38: 441–447
44 Dardenne M, Lepault F, Bendelac A, Bach JF (1989) Acceleration of the onset of diabetes in NOD mice by thymectomy at weaning. *Eur J Immunol* 19: 889–895
45 Herbelin A, Gombert JM, Lepault F, Bach JF, Chatenoud L (1998) Mature mainstream TCR alpha beta$^+$CD4$^+$ thymocytes expressing L-selectin mediate "active tolerance" in the nonobese diabetic mouse. *J Immunol* 161: 2620–2628
46 Hayakawa K, Lin BT, Hardy RR (1992) Murine thymic CD4$^+$ T cell subsets: a subset (Thy0) that secretes diverse cytokines and overexpresses the V beta 8 T cell receptor gene family. *J Exp Med* 176: 269–274
47 Arase H, Arase N, Ogasawara K, Good RA, Onoe K (1992) An NK1.1$^+$ CD4$^+$8$^-$ single-positive thymocyte subpopulation that expresses a highly skewed T-cell antigen receptor V beta family. *Proc Natl Acad Sci USA* 89: 6506–6510
48 Hanke T, Mitnacht R, Boyd R, Hunig T (1994) Induction of interleukin 2 receptor beta chain expression by self-recognition in the thymus. *J Exp Med* 180: 1629–1636
49 Bendelac A, Rivera MN, Park SH, Roark JH (1997) Mouse CD1-specific NK1 T cells: development, specificity, and function. *Annu Rev Immunol* 15: 535–562
50 Vicari AP, Mocci S, Openshaw P, O'Garra A, Zlotnik A (1996) Mouse gamma delta TCR$^+$NK1.1$^+$ thymocytes specifically produce interleukin-4, are major histocompatibility complex class I independent, and are developmentally related to alpha beta TCR$^+$NK1.1$^+$ thymocytes. *Eur J Immunol* 26: 1424–1429
51 Gombert JM, Herbelin A, Tancrede-Bohin E, Dy M, Carnaud C, Bach JF (1996) Early quantitative and functional deficiency of NK1$^+$-like thymocytes in the NOD mouse. *Eur J Immunol* 26: 2989–2998
52 Boden E, Tang Q, Bour-Jordan H, Bluestone JA (2003) The role of CD28 and CTLA4 in the function and homeostasis of CD4$^+$CD25$^+$ regulatory T cells. *Novartis Found Symp* 252: 55–63; discussion 63–66, 106–114
53 Kanagawa O, Militech A, Vaupel BA (2002) Regulation of diabetes development by regulatory T cells in pancreatic islet antigen-specific TCR transgenic nonobese diabetic mice. *J Immunol* 168: 6159–6164

54 Asano M, Toda M, Sakaguchi N, Sakaguchi S (1996) Autoimmune disease as a consequence of developmental abnormality of a T cell subpopulation. *J Exp Med* 184: 387–396

55 Suri-Payer E, Kehn PJ, Cheever AW, Shevach EM (1996) Pathogenesis of post-thymectomy autoimmune gastritis. Identification of anti-H/K adenosine triphosphatase-reactive T cells. *J Immunol* 157: 1799–1805

56 Suri-Payer E, Amar AZ, Thornton AM, Shevach EM (1998) $CD4^{+}CD25^{+}$ T cells inhibit both the induction and effector function of autoreactive T cells and represent a unique lineage of immunoregulatory cells. *J Immunol* 160: 1212–1218

57 Stephens LA, Mason D (2000) CD25 is a marker for $CD4^{+}$ thymocytes that prevent autoimmune diabetes in rats, but peripheral T cells with this function are found in both $CD25^{+}$ and $CD25^{-}$ subpopulations. *J Immunol* 165: 3105–3110

58 Seddon B, Mason D (1999) Regulatory T cells in the control of autoimmunity: the essential role of transforming growth factor beta and interleukin 4 in the prevention of autoimmune thyroiditis in rats by peripheral CD4(+)CD45RC- cells and CD4(+)CD8(–) thymocytes. *J Exp Med* 189: 279–288

59 Saoudi A, Seddon B, Fowell D, Mason D (1996) The thymus contains a high frequency of cells that prevent autoimmune diabetes on transfer into prediabetic recipients. *J Exp Med* 184: 2393–2398

60 Herman AE, Freeman GJ, Mathis D, Benoist C (2004) $CD4^{+}CD25^{+}$ T regulatory cells dependent on ICOS promote regulation of effector cells in the prediabetic lesion. *J Exp Med* 199: 1479–1489

61 Gavin M, Rudensky A (2003) Control of immune homeostasis by naturally arising regulatory $CD4^{+}$ T cells. *Curr Opin Immunol* 15: 690–696

62 Ramsdell F (2003) Foxp3 and natural regulatory T cells: key to a cell lineage? *Immunity* 19: 165–168

63 McHugh RS, Whitters MJ, Piccirillo CA, Young DA, Shevach EM, Collins M, Byrne MC (2002) CD4(+)CD25(+) immunoregulatory T cells: gene expression analysis reveals a functional role for the glucocorticoid-induced TNF receptor. *Immunity* 16: 311–323

64 Tone M, Tone Y, Adams E, Yates SF, Frewin MR, Cobbold SP, Waldmann H (2003) Mouse glucocorticoid-induced tumor necrosis factor receptor ligand is costimulatory for T cells. *Proc Natl Acad Sci USA* 100: 15059–15064

65 Nakamura K, Kitani A, Strober W (2001) Cell contact-dependent immunosuppression by CD4(+)CD25(+) regulatory T cells is mediated by cell surface-bound transforming growth factor beta. *J Exp Med* 194: 629–644

66 Zelenika D, Adams E, Humm S, Graca L, Thompson S, Cobbold SP, Waldmann H (2002) Regulatory T cells overexpress a subset of Th2 gene transcripts. *J Immunol* 168: 1069–1079

67 Lehmann J, Huehn J, de la Rosa M, Maszyna F, Kretschmer U, Krenn V, Brunner M, Scheffold A, Hamann A (2002) Expression of the integrin alpha Ebeta 7 identifies unique subsets of $CD25^{+}$ as well as CD25- regulatory T cells. *Proc Natl Acad Sci USA* 99: 13031–13036

68 Shevach EM (2002) $CD4^+$ $CD25^+$ suppressor T cells: more questions than answers. *Nat Rev Immunol* 2: 389–400
69 Piccirillo CA, Thornton AM (2004) Cornerstone of peripheral tolerance: naturally occurring $CD4^+CD25^+$ regulatory T cells. *Trends Immunol* 25: 374–380
70 Tang Q, Henriksen KJ, Bi M, Finger EB, Szot G, Ye J, Masteller EL, McDevitt H, Bonyhadi M, Bluestone JA (2004) *In vitro*-expanded antigen-specific regulatory T cells suppress autoimmune diabetes. *J Exp Med* 199: 1455–1465
71 Sakaguchi S (2004) Naturally arising $CD4^+$ regulatory t cells for immunologic self-tolerance and negative control of immune responses. Annu Rev Immunol 22: 531–562
72 Stephens LA, Mottet C, Mason D, Powrie F (2001) Human CD4(+)CD25(+) thymocytes and peripheral T cells have immune suppressive activity *in vitro*. *Eur J Immunol* 31: 1247–1254
73 Baecher-Allan C, Brown JA, Freeman GJ, Hafler DA (2001) $CD4^+$CD25high regulatory cells in human peripheral blood. *J Immunol* 167: 1245–1253
74 Dieckmann D, Plottner H, Berchtold S, Berger T, Schuler G (2001) Ex vivo isolation and characterization of CD4(+)CD25(+) T cells with regulatory properties from human blood. *J Exp Med* 193: 1303–1310
75 Baecher-Allan C, Viglietta V, Hafler DA (2004) Human $CD4^+CD25^+$ regulatory T cells. *Semin Immunol* 16: 89–98
76 Kukreja A, Cost G, Marker J, Zhang C, Sun Z, Lin-Su K, Ten S, Sanz M, Exley M, Wilson B et al. (2002) Multiple immuno-regulatory defects in type-1 diabetes. *J Clin Invest* 109: 131–140
77 Liu MF, Wang CR, Fung LL, Wu CR (2004) Decreased $CD4^+CD25^+$ T cells in peripheral blood of patients with systemic lupus erythematosus. *Scand J Immunol* 59: 198–202
78 Kriegel MA, Lohmann T, Gabler C, Blank N, Kalden JR, Lorenz HM (2004) Defective suppressor function of human $CD4^+$ $CD25^+$ regulatory T cells in autoimmune polyglandular syndrome type II. *J Exp Med* 199: 1285–1291
79 You S, Slehoffer G, Barriot S, Bach JF, Chatenoud L (2004) Unique role of $CD4^+CD62L^+$ regulatory T cells in the control of autoimmune diabetes in T cell receptor transgenic mice. *Proc Natl Acad Sci USA* 10152: 14580–14585
80 Lehuen A, Lantz O, Beaudoin L, Laloux V, Carnaud C, Bendelac A, Bach JF, Monteiro RC (1998) Overexpression of natural killer T cells protects Valpha14-Jalpha281 transgenic nonobese diabetic mice against diabetes. *J Exp Med* 188: 1831–1839
81 Wang B, Geng YB, Wang CR (2001) CD1-restricted NK T cells protect nonobese diabetic mice from developing diabetes. *J Exp Med* 194:313–320
82 Shi FD, Flodstrom M, Balasa B, Kim SH, Van Gunst K, Strominger JL, Wilson SB, Sarvetnick N (2001) Germ line deletion of the CD1 locus exacerbates diabetes in the NOD mouse. *Proc Natl Acad Sci USA* 98: 6777–6782
83 Hong S, Wilson MT, Serizawa I, Wu L, Singh N, Naidenko OV, Miura T, Haba T, Scherer DC, Wei J et al. (2001) The natural killer T-cell ligand alpha-galactosylceramide prevents autoimmune diabetes in non-obese diabetic mice. *Nat Med* 7: 1052–1056
84 Roelofs-Haarhuis K, Wu X, Gleichmann E (2004) Oral tolerance to nickel requires

CD4$^+$ invariant NKT cells for the infectious spread of tolerance and the induction of specific regulatory T cells. *J Immunol* 173: 1043–1050
85 Wilson SB, Kent SC, Patton KT, Orban T, Jackson RA, Exley M, Porcelli S, Schatz DA, Atkinson MA, Balk SP et al. (1998) Extreme Th1 bias of invariant Valpha24JalphaQ T cells in type 1 diabetes. *Nature* 391: 177–181
86 Lee PT, Putnam A, Benlagha K, Teyton L, Gottlieb PA, Bendelac A (2002) Testing the NKT cell hypothesis of human IDDM pathogenesis. J Clin Invest 110: 793–800
87 Oikawa Y, Shimada A, Yamada S, Motohashi Y, Nakagawa Y, Irie J, Maruyama T, Saruta T (2002) High frequency of valpha24(+) vbeta11(+) T-cells observed in type 1 diabetes. *Diabetes Care* 25: 1818–1823
88 Berzins SP, Kyparissoudis K, Pellicci DG, Hammond KJ, Sidobre S, Baxter A, Smyth MJ, Kronenberg M, Godfrey DI (2004) Systemic NKT cell deficiency in NOD mice is not detected in peripheral blood: implications for human studies. *Immunol Cell Biol* 82: 247–252
89 Andre I, Gonzalez A, Wang B, Katz J, Benoist C, Mathis D (1996) Checkpoints in the progression of autoimmune disease: lessons from diabetes models. *Proc Natl Acad Sci USA* 93: 2260–2263
90 Wu Q, Salomon B, Chen M, Wang Y, Hoffman LM, Bluestone JA, Fu YX (2001) Reversal of spontaneous autoimmune insulitis in nonobese diabetic mice by soluble lymphotoxin receptor. *J Exp Med* 193: 1327–1332
91 Berzins SP, Venanzi ES, Benoist C, Mathis D (2003) T-cell compartments of prediabetic NOD mice. *Diabetes* 52: 327–334
92 Gregori S, Giarratana N, Smiroldo S, Adorini L (2003) Dynamics of pathogenic and suppressor T cells in autoimmune diabetes development. *J Immunol* 171: 4040–4047
93 Prasad SJ, Goodnow CC (2002) Cell-intrinsic effects of non-MHC NOD genes on dendritic cell generation *in vivo*. *Int Immunol* 14: 677–684
94 Prasad SJ, Goodnow CC (2002) Intrinsic *in vitro* abnormalities in dendritic cell generation caused by non-MHC non-obese diabetic genes. Immunol Cell Biol 80: 198–206
95 Tarbell KV, Yamazaki S, Olson K, Toy P, Steinman RM (2004) CD25$^+$ CD4$^+$ T cells, expanded with dendritic cells presenting a single autoantigenic peptide, suppress autoimmune diabetes. *J Exp Med* 199: 1467–1477
96 Yamazaki S, Iyoda T, Tarbell K, Olson K, Velinzon K, Inaba K, Steinman RM (2003) Direct expansion of functional CD25$^+$ CD4$^+$ regulatory T cells by antigen-processing dendritic cells. *J Exp Med* 198: 235–247
97 von Herrath MG, Harrison LC (2003) Antigen-induced regulatory T cells in autoimmunity. *Nat Rev Immunol* 3: 223–232
98 Thornton AM, Shevach EM (1998) CD4$^+$CD25$^+$ immunoregulatory T cells suppress polyclonal T cell activation *in vitro* by inhibiting interleukin 2 production. *J Exp Med* 188: 287–296
99 Piccirillo CA, Letterio JJ, Thornton AM, McHugh RS, Mamura M, Mizuhara H, Shevach EM (2002) CD4(+)CD25(+) regulatory T cells can mediate suppressor function in

the absence of transforming growth factor beta1 production and responsiveness. *J Exp Med* 196: 237–246
100 Rapoport MJ, Jaramillo A, Zipris D, Lazarus AH, Serreze DV, Leiter EH, Cyopick P, Danska JS, Delovitch TL (1993) Interleukin 4 reverses T cell proliferative unresponsiveness and prevents the onset of diabetes in nonobese diabetic mice. *J Exp Med* 178: 87–99
101 Pennline KJ, Roque-Gaffney E, Monahan M (1994) Recombinant human IL-10 prevents the onset of diabetes in the nonobese diabetic mouse. *Clin Immunol Immunopathol* 71: 169–175
102 Wang B, Gonzalez A, Hoglund P, Katz JD, Benoist C, Mathis D (1998) Interleukin-4 deficiency does not exacerbate disease in NOD mice. *Diabetes* 47: 1207–1211
103 Chen W, Wahl SM (2003) TGF-beta: the missing link in CD4$^+$CD25$^+$ regulatory T cell-mediated immunosuppression. *Cytokine Growth Factor Rev* 14: 85–89
104 Bach JF (2001) Immunotherapy of insulin-dependent diabetes mellitus. *Curr Opin Immunol* 13: 601–605
105 Chatenoud L (2001) Restoration of self-tolerance is a feasible approach to control ongoing beta-cell specific autoreactivity: its relevance for treatment in established diabetes and islet transplantation. *Diabetologia* 44: 521–536
106 Elias D, Meilin A, Ablamunits V, Birk OS, Carmi P, Konen-Waisman S, Cohen IR (1997) Hsp60 peptide therapy of NOD mouse diabetes induces a Th2 cytokine burst and downregulates autoimmunity to various beta-cell antigens. *Diabetes* 46: 758–764
107 Tian J, Clare-Salzler M, Herschenfeld A, Middleton B, Newman D, Mueller R, Arita S, Evans C, Atkinson MA, Mullen Y et al. (1996) Modulating autoimmune responses to GAD inhibits disease progression and prolongs islet graft survival in diabetes-prone mice. *Nat Med* 2: 1348–1353
108 Tian J, Atkinson MA, Clare-Salzler M, Herschenfeld A, Forsthuber T, Lehmann PV, Kaufman DL (1996) Nasal administration of glutamate decarboxylase (GAD65) peptides induces Th2 responses and prevents murine insulin-dependent diabetes. *J Exp Med* 183: 1561–1567
109 Homann D, Holz A, Bot A, Coon B, Wolfe T, Petersen J, Dyrberg TP, Grusby MJ, von Herrath MG (1999) Autoreactive CD4$^+$ T cells protect from autoimmune diabetes *via* bystander suppression using the IL-4/Stat6 pathway. *Immunity* 11: 463-472
110 Zhang ZJ, Davidson L, Eisenbarth G, Weiner HL (1991) Suppression of diabetes in nonobese diabetic mice by oral administration of porcine insulin. *Proc Natl Acad Sci USA* 88: 10252–10256
111 Weiner HL (2001) The mucosal milieu creates tolerogenic dendritic cells and T(R)1 and T(H)3 regulatory cells. *Nat Immunol* 2: 671–672
112 Pozzilli P, Pitocco D, Visalli N, Cavallo MG, Buzzetti R, Crino A, Spera S, Suraci C, Multari G, Cervoni M et al (2000) No effect of oral insulin on residual beta-cell function in recent-onset type I diabetes (the IMDIAB VII). IMDIAB Group. *Diabetologia* 43: 1000–1004
113 Chaillous L, Lefevre H, Thivolet C, Boitard C, Lahlou N, Atlan-Gepner C, Bouhanick

B, Mogenet A, Nicolino M, Carel JC et al (2000) Oral insulin administration and residual beta-cell function in recent-onset type 1 diabetes: a multicentre randomised controlled trial. Diabete Insuline Orale group. *Lancet* 356: 545–549

114 Chatenoud L (2002) The use of monoclonal antibodies to restore self-tolerance in established autoimmunity. *Endocrinol Metab Clin North Am* 31: 457–475, ix

115 Chatenoud L, Thervet E, Primo J, Bach JF (1994) Anti-CD3 antibody induces long-term remission of overt autoimmunity in nonobese diabetic mice. *Proc Natl Acad Sci USA* 91: 123–127

116 Chatenoud L, Primo J, Bach JF (1997) CD3 antibody-induced dominant self tolerance in overtly diabetic NOD mice. *J Immunol* 158: 2947–2954

117 Bach JF, Chatenoud L (2001) Tolerance to islet autoantigens in type 1 diabetes. *Annu Rev Immunol* 19: 131–161

118 Chatenoud L, Salomon B, Bluestone JA (2001) Suppressor T cells – they're back and critical for regulation of autoimmunity! *Immunol Rev* 182: 149–163

119 Belghith M, Bluestone JA, Barriot S, Megret J, Bach JF, Chatenoud L (2003) TGF-beta-dependent mechanisms mediate restoration of self-tolerance induced by antibodies to CD3 in overt autoimmune diabetes. *Nat Med* 9: 1202–1208

120 Herold KC, Hagopian W, Auger JA, Poumian-Ruiz E, Taylor L, Donaldson D, Gitelman SE, Harlan DM, Xu D, Zivin RA et al (2002) Anti-CD3 monoclonal antibody in new-onset type 1 diabetes mellitus. *N Engl J Med* 346: 1692–1698

121 Yamamoto AM, Chernajovsky Y, Lepault F, Podhajcer O, Feldmann M, Bach JF, Chatenoud L (2001) The activity of immunoregulatory T cells mediating active tolerance is potentiated in nonobese diabetic mice by an IL-4-based retroviral gene therapy. *J Immunol* 166: 4973–4980

122 Goudy KS, Burkhardt BR, Wasserfall C, Song S, Campbell-Thompson ML, Brusko T, Powers MA, Clare-Salzler MJ, Sobel ES, Ellis TM et al (2003) Systemic overexpression of IL-10 induces $CD4^+CD25^+$ cell populations *in vivo* and ameliorates type 1 diabetes in nonobese diabetic mice in a dose-dependent fashion. *J Immunol* 171: 2270–2278

123 Moritani M, Yoshimoto K, Tashiro F, Hashimoto C, Miyazaki J, Ii S, Kudo E, Iwahana H, Hayashi Y, Sano T et al (1994) Transgenic expression of IL-10 in pancreatic islet A cells accelerates autoimmune insulitis and diabetes in non-obese diabetic mice. *Int Immunol* 6: 1927–1936

124 Wogensen L, Lee MS, Sarvetnick N (1994) Production of interleukin 10 by islet cells accelerates immune-mediated destruction of beta cells in nonobese diabetic mice. *J Exp Med* 179: 1379–1384

125 Goudy K, Song S, Wasserfall C, Zhang YC, Kapturczak M, Muir A, Powers M, Scott-Jorgensen M, Campbell-Thompson M, Crawford JM, et al. (2001) Adeno-associated virus vector-mediated IL-10 gene delivery prevents type 1 diabetes in NOD mice. *Proc Natl Acad Sci USA* 98: 13913–13918

126 Peng Y, Laouar Y, Li MO, Green EA, Flavell RA (2004) TGF-beta regulates *in vivo* expansion of Foxp3-expressing $CD4^+CD25^+$ regulatory T cells responsible for protection against diabetes. *Proc Natl Acad Sci USA* 101: 4572–4577

127 Steinman RM, Hawiger D, Nussenzweig MC (2003) Tolerogenic dendritic cells. *Annu Rev Immunol* 21: 685–711
128 Jonuleit H, Schmitt E, Schuler G, Knop J, Enk AH (2000) Induction of interleukin 10-producing, nonproliferating CD4(+) T cells with regulatory properties by repetitive stimulation with allogeneic immature human dendritic cells. *J Exp Med* 192: 1213–1222
129 Adorini L, Giarratana N, Penna G (2004) Pharmacological induction of tolerogenic dendritic cells and regulatory T cells. *Semin Immunol* 16: 127–134
130 Gregori S, Casorati M, Amuchastegui S, Smiroldo S, Davalli AM, Adorini L (2001) Regulatory T cells induced by 1 alpha,25-dihydroxyvitamin D3 and mycophenolate mofetil treatment mediate transplantation tolerance. *J Immunol* 167: 1945–1953
131 Gregori S, Giarratana N, Smiroldo S, Uskokovic M, Adorini L (2002) A 1alpha,25-dihydroxyvitamin D(3) analog enhances regulatory T-cells and arrests autoimmune diabetes in NOD mice. *Diabetes* 51: 1367–1374
132 Chilton PM, Rezzoug F, Fugier-Vivier I, Weeter LA, Xu H, Huang Y, Ray MB, Ildstad ST (2004) Flt3-ligand treatment prevents diabetes in NOD mice. *Diabetes* 53: 1995–2002
133 Mukherjee R, Chaturvedi P, Qin HY, Singh B (2003) $CD4^+CD25^+$ regulatory T cells generated in response to insulin B:9-23 peptide prevent adoptive transfer of diabetes by diabetogenic T cells. *J Autoimmun* 21: 221–237

The potential for targeting CD4⁺CD25⁺ regulatory T cells in the treatment of multiple sclerosis in humans

Clare Baecher-Allan, Vissia Viglietta and David A. Hafler

Laboratory of Molecular Immunology, Center for Neurologic Diseases, Brigham and Women's Hospital and Harvard Medical School, 77 Avenue Louis Pasteur, Boston, MA 02115, USA

Introduction

Comparative analyses of surface-marker expression and *in vitro* assays of function indicate that significant differences exist in the state of activation of the lymphoid cells in both the blood and cerebrospinal fluid (CSF) between patients with multiple sclerosis (MS) and healthy controls. Of specific interest is the finding that T cells recognizing central-nervous-system (CNS)-derived antigens are found at slightly higher frequencies and in a higher activation state in the circulation of patients with MS as compared to healthy controls [1–3]. This implicates that the autoreactive T cell and the loss of immune regulatory mechanisms are critical in the etiology of MS. Various therapeutic interventions have focused on reducing either the general state of activation of the entire immune system (such as high-dose steroids during acute episodes) or the specific activation of CNS-responsive T cells. The goals of T-cell-directed therapies have been to induce T-cell anergy, to alter the T-cell response in the patient from Th1 (inflammatory) to Th2 (anti-inflammatory), or induce the peripheral deletion of autoreactive T cells. The recently identified population of T cells known as $CD4^+CD25^+$ regulatory T cells (T_R cells) that appears to have an increased propensity for recognition of self-antigens and act to down-modulate the immune response was found to be functionally deficient in patients with MS as compared to those isolated from healthy subjects [4]. Thus the $CD4^+CD25^+$ T_R population may be an ideal target for the design of future therapies focused on augmenting their inhibitory activity in patients with MS. In this chapter we will discuss what is known about the state of the immune system in patients with MS, and how the various current and future therapies may affect the interaction between inflammatory and regulatory T-cell subsets and thus regulate the disease.

MS, the diagnosis

MS is generally categorized as being either relapsing-remitting MS (RRMS) or primary progressive MS (PPMS) in onset. RRMS, observed in approximately 85% of

Regulatory T Cells in Inflammation, edited by Leonie S. Taams, Arne N. Akbar and Marca H.M. Wauben

patients, is characterized by a series of exacerbations that result in varying degrees of disability from which the patients recover partially or completely followed by a remission period of variable duration. Later in disease the extent of recovery from attacks is often decreased and baseline neurological disability increases as approximately 40% of RRMS patients ultimately change to a progressive form of the disease known as secondary progressive MS (SPMS) in which there is a relative absence of attacks while there is a gradual clinical decline [5]. In contrast, PPMS may represent a very different disease from RRMS and SPMS as it is characterized by the absence of acute attacks from the onset, involves a gradual clinical decline, and is resistant to any form of immunotherapy [6].

The hallmark of MS is the inflammatory perivascular demyelinated plaque with the formation of a reactive glial scar although there is great variability in the clinical presentation and course of illness [7]. Inflammatory lesions may develop at any site along myelinated CNS white-matter tracts. Some degree of spontaneous improvement in symptoms following an acute attack occurs in most patients and has been attributed to the resolution of inflammation at the site of the responsible lesion rather than the reversal of the demyelination that may persist even in the absence of symptoms [7]. Transition to the SPMS disease state is associated with significantly fewer lesions and a decrease in brain parenchymal volume [8, 9]. Due to the heterogeneity in symptom presentation, the diagnosis of MS requires monitoring disease activity over time throughout the CNS. Thus, part of the criteria for the diagnosis of clinically definite MS (CDMS) is that there must be a relapsing episode after the initial presentation in which CNS involvement was shown to exist in at least two separate white-matter sites as demonstrated by magnetic resonance imaging (MRI). These MRI-detected lesions, demonstrating regions of enhanced contrast following intravenous administration of gadolinium chelate, reflect a focal breakdown of the blood-brain barrier (BBB). Thus, the MRI is useful to establish the multifocal nature of the disease and to assess disease activity and burden [10]. Relapses can persist for days to several weeks and are likely caused by the traffic of activated, myelin reactive T cells into the CNS causing acute inflammation. Over time, the extent of recovery from attacks decreases and the baseline neurological disability accrues. In addition, although often absent early in the disease, over 90% of patients with CDMS proceed to present oligoclonal bands, which are immunoglobulin proteins found in the CSF that are derived from a clonal population of activated B cells and are detected by their limited set of charges [11]. Importantly, the identification of oligoclonal bands alone is not informative as they are often found in other inflammatory infectious diseases of the CNS such as human T-cell lymphotrophic virus type 1 (HTLV-1) [12] or varicella zoster virus [13]. Regardless, although the antigen specificity of these MS-associated oligoclonal Igs remains unknown, their presence indicates CNS B-cell activation.

Linkage-analysis studies of multiplex families with more than one affected member have revealed an association of the MHC alleles with increased risk of developing the illness [14]. Although, class II direct repeat 2 (DR2) confers increased susceptibility [15], other HLA alleles, the T-cell receptor (TCR), Ig-encoding regions, myelin basic protein (MBP) and tumor necrosis factor (TNF) have also been weakly associated with disease [16, 17].

Central role of T cells in MS

Studies in experimental autoimmune encephalomyelitis (EAE), the most commonly used animal model of MS, have provided important insights into the mechanism of the disease and strongly suggest a central role for T cells in the initiation and propagation of the disease. Upon immunization with brain-derived proteins, sensitive strains of mice develop CNS lesions that are characterized by multifocal plaques and a perivascular infiltrate primarily consisting of T cells and monocytes. The MHC background is an important component of susceptibility to EAE as it influences the TCR repertoire and the specific encephalitogenic peptides of the myelin proteins that are presented by the MHC [18]. Importantly, it was shown that EAE can be induced by the transfer of $CD4^+$ T cells that specifically react to these CNS autoantigens in the context of class II [19] and can be induced in several species of animal by injection of homogenized whole myelin or myelin components such as MBP, proteolipid protein (PLP) or myelin oligodendrocyte glycoprotein (MOG). Although it appears that EAE is a T-cell-mediated disease, it is also modified by the presence of B cells possibly as a result of their production of antibodies or ability to function as antigen-presenting cells (APCs) [20]. In the EAE model, $CD4^+$ T cells that produce interferon (IFN)-γ, interleukin (IL)-12, and TNF-α, which have been defined as Th1 (inflammatory) cells, have been shown to be pathogenic while Th2 (anti-inflammatory) $CD4^+$ T cells have a protective role. The neutralization of IL-12, which is strongly induced by IFN-γ and potentiates Th1 cell differentiation, inhibits the development of EAE [21]. In contrast, the $CD4^+$ T cells which actually promote humoral responses and produce IL-4, IL-5, IL-10 or IL-13 have been shown to be protective in EAE [22, 23].

The analyses of human clinical samples suggest that there is a dysregulation of autoreactive T cells in MS. Self-antigen-specific T cells can be found in the circulation of patients as well as of healthy individuals. Using established cloning techniques the frequency of those T cells that can expand *in vitro* and whose TCRs specifically recognize the different exogenously provided, brain-derived antigens in the context of MHC, has been calcualated. Performed in this manner, several studies have demonstrated that MBP- or MOG-specific T cells in the circulation of patients exist at similar or slightly higher frequencies than in normal controls [24, 25]. PLP-reactive T cells were also found increased in the blood and CSF of MS patients compared to normal controls [26]. In contrast, there are some reports in which no difference was observed in the frequency of isolating brain-reactive cells from the circulation of patients as compared to controls [27]. However, additional studies indicated that although limiting dilution culture conditions gave rise to roughly equal numbers of MBP-reactive cells from the PBMCs from both patients with MS and controls, the patient-derived circulating myelin-reactive T cells were in an enhanced state of activation that became highly apparent when the cells were pre-cultured in exogenous IL-2 [28]. Thus the myelin-reactive cells isolated from patients differed from those found in healthy individuals as they exhibit an increased baseline of activation indicated by a higher expression of CD25. Experimentally, if the cells were first pre-expanded in IL-2 before being cultured against myelin antigens, much higher frequencies of MBP- and PLP-reactive $CD4^+$ T cells were observed in patient-derived samples. Furthermore, using a similar IL-2 pre-culture, MBP-reactive T cells could be isolated from the CSF of MS patients but not of controls. These data indicate that the circulation of healthy individuals contains self-reactive T cells in a non-activated state, while those in the circulation of patients with MS are in an activated state, suggesting a loss of regulation in the patients. This finding indicates that all self-reactive T cells are not deleted during T-cell development and that it may be the loss of the ability to regulate these circulating self-reactive cells that contributes to the development of MS.

Histological analyses of brain sections also provide insight into the cellular composition and the nature of the T cells in the MS lesion. Active lesions exhibit a perivascular infiltrate consisting of $CD4^+$, $CD8^+$ and γ/δ T cells, monocytes and occasional B cells [29–33] while macrophages containing myelin debris are usually found in the center of the plaque [34]. In contrast, chronic-active lesions exhibit a reduced cellular infiltrate that is basically restricted to the edge of the plaque. Furthermore, increased frequencies of $CD4^+CXCR3^+$ Th1 cells in the circulation of patients with active disease were found to correlate with the number of MRI lesions. Thus, similar to the data resulting from the study of EAE, the presence of MS appears to be highly correlated with a deleterious Th1 response.

Currently, it is hypothesized that Th1 cells may be pathogenic in MS. Although there are reports that patient-derived MBP-reactive T cells can produce either Th1 or Th2 cytokines depending upon culture conditions [35] a number of reports have associated disease severity with Th1 responses and favorable therapies with Th2 responses [36, 37]. Furthermore, it has been reported that the cytokine profiles of PLP-reactive T-cell clones generated from MS patients correlated with different clinical stages of disease in that acute attacks were associated with Th1-like patterns, while periods of remission were characterized by Th0, Th1 and Th2 responses [38]. Another study found an increased production of IL-12 upon monocyte activation and an increased frequency of IFN-γ- and TNF-α-secreting T cells in the circulation of untreated patients with MS as compared to controls [39]. Furthermore, IL-12 and TNF-α have been detected in MS plaques [40] although the expression of TNF-α was shown to be produced primarily by astrocytes and macrophages within the plaque [41]. Thus, although it is not clear cut, in general the skewing of the cytokine response toward Th2 or away from Th1 is considered a beneficial prognostic indicator.

Features of the $CD4^+CD25^+$ subset of T_R cells

The relatively recently described $CD4^+CD25^+$ T_R subset is one of several different types of regulatory T cell that have a down-modulatory effect on the immune response. Defined by their production of highly restricted sets of specific biological response modifiers (i.e. cytokines) during an immune response, the Th1 or Th2 classification of T cells is the most basic and may possibly represent the most influential type of regulatory T cell. Keeping this in mind, a number of subsets of T cells have been defined as having the ability to modulate Th1 and/or Th2 responses through different modes of action. In general the three most prevalently studied $CD4^+$ regulatory T-cell subsets are the T_R1 cells that inhibit T-cell proliferation primarily *via* secretion of IL-10 and TGF-β, the Th3 subset that appears to suppress activation *via* the secretion of TGF-β and the $CD4^+CD25^+$ T_R cells that inhibit immune cell activation through an as yet undefined mechanism that appears [42] to require cell contact but is generally believed to be independent of cytokines.

The importance and function of the $CD4^+CD25^+$ subset of regulatory T cells were first indicated by studies performed in the mouse. In a mouse model of autoimmunity in which the thymus was removed at neonatal day 3, $CD4^+CD25^+$ cells were absent and the mice developed multiple-organ autoimmune disease [42]. Importantly, it was shown that the adoptive transfer of $CD4^+CD25^+$ cells could abrogate the development of anti-self responses in this as well as other models of autoimmune disease [43]. Additional mouse studies by Jordan et al. [44] indicated that $CD4^+CD25^+$ T_R cells had an increased propensity for the recognition of self-antigen as they were positively selected for the ability to recognize self-antigens during their development in the thymus. In these studies, increased numbers of $CD4^+CD25^+$ T_R cells were observed in peripheral lymphoid organs if $CD4^+CD25^+$ T_R cells were able to interact with self-antigen in the thymus as demonstrated in mouse model where offspring of crosses between a TCR transgenic mouse in which all the T cells express a single TCR recognizing the influenza hemagglutinin (HA) peptide and an HA transgenic mouse in which HA is expressed as a self-antigen. $CD4^+CD25^+$ T_R cells also appear to require constant low-level co-stimulation in order to maintain their normal homeostasis as NOD mice, modified to lack expression of B7-1/B7-2 co-stimulatory molecules, exhibited increased disease severity and a deficiency in the $CD4^+CD25^+$ T_R population [45]. Together, these findings suggest that fully functional $CD4^+CD25^+$ T_R cells may require frequent interaction with self-antigen in the context of a low-level co-stimulation.

Despite the impossibility to perform similar *in vivo* experiments, *in vitro* assays have demonstrated the existence of a functionally similar subset of $CD4^+CD25^+$ T_R

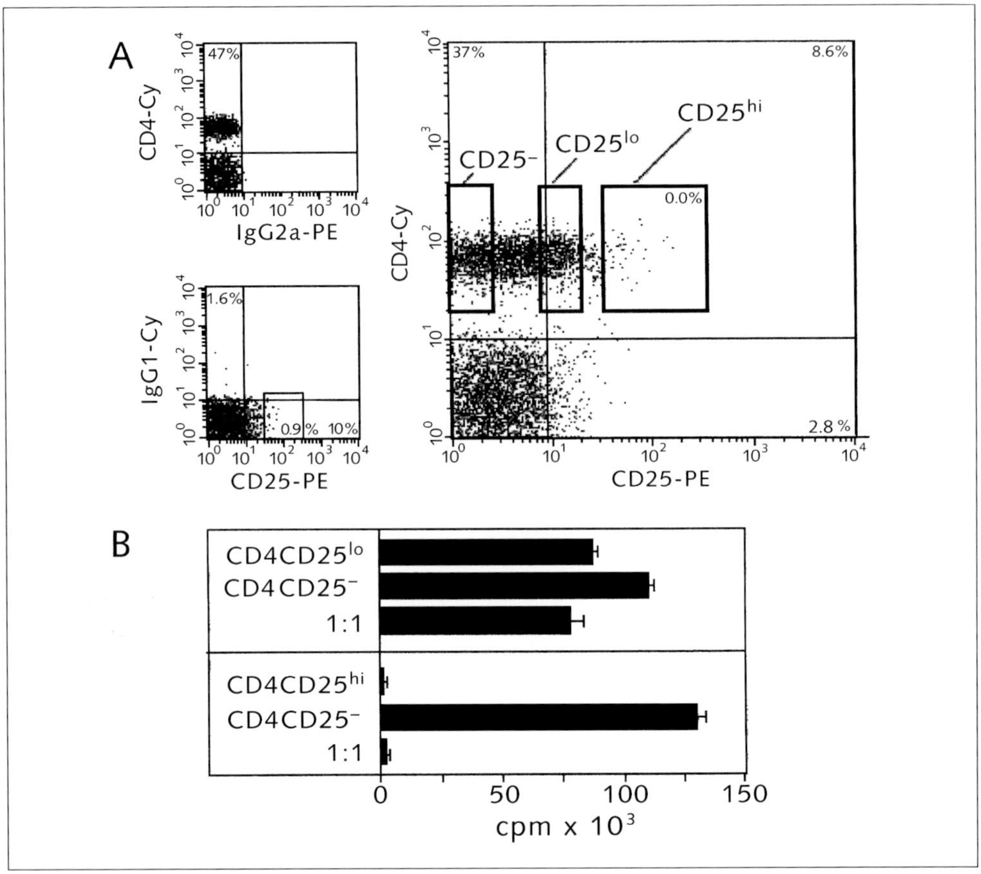

Figure 1
(A) Whole PBMCs were stained with CD4, CD25 and a cocktail of monocyte markers to remove potential CD4^{+} *monocytic cells from the isolation. Three populations were isolated via high-speed FACS sorting: CD4*$^{+}$*CD25*$^{-}$ *(T responder), CD4*$^{+}$*CD25*^{high} *(*T_R*) and CD4*$^{+}$*CD25*^{low} *cells. (B) Co-cultures were established with T-responder cells (CD4*$^{+}$*CD25*$^{-}$*) in the presence or absence of CD4*$^{+}$*CD25*^{high} T_R *cells or CD4*$^{+}$*CD25*^{low} *cells at a 1:1 ratio under stimulation with anti-CD3 and T-depleted accessory cells. [*^{3}*H]Thymidine (1 µCi/well) was added at day 5 of the culture and harvested after an additional 16 h.*

cells as the one described in mice [46, 47]. As the study of human samples cannot take advantage of transgenic technology in which practically all T cells express a single TCR recognizing a known antigen, the assays of human T cells usually utilize specific anti-CD3 monoclonal antibodies (mAbs) that stimulate through the TCR to induce T-cell activation and proliferation. Although stimulation *via* different plate-

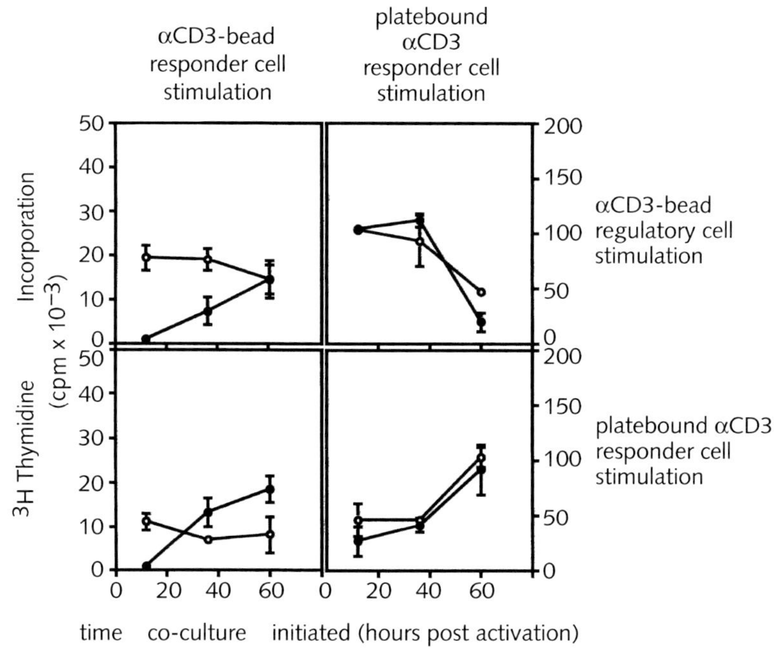

Figure 2
CD4⁺CD25⁻ T-responder cells or CD4⁺CD25^high^ T_R cells were stimulated separately under conditions of strong stimulation (plate-bound anti-CD3 at 5 μg/ml) or weak stimulation (anti-CD3 beads) in the presence of T-depleted irradiated PBMCs (accessory cells). The cultures were then combined at varying times post-stimulation (•). Similarly stimulated cultures of accessory cells only were also combined with activated T-responder cultures at the same time (○) to determine what effect the addition of activated T_R/accessory cells has on T-responder proliferation as compared to the addition of activated accessory cells only.

bound anti-CD3 concentrations is highly artificial, it serves as a model for different strengths of potential TCR engagement that could normally occur *in vivo*. The data in Figure 1 demonstrate that in humans the T_R-cell subset is characterized by a slightly more restrictive surface phenotype as, upon anti-CD3 mAb stimulation, the greatest suppressive activity *in vitro* is a feature only of those $CD4^+$ cells that express the highest levels of CD25. These cells represent 2–5% of the CD4 T cells derived from blood. By varying the concentration of the anti-CD3 stimulus in the co-cultures, it was further shown that inhibition of responder T-cell ($CD4^+CD25^-$) proliferation by $CD4^+CD25^{high}$ T_R cells is inversely related to the strength of the signal provided to the TCR (Fig. 2). Furthermore, enhancing T-cell signals *via* CD28 co-stimulation or

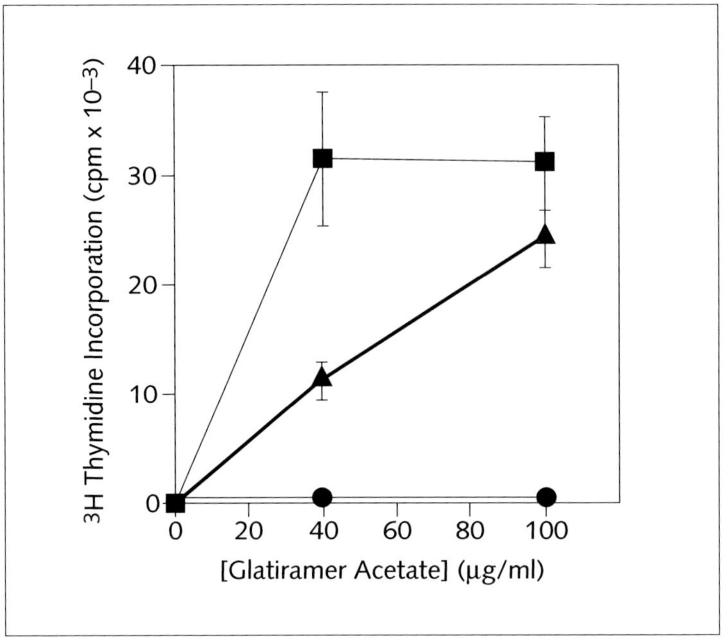

Figure 3
$CD4^+CD25^-$ T-responder cells were stimulated alone (■) or in combination with $CD4^+CD25^{high}$ T_R cells (▲) with different concentrations of glatiramer acetate (Copaxone) and irradiated T-depleted accessory cells. The data represent the proliferation of 1×10^4 T-responder cells per well in the absence or presence of 5×10^3 T_R cells per well. $CD4^+CD25^{high}$ T_R cells were also stimulated under the same conditions in the presence of antigen and accessory cells only (•).

exogenous IL-2, a T-cell growth factor, also abrogates suppression by $CD4^+CD25^{high}$ T_R cells. Thus, these data showing loss of suppression under strong TCR stimulation may be highly relevant to the development of appropriate suppression of self-reactive T cells in the healthy individual as these cells were not deleted in the thymus due to their recognition of self-antigen-MHC complexes *via* weak signals. Thereby, suppression by $CD4^+CD25^{high}$ T_R cells would not occur under conditions of strong TCR signaling, such as during a pathogen response, or upon providing additional co-stimulation *via* CD28 signal or exogenous IL-2 [47, 48].

The mechanism by which human $CD4^+CD25^{high}$ T_R cells exert their suppressive function is unclear. The mouse studies have indicated that the activation of T_R-cell effector function is antigen-specific while the suppression does not need to be. There are a few reports demonstrating suppression of specific antigen responses *in vitro* by the co-culture of $CD4^+CD25^{high}$ cells with $CD4^+CD25^-$ cells in the presence of antigen (MOG or Copaxone®) and autologous APCs [47, 49]. We have shown that cells derived from healthy controls can inhibit Copaxone-induced T-cell proliferation (Fig. 3) [48]. In contrast, there are many more reports of "antigen-specific suppression" in human cells in which the $CD4^+$ T-cell response against antigens was increased upon depletion of $CD25^+$ cells [50]. Furthermore, the potential involvement of the inhibitory surface molecules CTLA-4 and membrane-bound TGF-β is controversial as blocking these molecules leads to the abrogation of suppression only in a subset of studies. However, all studies agree that $CD4^+CD25^+$ T_R cells must physically contact the responding target T cells in order to inhibit their response.

$CD4^+CD25^+$ T_R cells derived from patients with MS exhibit functional impairment

Although the analysis of PBMC samples indicated no difference in the frequency of $CD4^+CD25^{high}$ T_R cells in the circulation of patients with MS as compared to healthy controls, $CD4^+CD25^{high}$ T_R cells isolated from patients with MS were significantly less suppressive compared to those from healthy donors.

Utilizing the established *in vitro* co-culture assay, the function of $CD4^+CD25^{high}$ regulatory cells isolated from patients with MS or normal healthy controls was compared under multiple culture conditions that encompassed different strengths of TCR signaling (*via* different doses of plate-bound anti-CD3 mAb) and different types of signal (either plate-bound anti-CD3 or soluble anti-CD3 and anti-CD28). As before, the cell populations were isolated *via* FACS cell sorting in which monocytes were effectively removed by virtue of their expression of CD14, CD32 and/or CD116, while the T_R and $CD4^+CD25^-$ of subsets were isolated by virtue of their expression of CD4 and the presence of either high or no CD25, respectively. Testing these two T-cell subsets for activity upon co-culture in the same well at varying ratios with different strengths of TCR signaling delivered by multiple doses of anti-CD3 mAb (0, 0.1, 0.5 and 2.5 μg/ml) demonstrated that patient-derived $CD4^+CD25^{high}$ T cells were significantly less able to suppress the proliferation of responder T cells than the co-cultures initiated with cell subsets isolated from healthy controls (Fig. 4). Furthermore, secretion of the Th1 cytokine IFN-γ, postulated to be produced by destructive, autoreactive T cells, was suppressed in co-cultures of $CD4^+CD25^{high}$ T cells from healthy controls, but not in co-cultures established with cells isolated from patients with MS. Importantly, the $CD4^+CD25^-$ responder cells from both patients with multiple sclerosis and healthy individuals gave similar dose-response proliferation to the plate-bound anti-CD3, while the $CD4^+CD25^{high}$ T cells isolated from both groups were similarly anergic to plate-bound anti-CD3 stimulation. Although, there was no detectable secretion of the Th2 cytokine IL-13 at the low cell numbers used in these experiments, and there was variable secretion of IL-10 by both target cells and $CD4^+CD25^{high}$ T_R cells. Thus,

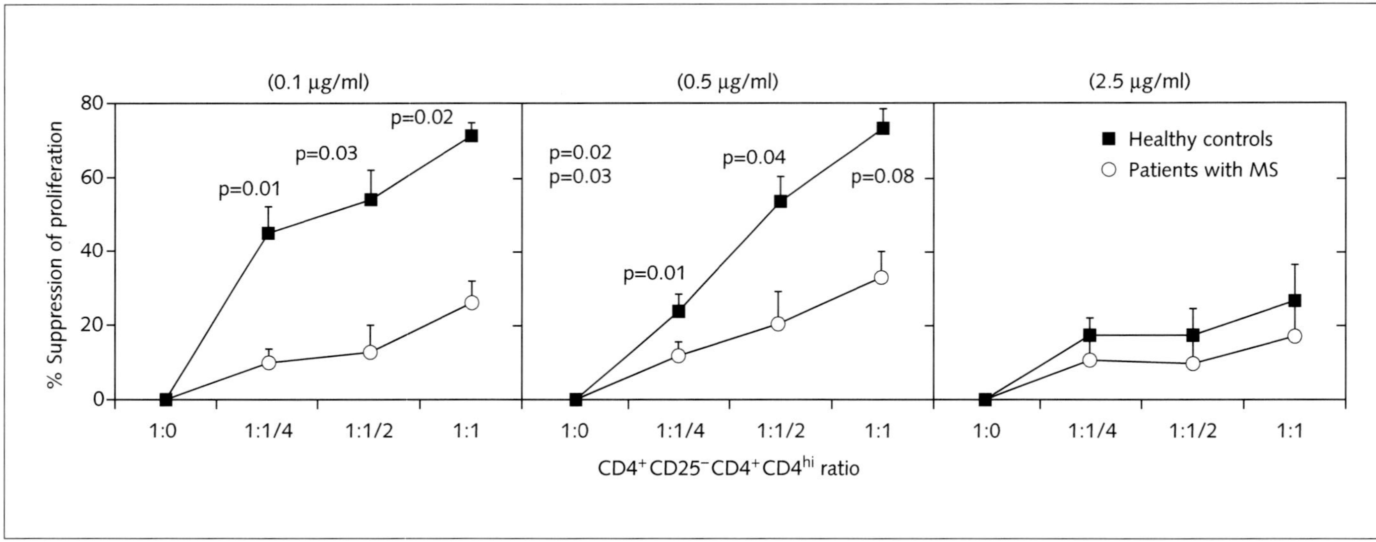

Figure 4

Summary of CD4+CD25hi T-cell regulatory function that is altered in patients with MS. The mean percentage inhibition of the proliferative response by CD4+CD25hi derived from 15 MS patients (○) and 21 healthy controls (■) was calculated. CD4+CD25− and CD4+CD25hi populations were stimulated with plate-bound anti-CD3 mAb, alone or co-cultured at varying ratios. The proliferative response was inhibited upon addition of CD4+CD25hi cells to the CD4+CD25− responder cells at a 1:1 ratio in normal controls. Regulatory T cells from patients with MS exhibited significantly less suppressor activity. Decreasing the number of CD4+CD25hi T cells (responder/suppressor ratio 1:1/2, 1:1/4) resulted in less suppression in all the conditions examined.

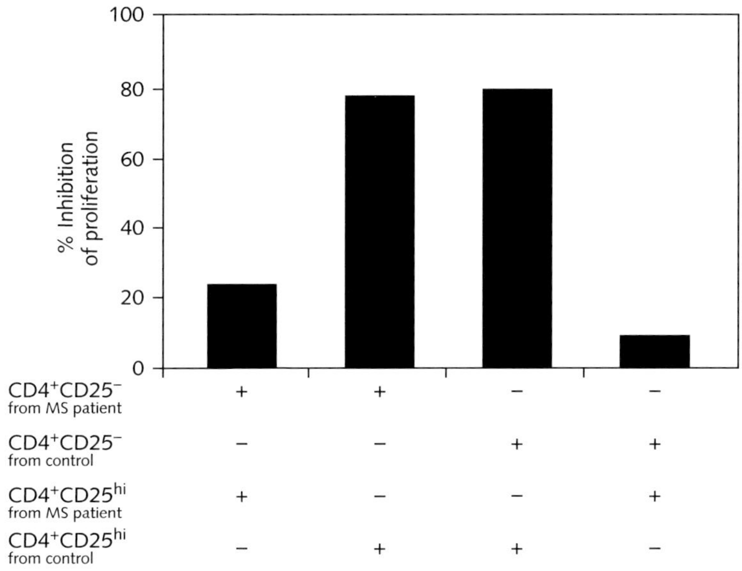

Figure 5
$CD4^+CD25^{hi}$ T cells from patients with MS do not inhibit proliferation of responder T cells isolated from either the autologous individual or healthy donors. $CD4^+CD25^-$ responder T cells and $CD4^+CD25^{hi}$ from MS patients and normal controls were stimulated with plate-bound anti-CD3 at 0.5 μg/ml. 2.5×10^3/well responder T cells from MS patients were co-cultured with the same number of autologous $CD4^+CD25^{hi}$ regulatory T cells (first column on the left) or with regulatory cells isolated from healthy donors (second column). Conversely, responder T cells from healthy donors were co-cultured either with $CD4^+CD25^{hi}$ derived from the same subject (third column), or with regulatory cells isolated from patients with MS (fourth column).

there was a striking decrease in the ability of MS patient-derived T_R cells to suppress Th1 cytokines and proliferation [4, 51].

Mixing experiments in which $CD4^+CD25^{high}$ T_R cells from either healthy subjects or patients with MS were co-cultured with the autologous or the converse (allogenic) target T cells, demonstrated that the primary defect in the inhibition of co-culture proliferation in the analysis of patient samples, is due to a decrease in the function of the T_R cells. As shown in Figure 5, T_R cells from MS patients could not suppress the proliferative response of patient or control target T cells, while in the reciprocal experiment, $CD4^+CD25^{high}$ T_R cells from healthy controls could strongly suppress the proliferative response of target $CD4^+CD25^-$ T cells from both patients and controls. These data indicate that the primary defect is in the function of regulatory T cells isolated from the circulation of patients with MS and not due to the more trivial explanation that the responder T cells derived from patients are in a more activated state and thus are more difficult to inhibit [4, 51].

The possibility still had to be addressed of whether the deficiency in $CD4^+CD25^{high}$ T_R effector function was a result of the dilution of T_R cells with activated responder type T cells especially when isolated from a person with an activated immune system such as a patient with MS. This possibility is a direct result of the fact that, upon activation, basically all T cells express CD25 – that is, the α chain of IL-2 – the major growth factor for T cells. Thus additional experiments were performed in which the isolation of patient and control $CD4^+CD25^{high}$ T_R cell populations was further restricted to include only those cells that also expressed high levels of CD62L, as they are believed to represent non-activated T cells. As can be seen in Figure 6, comparing the functional outcome of patient- and control-derived $CD4^+CD25^{high}CD62L^{high}$ cells produced an even greater disparity in T_R function [4, 51]. Thus although the underlying cause for this functional T_R deficiency is as yet unknown, the observed decrease in *in vitro* T_R activity in MS patient samples is not explained by dilution of $CD4^+$CD25high T_R cells with activated responder-type T cells.

A potential role for $CD4^+CD25^+$ T_R cells in regulating co-stimulation

It is generally accepted that optimal $CD4^+$ T-cell activation requires two signals, one through the TCR recognition of antigen-MHC class II complex and the other *via* interaction with co-stimulatory molecules on APCs. Interaction of APC B7-1 and B7-2 (also known as CD80 and CD86) molecules with CD28 on the surface of the T cell delivers very strong signals that enable T-cell responses to be enormously amplified and highly responsive. Evolutionarily speaking, co-stimulation may have conferred greater selection and survival by augmenting the immune responsiveness to pathogen-derived antigens. Unfortunately, the development of auto-immunity may be a negative side effect of co-stimulation due to inadvertent amplification of weak anti-self TCR signals in certain individuals.

The regulation of co-stimulation may be important in the etiology of MS. How T cells respond to specific co-stimulation and whether they even require co-stimulatory signals is greatly influenced by the strength of the signal delivered through the TCR and the activation state of the T cell itself. Thus, while expansion of MBP-reactive T cells derived from the peripheral blood of healthy controls has been shown to require co-stimulation through APC B7-1/B7-2 molecules, MBP-reactive T cells derived from patients with MS were able to respond in the absence of co-stimula-

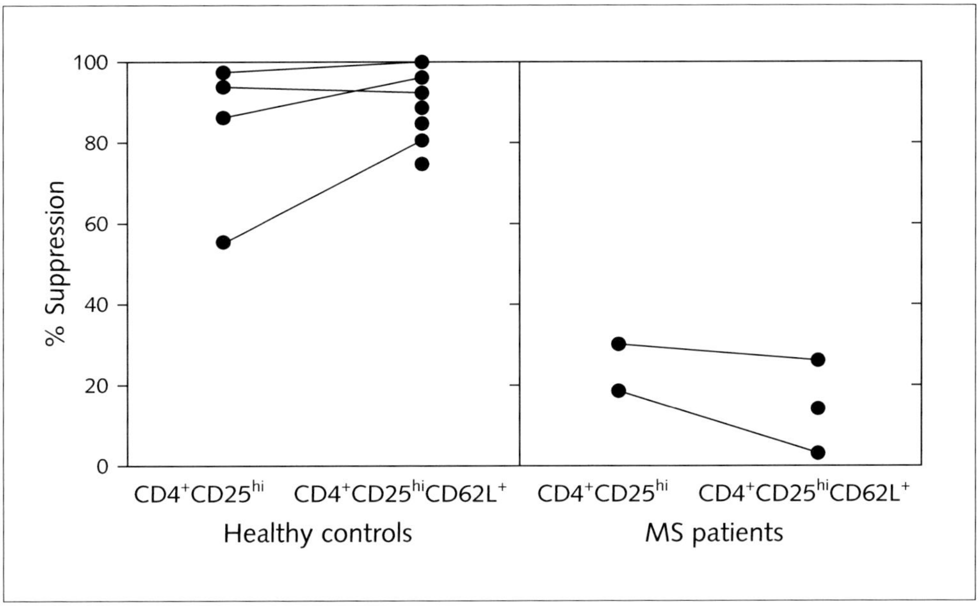

Figure 6
The CD62L$^+$ subset of CD4$^+$CD25^{hi} regulatory T cells exhibits decreased suppressive function in patients with MS. CD4$^+$CD25^{hi}CD62L$^+$ T cells were isolated from seven healthy individuals (left-hand panel) and three patients with MS (right-hand panel). Each symbol is representative of a different individual. Total CD4$^+$CD25^{hi} T cells were also isolated from four of the seven healthy subjects and two of the three patients. The percentage suppression of co-culture proliferation in response to plate-bound anti-CD3 was calculated in each individual. The CD4$^+$CD25^{hi} regulatory T cells derived from healthy controls were able to induce strong inhibition of the proliferative response (55–97% suppression) as shown previously. The CD62L$^+$ subset of CD4$^+$CD25^{hi} T cells derived from the same individuals exhibit enhanced suppressive capacity. In contrast, the regulatory cells derived from patients, although depleted of CD62L-activated T cells, show decreased inhibitory function (4–29% suppression) as compared to the suppression observed in healthy individuals.

tion [1–3]. Importantly, these data provide support for the idea that MBP-reactive T cells in patients with MS are in a more activated state than those isolated from the circulation of healthy controls. Thus it is postulated that dysregulation of the co-stimulatory activity of astrocytes and microglial cells, the resident APCs of the CNS, may be involved in the initiation and/or the augmentation of MS as these two cell types have been found in chronic active lesions and are able to express not only HLA class II, but also the co-stimulatory molecules, B7-1 and B7-2.

Importantly, a number of reports in both mouse and human systems have suggested that $CD4^+CD25^+$ T_R cells may down-modulate the expression of co-stimulatory molecules on APCs [52]. Thus, in addition to inhibiting the activation of potentially autoreactive T cells directly, $CD4^+CD25^+$ T_R cells may indirectly suppress undesirable low-level TCR signaling by inhibiting expression of co-stimulatory molecules.

Potential role for T_R cells in current and future therapies for MS

There are currently three different immunomodulatory therapies that have been approved for the treatment of RRMS in the USA, as they all have been found to impact clinical and/or MRI activities in relapsing-remitting disease. These are recombinant interferon B-1b (rhIFNB-1b; Betaseron and Betaferon), recombinant interferon B-1a (rhIFNb-1a; Avonex and Rebif) and glatiramer-acetate (GA; copolymer-1 or Copaxone). The methods by which these drugs alter the immune system and potentially ameliorate the disease are unknown. As it is difficult to compare the efficacies of the different drugs when tested in separate clinical trials, some studies have compared disease progression and MRI activities among different treatment groups. An evaluation of IFNb-1a (Avonex) treatment found the drug decreasing the percentage of RRMS patients who progressed to sustained disability by the end of the trial from 34.9% in the placebo group to 21.9% in the treated group. There was also a 32% decrease in the annual exacerbation rate and a 40% reduction in the mean MRI load in treated patients as compared to placebo group [53]. Studies on the effects of rhIFNb-1b (Betaseron) demonstrated a significant decrease in the accumulation of MRI lesions [54, 55]. It is currently under investigation whether these treatments may alter T_R suppressive capabilities over time, as it may contribute to observed efficacy.

Increasing the number or functional capacity of human $CD4^+CD25^+$ T_R cells

A potential therapy, albeit less attractive due to the magnitude of the idea, would be enhanced generation or activation *ex vivo* with subsequent re-introduction of the autologous T_R cells after expansion. This possibility arises from the described ability to expand these cells *in vitro* under various growth conditions [4, 56]. We have found that the T_R cells isolated from patients with MS expand at a greatly reduced frequency compared with those isolated from healthy controls [4]. A more attractive approach would be to determine what drugs/treatments might be able to induce expansion of these cells *in vivo*. In this regard, a recent study describing patients with juvenile idiopathic arthritis treated with TNF-α demonstrated an enhanced peripheral frequency of T_R cells in patients with the best prognosis [57].

Conclusion

As the different types or progressive stages of MS, from RRMS to SPMS, which correlate with marked decreases in inflammation, lesion burden, and response to immunotherapies, indicate a conversion from early inflammatory disease to a state of neurodegeneration which is refractory to immunologic intervention, the modulation of disease by specific immunotherapies would be expected to be most effective early in disease. Yet it is often difficult to diagnose the disease in its early stages as the symptoms of MS are fairly heterogeneous and often depend upon recurrence over time, and therefore may make early treatment more difficult. Thus, it is also likely that different treatment modalities will prove efficacious for the treatment of different stages or classifications of MS.

References

1 Scholz C, Patton KT, Anderson DE, Freeman GJ, Hafler DA (1998) Expansion of autoreactive T cells in multiple sclerosis is independent of exogenous B7 costimulation. *J Immunol* 160: 1532–1538

2 Viglietta V, Kent SC, Orban T, Hafler DA (2002) GAD65-reactive T cells are activated in patients with autoimmune type 1a diabetes. *J Clin Invest* 109: 895–903

3 Lovett-Racke AE, Trotter JL, Lauber J, Perrin PJ, June CH, Racke MK (1998) Decreased dependence of myelin basic protein-reactive T cells on CD28-mediated costimulation in multiple sclerosis patients. A marker of activated/memory T cells. *J Clin Invest* 101: 725–730

4 Viglietta V, Baecher-Allan C, Weiner HL, Hafter DA (2004) Loss of functional suppression by $CD4^{+}CD25^{+}$ regulatory T cells in patients with multiple sclerosis. *J Exp Med* 199: 971–979

5 Confavreux C, Vukusic S, Moreau T, Adeleine P (2000) Relapses and progression of disability in multiple sclerosis. *N Engl J Med* 343: 1430–1438

6 Hohol MJ, Olek MJ, Orav EJ, Stazzone L, Hafler DA, Khoury SJ, Dawson DM, Weiner HL (1999) Treatment of progressive multiple sclerosis with pulse cyclophosphamide/methylprednisolone: response to therapy is linked to the duration of progressive disease. *Mult Scler* 5: 403–409

7 Davis RL, Robertson DM (eds) (1997) *Textbook of Neuropathology*, 3rd ed. Williams & Wilkins, Baltimore

8 Khoury SJ, Guttmann CR, Orav EJ, Hohol MJ, Ahn SS, Hsu L, Kikinis R, Mackin GA, Jolesz FA, Weiner HL (1994) Longitudinal MRI in multiple sclerosis: correlation between disability and lesion burden. *Neurology* 44: 2120–2124

9 Filippi M, Paty DW, Kappos L, Barkhof F, Compston DA, Thompson AJ, Zhao GJ, Wiles CM, McDonald WI, Miller DH (1995) Correlations between changes in disabili-

ty and T2-weighted brain MRI activity in multiple sclerosis: a follow-up study. *Neurology* 45: 255–260

10 Genain CP, Cannella B, Hauser SL, Raine CS (1999) Identification of autoantibodies associated with myelin damage in multiple sclerosis. *Nat Med* 5: 170–175

11 McLean BN, Luxton RW, Thompson EJ (1990) A study of immunoglobulin G in the cerebrospinal fluid of 1007 patients with suspected neurological disease using isoelectric focusing and the Log IgG-index: A comparison and diagnostic applications. *Brain* 113: 1269–1289

12 Matsui M, Nagumo F, Tadano J, Kuroda Y (1995) Characterization of humoral and cellular immunity in the central nervous system of HAM/TSP. *J Neurol Sci* 130: 183–189

13 Vartdal F, Vandvik B, Norrby E (1982) Intrathecal synthesis of virus-specific oligoclonal IgG, IgA and IgM antibodies in a case of varicella-zoster meningoencephalitis. *J Neurol Sci* 57: 121–132

14 Jersild C, Fog T, Hansen GS, Thomsen M, Svejgaard A, Dupont B (1973) Histocompatibility determinants in multiple sclerosis, with special reference to clinical course. *Lancet* 2: 1221–1225

15 Hauser SL, Fleischnick E, Weiner HL, Marcus D, Awdeh Z, Yunis EJ, Alper CA (1989) Extended major histocompatibility complex haplotypes in patients with multiple sclerosis. *Neurology* 39: 275–277

16 Dyment DA, Sadovnick AD, Ebers GC, Sadnovich AD (1997) Genetics of multiple sclerosis. *Hum Mol Genet* 6: 1693–1698

17 Oksenberg JR, Seboun E, Hauser SL (1996) Genetics of demyelinating diseases. *Brain Pathol* 6: 289–302

18 Fritz RB, McFarlin DE (1989) Encephalitogenic epitopes of myelin basic protein. *Chem Immunol* 46: 101–125

19 Mokhtarian F, McFarlin DE, Raine CS (1984) Adoptive transfer of myelin basic protein-sensitized T cells produces chronic relapsing demyelinating disease in mice. *Nature* 309: 356–358

20 Wolf SD, Dittel BN, Hardardottir F, Janeway CA Jr (1996) Experimental autoimmune encephalomyelitis induction in genetically B cell-deficient mice. *J Exp Med* 184: 2271–2278

21 Leonard JP, Waldburger KE, Goldman SJ (1995) Prevention of experimental autoimmune encephalomyelitis by antibodies against interleukin 12. *J Exp Med* 181: 381–386

22 Khoury SJ, Hancock WW, Weiner HL (1992) Oral tolerance to myelin basic protein and natural recovery from experimental autoimmune encephalomyelitis are associated with downregulation of inflammatory cytokines and differential upregulation of transforming growth factor beta, interleukin 4, and prostaglandin E expression in the brain. *J Exp Med* 176: 1355–1364

23 Antel J (1999) Multiple sclerosis--emerging concepts of disease pathogenesis. *J Neuroimmunol* 98: 45–48

24 Ota K, Matsui M, Milford EL, Mackin GA, Weiner HL, Hafler DA (1990) T-cell recog-

nition of an immunodominant myelin basic protein epitope in multiple sclerosis. *Nature* 346: 183–187

25 Kerlero de Rosbo N, Milo R, Lees MB, Burger D, Bernard CC, Ben-Nun A (1993) Reactivity to myelin antigens in multiple sclerosis. Peripheral blood lymphocytes respond predominantly to myelin oligodendrocyte glycoprotein. *J Clin Invest* 92: 2602–2608

26 Markovic-Plese S, Fukaura H, Zhang J, al-Sabbagh A, Southwood S, Sette A, Kuchroo VK, Hafler DA (1995) T cell recognition of immunodominant and cryptic proteolipid protein epitopes in humans. *J Immunol* 155: 982–992

27 Jingwu Z, Medaer R, Hashim GA, Chin Y, van den Berg-Loonen E, Raus JC (1992) Myelin basic protein-specific T lymphocytes in multiple sclerosis and controls: precursor frequency, fine specificity, and cytotoxicity. *Ann Neurol* 32: 330–338

28 Zhang J, Markovic-Plese S, Lacet B, Raus J, Weiner HL, Hafler DA (1994) Increased frequency of interleukin 2-responsive T cells specific for myelin basic protein and proteolipid protein in peripheral blood and cerebrospinal fluid of patients with multiple sclerosis. *J Exp Med* 179: 973–984

29 Wucherpfennig KW, Newcombe J, Li H, Keddy C, Cuzner ML, Hafler DA (1992) T cell receptor V alpha-V beta repertoire and cytokine gene expression in active multiple sclerosis lesions. *J Exp Med* 175: 993–1002

30 Traugott U, Reinherz EL, Raine CS (1983) Multiple sclerosis: distribution of T cell subsets within active chronic lesions. *Science* 219: 308–310

31 Hauser SL, Bhan AK, Gilles F, Kemp M, Kerr C, Weiner HL (1986) Immunohistochemical analysis of the cellular infiltrate in multiple sclerosis lesions. *Ann Neurol* 19: 578–587

32 Wucherpfennig KW, Newcombe J, Li H, Keddy C, Cuzner ML, Hafler DA (1992) Gamma delta T-cell receptor repertoire in acute multiple sclerosis lesions. *Proc Natl Acad Sci USA* 89: 4588–4592

33 Prineas JW, Wright RG (1978) Macrophages, lymphocytes, and plasma cells in the perivascular compartment in chronic multiple sclerosis. *Lab Invest* 38: 409–421

34 Prineas J (1975) Pathology of the early lesion in multiple sclerosis. *Hum Pathol* 6: 531–554

35 Windhagen A, Anderson DE, Carrizosa A, Balashov K, Weiner HL, Hafler DA (1998) Cytokine secretion of myelin basic protein reactive T cells in patients with multiple sclerosis. J *Neuroimmunol* 91: 1–9

36 Sorensen TL, Tani M, Jensen J, Pierce V, Lucchinetti C, Folcik VA, Qin S, Rottman J, Sellebjerg F, Strieter RM et al. (1999) Expression of specific chemokines and chemokine receptors in the central nervous system of multiple sclerosis patients. *J Clin Invest* 103: 807–815

37 Balashov KE, Rottman JB, Weiner HL, Hancock WW (1999) *CCR5(+)* and *CXCR3(+)* T cells are increased in multiple sclerosis and their ligands MIP-1alpha and IP-10 are expressed in demyelinating brain lesions. *Proc Natl Acad Sci USA* 96: 6873–6878

38 Correale J, Gilmore W, McMillan M, Li S, McCarthy K, Le T, Weiner LP (1995) Pat-

terns of cytokine secretion by autoreactive proteolipid protein-specific T cell clones during the course of multiple sclerosis. *J Immunol* 154: 2959–2968
39 Comabella M, Balashov K, Issazadeh S, Smith D, Weiner HL, Khoury SJ (1998) Elevated interleukin-12 in progressive multiple sclerosis correlates with disease activity and is normalized by pulse cyclophosphamide therapy. *J Clin Invest* 102: 671–678
40 Windhagen A, Newcombe J, Dangond F, Strand C, Woodroofe MN, Cuzner ML, Hafler DA (1995) Expression of costimulatory molecules B7-1 (CD80), B7-2 (CD86), and interleukin 12 cytokine in multiple sclerosis lesions. *J Exp Med* 182: 1985–1996
41 Hofman FM, Hinton DR, Johnson K, Merrill JE (1989) Tumor necrosis factor identified in multiple sclerosis brain. *J Exp Med* 170: 607–612
42 Sakaguchi S, Fukuma K, Kuribayashi K, Masuda T (1985) Organ-specific autoimmune diseases induced in mice by elimination of T cell subset. I. Evidence for the active participation of T cells in natural self-tolerance; deficit of a T cell subset as a possible cause of autoimmune disease. *J Exp Med* 161: 72–87
43 Suri-Payer E, Amar AZ, Thornton AM, Shevach EM (1998) $CD4^+CD25^+$ T cells inhibit both the induction and effector function of autoreactive T cells and represent a unique lineage of immunoregulatory cells. *J Immunol* 160: 1212–1218
44 Jordan MS, Boesteanu A, Reed AJ, Petrone AL, Holenbeck AE, Lerman MA, Naji A, Caton AJ (2001) Thymic selection of $CD4^+CD25^+$ regulatory T cells induced by an agonist self-peptide. *Nat Immunol* 2: 301–306
45 Salomon B, Rhee L, Bour-Jordan H, Hsin H, Montag A, Soliven B, Arcella J, Girvin AM, Padilla J, Miller SD, Bluestone JA (2001) Development of spontaneous autoimmune peripheral polyneuropathy in B7-2-deficient NOD mice. *J Exp Med* 194: 677–684
46 Thornton AM, Shevach EM (1998) $CD4^+CD25^+$ immunoregulatory T cells suppress polyclonal T cell activation *in vitro* by inhibiting interleukin 2 production. *J Exp Med* 188: 287–296
47 Baecher-Allan C, Brown JA, Freeman GJ, Hafler DA (2001) $CD4^+$CD25high regulatory cells in human peripheral blood. *J Immunol* 167: 1245–1253
48 Baecher-Allan C, Viglietta V, Hafler DA (2002) Inhibition of human CD4(+) CD25(+high) regulatory T cell function. *J Immunol* 169: 6210–6217
49 Wing K, Lindgren S, Kollberg G, Lundgren A, Harris RA, Rudin A, Lundin S, Suri-Payer E (2003) CD4 T cell activation by myelin oligodendrocyte glycoprotein is suppressed by adult but not cord blood $CD25^+$ T cells. *Eur J Immunol* 33: 579–587
50 Taylor PA, Noelle RJ, Blazar BR (2001) CD4(+)CD25(+) immune regulatory cells are required for induction of tolerance to alloantigen *via* costimulatory blockade. *J Exp Med* 193: 1311–1318
51 Baecher-Allan C, Viglietta V, Hafler DA (2004) Human $CD4^+CD25^+$ regulatory T cells. *Semin Immunol* 16: 89–98
52 Misra N, Bayry J, Lacroix-Desmazes S, Kazatchkine MD, Kaveri SV (2004) Cutting edge: human $CD4^+CD25^+$ T cells restrain the maturation and antigen-presenting function of dendritic cells. *J Immunol* 172: 4676–4680
53 Logan-Clubb L, Stacy M (1995) An open-labelled assessment of adverse effects associ-

ated with interferon 1-beta in the treatment of multiple sclerosis. *J Neurosci Nurs* 27: 344–347

54 The IFNB Multiple Sclerosis Study Group and The University of British Columbia MS/MRI Analysis Group (1995) Interferon beta-1b in the treatment of multiple sclerosis: final outcome of the randomized controlled trial. *Neurology* 45: 1277–1285

55 The IFNB Multiple Sclerosis Study Group (1993) Interferon beta-1b is effective in relapsing-remitting multiple sclerosis. I. Clinical results of a multicenter, randomized, double-blind, placebo-controlled trial. Neurology 43: 655–661

56 Levings MK, Sangregorio R, Sartirana C, Moschin AL, Battaglia M, Orban PC, Roncarolo MG (2002) Human $CD25^+CD4^+$ T suppressor cell clones produce transforming growth factor beta, but not interleukin 10, and are distinct from type 1 T regulatory cells. *J Exp Med* 196: 1335–1346

57 de Kleer IM, Wedderburn LR, Taams LS, Patel A, Varsani H, Klein M, de Jager W, Pugayung G, Giannoni F, Rijkers G et al. (2004) $CD4^+$CD25(bright) regulatory T cells actively regulate inflammation in the joints of patients with the remitting form of juvenile idiopathic arthritis. *J Immunol* 172: 6435–6443

Immunotherapy of rheumatoid arthritis using CD4$^+$CD25$^+$ regulatory T cells

Jocea M. van Amelsfort[1], *Johannes W. J. Bijlsma*[1] *and Leonie S. Taams*[2]

[1]Department of Rheumatology & Clinical Immunology, University Medical Center Utrecht, 3508 GA Utrecht, The Netherlands; [2]Infection & Immunity Research Group, Franklin-Wilkins Building, King's College London, 150 Stamford Street, London SE1 9NN, UK

Introduction to rheumatoid arthritis (RA)

RA is a disease characterized by a systemic polyarthritis of primarily the small joints of hand and feet. Although spontaneous remission occurs, the arthritis becomes chronic in most cases, resulting in progressive damage to the cartilage and destruction of the bone in the joints with an increased physical disability for the patient. Besides in the joints, inflammation occurs systemically, leading to general symptoms such as fatigue and anaemia. Furthermore, extra-articular manifestations of inflammation are found in patients with RA, such as rheumatoid nodules, vasculitis and pericarditis [1].

The incidence of RA in the general population is 0.5–1% [2–4]. Both genetics and environment are important in developing RA. Studies in relatives of RA patients showed a modest increased risk for relatives to develop RA [4]. Furthermore, twin studies show a 4-fold increased concordance in monozygotic twins compared to dizygotic twins, suggesting a role for the shared genetics in the twins [4, 5]. The main genetic risk factor of RA is the presence of the HLA DRB1 allele and this has been shown consistently in many populations throughout the world [4]. All HLA DRB1 alleles associated with RA encode a conserved amino acid sequence, which is commonly called the shared epitope. Environmental factors have also been suggested to be associated with RA, such as female sex hormones, infections such as Epstein-Barr virus and parvovirus, and more recently cigarette smoking [4].

The underlying mechanisms resulting in the development of the chronic inflammation remain largely unknown. RA is generally considered an autoimmune disease, since chronic inflammation with autoimmunity to many self-antigens is a key characteristic of the disease without the presence of a foreign antigen. However, the disease-causing self-antigen has yet to be discovered. Several joint-specific antigens have been proposed as likely candidates, such as collagen type II and chondrocyte gp39 [6–8], but ubiquitous antigens such as BiP, HSP60 and glucose-6-phosphate isomerase (GPI) have also been suggested to be autoantigens in RA [9–12].

Regulatory T Cells in Inflammation, edited by Leonie S. Taams, Arne N. Akbar and Marca H.M. Wauben

Based on the association with HLA-DR, the concept of RA as a $CD4^+$ T-cell-mediated disease was established. This was supported by evidence from several experimental arthritis models, including collagen-induced arthritis and adjuvant arthritis, which showed that the disease could be transferred by autoreactive $CD4^+$ T cells [13–15]. In humans, $CD4^+$ T-cell infiltrations have been observed in the synovial tissue of RA patients [16] and $CD4^+$ T cells are found in synovial fluid. Based on their cytokine profile, these $CD4^+$ T cells display a predominant Th1 activity and a lack of Th2 activity. This imbalance has been suggested to contribute to the activation of macrophages and B cells [17, 18].

The $CD4^+$ T cells in the synovium have several interesting features. First, they have a high expression of CD45RO and low expression of CD45RB, which characterizes them as highly differentiated memory T cells and indicates that these cells have been exposed to antigens *in vivo* [19]. T cells with such a highly differentiated phenotype are usually prone to apoptosis. It has been suggested however that synovial T cells might have a prolonged survival due to the presence of type I interferons, thus contributing to the chronic inflammation [20]. Secondly, synovial T cells express CD69, an early marker of activation. Because T cells in the blood of RA patients rarely express CD69, activation in the synovium is likely to have occurred, possibly as a consequence of antigen recognition, passage through endothelial cells or exposure to cytokines present in the tissue environment. A correlation between CD69 expression on synovial T cells and disease severity has been reported [21]. Third, synovial T cells produce a number of molecules that contribute to the inflammation, such as interferon (IFN)-γ and CD40 ligand. Interaction of CD40 ligand and CD40 on B cells induces B-cell proliferation and immunoglobulin production. CD40 engagement also induces monocyte activation and dendritic cell differentiation and vascular endothelial growth factor production by synovial fibroblasts [22–24].

The efficacy of anti-CD4 monoclonal antibody (mAb) therapy in several animal arthritis models [25–27] has provided further evidence for the role of $CD4^+$ T cells in the pathogenesis of arthritis [28]. In clinical trials with RA patients, however, no significant improvement of symptoms was observed upon anti-CD4 mAb treatment, suggesting that perhaps T cells were not key participants in the chronic inflammation [29–31]. However, more recent studies with CTLA-4 Ig in RA patients do point to a role for T cells. CTLA-4 Ig prevents T-cell activation and was shown to be clinically effective in RA patients [32, 33].

Besides T cells, several other cell types are considered to be important for initiating and sustaining the inflammation in the joints. In many RA patients the so-called rheumatoid factor (RF) is present, which is an autoantibody that recognizes the Fc region of self-IgG. The presence of these autoantibodies suggests a role for B cells. Immune complexes containing RF can activate the complement cascade and enhance the chronic inflammatory process. There is as yet little evidence that antibodies are directly involved in tissue destruction, but the presence of the antibodies is a prognostic factor, because RF positive patients have a more erosive and a more destructive disease [34]. In addition, recent studies by Edwards et al. [35] showed that B-cell-targeted therapy with an anti-CD20 mAb (Rituximab) led to a significant improvement in disease symptoms.

The efficacy of biological therapies aimed at blocking cytokines that are predominantly produced by macrophages, such as tumour necrosis factor (TNF)-α and interleukin (IL)-1, suggests an important role for macrophages in the ongoing inflammation and joint destruction in RA. Activated macrophages that express high levels of MHC-II and produce pro-inflammatory cytokines and growth factors, such as IL-1, TNF-α, IL-6, granulocyte-macrophage colony-stimulating factor (GM-CSF), chemokines such as IL-8 and macrophage inflammatory protein (MIP), and metalloproteinases, are abundantly present in the synovial membrane and at the cartilage-pannus junction [36]. Furthermore, it was shown that there is a correlation between the degree of synovial macrophage infiltration and the radiological progression of permanent joint damage, the disease feature that ultimately determines quality of life [37].

Therapeutic strategies in RA

The main goal in the treatment of RA is to stop the ongoing inflammation and with that prevent permanent joint damage. However, so far no treatment is available that can actually cure RA. Medications that are used to treat RA can be divided into three main groups: non-steroidal anti-inflammatory drugs (NSAIDs), corticosteroids and disease-modifying antirheumatic drugs (DMARDs). NSAIDs are particularly helpful in the first few weeks after diagnosis, because they provide relief of pain and stiffness; however, they have not been shown to slow the progression of the disease [38, 39]. Corticosteroids are potent suppressors of the inflammatory response in RA, but have significant dose-dependent side effects and controversy continues about when and how these medications should be used in the treatment of RA [40, 41]. The third group, the DMARDs, are defined by their ability to slow the progression of joint damage in RA. The most commonly used DMARD is methotrexate (MTX), which has been demonstrated to be effective and relatively safe in long-term therapies [42, 43]. Other DMARDs are antimalarial drugs, such as chloroquine and hydroxychloroquine, sulfasalazine, intramuscular gold and leflonumide. The mechanism by which DMARDs work is poorly understood. For MTX, several mechanisms have been described, such as promotion of adenosine release at the inflamed site and inhibition of transmethylation reactions [44, 45]. In the last decade, treatment with anti-TNF-α mAb or TNF-α-soluble receptors has proven to be very effective in the treatment of RA [46, 47]. Also several other therapies with clinical success have been developed targeting other pro-inflammatory cytokines, such as IL-1 and IL-6 [48, 49].

Treatment with anti-TNF-α mAb leads to a down-regulation of pro-inflammatory cytokines, such as IL-6 and IL-1 [47]. Furthermore, the expression of adhesion molecules such as E-selectin, intracellular cell-adhesion molecule 1 (ICAM-1) and vascular cell-adhesion molecule 1 (VCAM-1) was diminished in the synovium after treatment with anti-TNF-α mAb, suggesting a reduction of leukocyte trafficking into the joints [47]. Importantly though, the beneficial effects of anti-TNF-α therapy disappear upon cessation of antibody administration, indicating that the immune reactions leading to the inflammatory manifestations of the disease continue and are only transiently suppressed. Furthermore, a significant group of patients (30–40%) responds inadequately to treatment with anti-TNF-α mAb [47].

$CD4^+CD25^+$ regulatory T cells

As described in the previous section, most therapies in RA are aimed at general suppression of the inflammatory response. With the development of biologicals such as anti-TNF-α mAbs a more specific immune suppression is obtained. However, current therapies in RA are only temporarily beneficial and lead to general suppression of the whole immune system and even to suppression of other bone marrow functions, resulting in severe side effects. Since RA might be the result of an imbalance in immunoregulation, a different approach could be to stimulate immunoregulatory events, with the naturally occurring regulatory T-cell (T_R cell) population being a target for such an approach. These T_R cells, which can be identified by their expression of CD4 and CD25 (the IL-2 receptor α chain), constitute approximately 10% of peripheral blood $CD4^+$ T cells [50]. They are found in peripheral blood [51], lymphoid tissue such as tonsil, spleen, lymph node and thymus [52, 53], and many other organs and tissues, for example lung, gut and skin [54–56].

The spontaneous induction of autoimmunity in mice upon depletion of $CD4^+CD25^+$ T cells was the first indication that these cells are actively involved in maintaining immune tolerance to self-antigens [50, 57]. It is now well established that $CD4^+CD25^+$ T_R-cell depletion either *in vitro* or *in vivo* leads to increased antibody titres and T-cell responses (for review see [58]). In particular the suppressive effects on T cells have been investigated. It was shown that upon interaction with $CD4^+CD25^+$ T_R cells, both $CD4^+$ and $CD8^+$ T cells are impaired in their capacity to proliferate and produce cytokines [51, 53, 59, 60]. Cell contact between the T_R cells and responder T cells is required for the suppressive effects, and both surface-expressed CTLA-4 and glucocorticoid-induced TNF receptor (GITR) have been suggested as candidates to mediate these cell-contact-dependent effects [61–63]. However, *in vivo* transforming growth factor (TGF)-β and IL-10 might play a more important role in suppression than has been deduced so far from *in vitro* studies [64, 65]. TGF-β in particular might function *via* the induction of the *forkhead*/winged helix transcription factor Foxp3 [66, 67], which has recently been suggested to be the crucial molecule for functional T_R cells [68].

Importantly, the suppressive effects of $CD4^+CD25^+$ T_R cells are not confined to the adaptive immune system. Recent data indicate that T_R cells are also capable of down-regulating cells from the innate immune system. The initial data came from mouse studies, demonstrating that spleen-derived dendritic cells (DCs) displayed down-regulated expression of the co-stimulatory molecules CD80 and CD86 upon interaction with $CD4^+CD25^+$ T_R cells [69]. This has recently been confirmed in humans, showing that co-culture with $CD4^+CD25^+$ T_R cells led to down-regulated expression of CD40, CD80, CD86 and HLA class II on monocyte-derived DCs and a subsequent impaired capacity of these DCs to stimulate responder T cells [70]. This effect is not restricted to DCs, since we found that monocytes/macrophages are equally susceptible to the suppressive effects of $CD4^+CD25^+$ T_R cells (L.S. Taams et al. *in press, Human Immunology*, 2005). $CD4^+CD25^+$ T_R-monocyte interaction leads to down-modulated CD86 expression and impaired up-regulation of CD40, CD80 and HLA class II on monocytes compared to untreated monocytes or monocytes pre-cultured with non-regulatory T cells. When these monocytes are re-purified and used as antigen-presenting cells (APCs) in a T-cell-stimulation assay, they have a strongly reduced capacity to induce an antigen-specific T-cell response. Moreover, these monocytes display an impaired capacity to produce TNF-α and IL-6 in response to lipopolysaccharide (LPS) relative to control monocytes. These data indicate that $CD4^+CD25^+$ T_R cells can inhibit the activation and function of monocytes/macrophages, thus reducing their pro-inflammatory potential. This is of particular relevance to RA in which, as previously discussed, monocytes/macrophages together with $CD4^+$ T cells and B cells are key players in the inflammatory process in the joint. Specific targeting of these $CD4^+CD25^+$ T_R cells might therefore lead to a restoration of the immunological balance in the inflamed sites. However, before this could be achieved, the presence and function of $CD4^+CD25^+$ T_R cells in RA need to be identified.

$CD4^+CD25^+$ T_R cells in experimental arthritis models and RA patients

The initial studies by Sakaguchi and co-workers [50] demonstrated that depletion of $CD4^+CD25^+$ T_R cells in mice predominantly led to organ-specific autoimmunity, including some polyarthritis. In a study using a proteoglycan-induced arthritis model Bardos et al. [71] argued that $CD4^+CD25^+$ T_R cells do not protect against the induction of arthritis. In their model, CD25-depleted splenocytes from arthritic mice were injected into SCID mice in the presence of proteoglycan. Simultaneous transfer of $CD25^+$ T_R cells or $CD25^-$ T cells from naïve BALB/c mice did not alter disease severity. Also, no differences were observed when total splenic cells or CD25-depleted splenic cells were used from arthritic mice. Recent work by Toes and co-

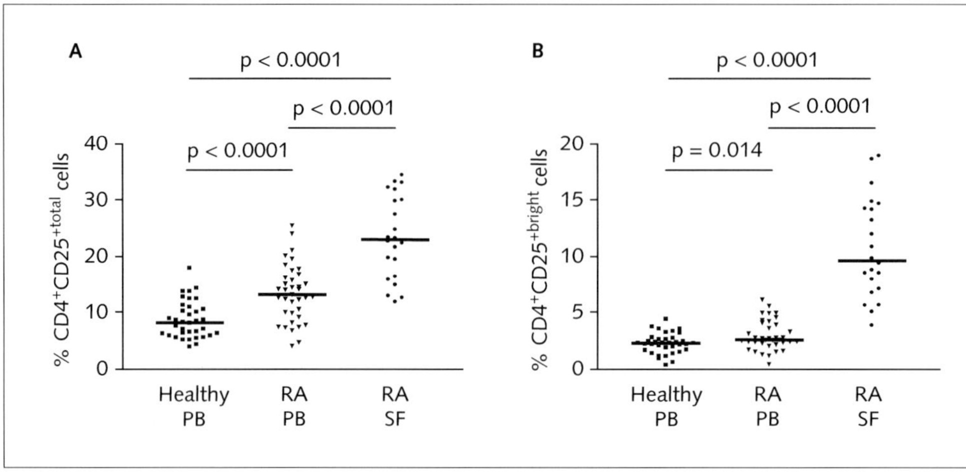

Figure 1.
The percentage of $CD4^+CD25^{+total}$ and $CD4^+CD25^{+bright}$ T cells is increased in both peripheral blood (PB) and synovial fluid (SF) of RA patients compared to healthy controls. The percentage $CD4^+CD25^+$ T cells was measured via *flow cytometry in healthy controls (n = 36) and PB (n = 37) and SF (n = 22) of RA patients. $CD4^+CD25^{+bright}$ cells were defined by a high expression of CD25 and a lower expression of CD4.*

workers [72] however supports the notion that $CD4^+CD25^+$ T_R cells do play a role in the regulation of arthritis. Using a model of collagen-induced arthritis (CIA) they showed that depletion of $CD4^+CD25^+$ T cells 14 days before immunization with collagen type II in Freund's complete adjuvant (CFA) led to an increase in the severity of CIA as well as increased collagen-specific antibody titres compared to control mice [72]. Adoptive transfer of $CD4^+CD25^+$ T_R cells reversed these effects. The role of T_R cells in prevention of experimental arthritis thus seems to be controversial.

Recently, a number of studies have described the presence and function of $CD4^+CD25^+$ T cells in patients with RA. We and others have shown that, in the synovial fluid (SF), the most easily accessible derivative of the target organ of RA, the percentage of $CD4^+CD25^+$ T cells was increased to approximately 20% compared to 15% in peripheral blood (PB) from RA patients (see Fig. 1) [73, 74]. Interestingly, as consistent as these studies were in their observation that the percentages of $CD4^+CD25^+$ T cells were elevated in SF of RA patients, in PB of RA patients the data were much more diverse. Focus has recently shifted towards the $CD4^+CD25^{+bright}$ population, $CD4^+$ T cells that express high levels of CD25. Baecher-Allan et al. [75] showed that the suppressive cells mainly reside within this population and not in the $CD4^+CD25^{+intermediate}$ population, which expresses only moderate levels of CD25. It was described that the percentage of $CD4^+CD25^{+bright}$ cells in PB of arthritis patients was equal or decreased compared to PB of healthy controls [73, 76]. However, we investigated the total population of $CD4^+CD25^+$ T cells and found an increased percentage in PB compared to healthy controls [74]. It could be argued that these are in part activated T cells; however, the $CD4^+CD25^+$ T cell population in PB of RA patients had the same non-activated phenotype as in healthy controls. Furthermore, when we gated for $CD4^+CD25^{+bright}$ cells, we still found a small but significant increase in PB of RA patients compared to healthy controls (mean ± S.E.M., 2.96 ± 0.2% *versus* 2.26 ± 0.2%; $p = 0.014$) (see Fig. 1). The differences between these studies can possibly be explained by differences in staining method and the definition of $CD4^+CD25^{+bright}$ cells. Furthermore, RA patients are a very diverse population with regard to disease duration, disease activity and use of medication, possibly resulting in differences in $CD4^+CD25^+$ T cells.

Are $CD4^+CD25^+$ T cells from SF of RA patients recently activated or T_R cells?

Since $CD4^+CD25^+$ T cells in SF are derived from an inflammatory milieu, it is possible that at least some of these cells are recently activated T cells that contribute to the inflammation, rather than T_R cells. We therefore examined the phenotype of both $CD4^+CD25^-$ and $CD4^+CD25^+$ T cells from SF. We and others found that approximately 50% of both $CD25^-$ and $CD25^+$ T-cell populations display an activated phenotype in SF compared to PB (Fig. 2). That is, they express increased levels of CD69, MHC II (Fig. 2) and CD45RO (not shown) [74, 76]. Interestingly, $CD4^+CD25^+$ from SF (and PB) expressed significantly higher levels of regulatory molecules such as CTLA-4, GITR and OX-40 compared to the $CD4^+CD25^-$ T cells from SF [74]. Thus, $CD4^+CD25^+$ T cells from SF have phenotypical features of both recently activated and highly regulatory T cells.

As described previously, the suppressive function of $CD4^+CD25^+$ T_R cells primarily resides in the $CD4^+CD25^{+bright}$ population. The phenotype of this population was shown to be different from the $CD4^+CD25^{+intermediate}$ population in PB of arthritis patients. $CD4^+CD25^{+bright}$ cells had higher levels of CTLA-4, CD45RO, HLA-DR, CD71, CD58, CD122 and CD62L compared to their $CD4^+CD25^{+intermediate}$ counterparts [73, 76]. Interestingly Cao et al. [73] showed that the $CD4^+CD25^{+bright}$ cells from SF displayed a similar phenotype as $CD4^+CD25^{+bright}$ cells derived from PB of RA patients and healthy controls. In contrast, we found that the $CD4^+CD25^{+bright}$ population from SF had even higher levels of markers such as MHC II, CTLA-4, GITR, CD69 and OX-40 compared to the $CD4^+CD25^{+bright}$ population in PB of RA (Fig. 2). Although these markers are normally up-regulated upon T-cell activation, it has also been shown that these molecules are important in the suppressive function of $CD4^+CD25^+$ T cells [51, 61–63, 77]. Similar results were found in the $CD4^+CD25^{+bright}$ cells from SF of juvenile idiopathic arthritis (JIA) patients, which is the most frequent rheumatic disease of childhood and differs from

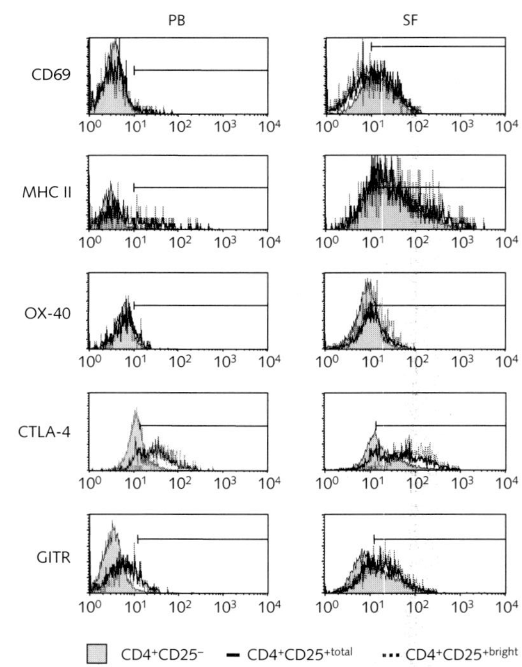

	$CD25^-$	$CD25^{+total}$	$CD25^{+intermediate}$	$CD25^{+bright}$
MFI $CD69^+$ PB	3.3 ± 0.2	3.7 ± 0.2	3.9 ± 0.2	3.0 ± 0.2
MFI $CD69^+$ SF	9.1 ± 1.7	10.4 ± 3.0	10.9 ± 2.2	6.9 ± 1.3
MFI MHC II^+ PB	3.8 ± 0.3	5.0 ± 0.3	4.7 ± 0.2	7.0 ± 0.9
MFI MHC II^+ SF	15.4 ± 3	17.3 ± 3	13.3 ± 2	26.6 ± 6
MFI OX-40 PB	6.0 ± 0.5	6.9 ± 0.2	7.3 ± 0.3	6.3 ± 0.2
MFI OX-40 SF	7.6 ± 0.9	10.3 ± 1.6	9.5 ± 1.5	11.9 ± 2.0
MFI CTLA-4 PB	16.0 ± 5.9	24.9 ± 6.3	20.9 ± 6.1	35.8 ± 7.4
MFI CTLA-4 SF	33.2 ± 7.6	60.9 ± 10.1	36.7 ± 8.5	123.3 ± 9.7
MFI GITR PB	3.2 ± 0.1	4.8 ± 0.5	4.7 ± 0.5	4.9 ± 0.6
MFI GITR SF	7.2 ± 1.4	12.9 ± 3.2	10.4 ± 2.2	16.9 ± 4.9

Figure 2.
Expression of cell-surface markers was measured by flow cytometry on $CD4^+CD25^-$, $CD4^+CD25^{+total}$, $CD4^+CD25^{+intermediate}$ and $CD4^+CD25^{+bright}$ T cells in both peripheral blood (PB) and synovial fluid (SF) of RA patients. Markers were set based on isotype control staining. Expression is shown as the mean fluorescence intensity (MFI). One representative example out of five paired samples is shown. In PB, both $CD25^{+total}$ and $CD25^{+bright}$ T cells showed an increased expression of MHC II, and particularly of CTLA-4 and GITR compared to $CD25^-$ T cells. In SF, all populations displayed a higher expression of the markers compared to PB. Notably, increased expression of OX-40, CTLA-4 and GITR was observed in both $CD25^{+total}$ and $CD25^{+bright}$ T cells relative to $CD25^-$ T cells.

adult RA in the fact that various subtypes exist with differences in disease severity and outcome. This study also showed that the Foxp3 levels were increased in the $CD4^+CD25^{+bright}$ T cells [76]. Interestingly, in the $CD4^+CD25^{+intermediate}$ cells from SF of these patients high levels of Foxp3 were also found, suggesting that in SF of JIA patients the T_R cells do not only reside within the $CD4^+CD25^{+bright}$ but also in the $CD4^+CD25^{+intermediate}$ population [76]. Figure 2 furthermore shows that $CD4^+CD25^{+intermediate}$ cells indeed display high levels of regulatory markers as GITR and CTLA-4, almost comparable to the levels of the $CD4^+CD25^{+bright}$ T cell population in PB. In conclusion, in SF of arthritis patients, the distinction between $CD4^+CD25^{+bright}$ and $CD4^+CD25^{+indermediate}$ as an indication of suppressive activity might not be suitable.

The question whether $CD4^+CD25^+$ T cells from SF are activated aggressor cells or regulatory cells can only be answered by functional analysis. It was found that $CD4^+CD25^{+bright}$ T cells from SF were able to suppress $CD4^+CD25^-$ T-cell proliferation [73, 76]. Furthermore, we demonstrated that the total population of $CD4^+CD25^+$ T cells in SF had a higher suppressive capacity than those cells from PB, resulting in an average 76% inhibition of T cell proliferation in SF compared to 41% in PB of RA patients [74]. This could be due to the fact that in SF the $CD4^+CD25^+$ T-cell population contains more $CD4^+CD25^{+bright}$ cells. However de Kleer et al. [76] showed that in JIA also the $CD4^+CD25^{+intermediate}$ T cells from SF were capable of inhibiting T-cell proliferation, which is in agreement with their increased Foxp3 levels. In summary, in the synovial fluid of RA patients an increased population of activated, highly suppressive $CD4^+CD25^+$ T cells is present. What is the exact function of these cells *in vivo* and why do patients still suffer chronic inflammation despite the presence of these regulatory cells?

Why does joint inflammation persist despite the presence of $CD4^+CD25^+$ T_R cells?

Although isolated $CD4^+CD25^+$ T cells from SF are highly suppressive *in vitro*, it is not clear whether their function is effectively maintained *in vivo*. It was shown that the suppressive effect of $CD4^+CD25^+$ T_R cells is not only dependent on the suppressor cells themselves, but also on the activation status of the responder/aggressor population, the $CD4^+CD25^-$ T cells [78]. Indeed we and others demonstrated that $CD4^+CD25^-$ T cells from SF displayed an activated phenotype and that this activation status led to a decreased susceptibility to $CD4^+CD25^+$ T-cell-mediated suppression [74, 76]. In addition, factors like IL-2 and co-stimulation through CD28 can break the anergic and suppressive state of $CD4^+CD25^+$ T cells [51, 60, 79, 80]. Thus the inflammatory milieu in the joint might inhibit the suppressive function of $CD4^+CD25^+$ T_R cells *via* the presence of pro-inflammatory cytokines/mediators and co-stimulation by APCs, such as activated macrophages, B cells and DCs. Indeed, recent data from our laboratory show that $CD4^+CD25^+$ T_R cell-mediated suppression is decreased by the addition of anti-CD28 mAb or pro-inflammatory cytokines such as TNF-α and IL-7, which are also increased in RA patients [81] (J.M. van Amelsfort et al., unpublished observations).

Another consideration is the presence of $CD4^+CD25^+$ T_R cells in the *synovial tissue*. Although the presence of $CD25^+$ T cells has been demonstrated in the synovial tissue [16, 82] and the presence of $CD4^+CD25^+$ T cells with regulatory capacities in the SF makes it likely that $CD25^+$ T cells in the synovial tissue are also T_R cells, no data are available on their suppressive capacities or Foxp3 expression. Since synovial tissue, rather than SF, is the actual site of inflammation it is essential that future work will dissect the functional properties of $CD4^+CD25^+$ T cells within the synovial tissue.

How does the presence of $CD4^+CD25^+$ T cells influence disease activity and response to treatment in arthritis patients?

As described previously JIA, the juvenile form of arthritis, is a disease that has different subtypes. de Kleer et al. [76] showed that patients with a form of JIA that is self-limiting (persistent JIA) had a higher percentage of $CD4^+CD25^{+bright}$ and $CD4^+CD25^{+total}$ cells than patients with a form of JIA that is more severe and chronic (extended JIA). These findings suggested that higher numbers of $CD4^+CD25^+$ T cells are beneficial for the clinical outcome. In our patient group with (adult) RA, we also investigated the relation between $CD4^+CD25^+$ T cells and disease activity and found a significant correlation between the percentage of $CD4^+CD25^+$ T cells in PB and the erythrocyte sedimentation rate (ESR) levels in blood, an important measure for inflammation [74]. However, this correlation was not found for $CD4^+CD25^+$ T cells in SF and no correlations with other disease parameters, such as C-reactive protein (CRP) levels and joint scores, were found. This is consistent with data from Cao et al. [83] who found no correlations between parameters for disease activity/severity and percentage $CD4^+CD25^+$ T cells in SF from different types of arthritis.

The influence of medication used in arthritis patients on the percentage $CD4^+CD25^+$ T cells was also investigated. Usage of medications such as MTX [74, 76], anti-TNF-α mAb [74] or intra-articular corticosteroids [83] had no influence on the percentage of $CD4^+CD25^+$ T cells in PB of RA patients. More interesting is the question whether treatment that resulted in a clinical improvement correlates with a change in presence and/or function of $CD4^+CD25^+$ T cells. Ehrenstein et al. [84] recently demonstrated that in active RA patients, before start of treatment with anti-TNF-α, $CD4^+CD25^+$ T cells were capable of inhibiting T-cell proliferation, but could not inhibit cytokine production by T cells and monocytes. This inability to suppress cytokine production might underscore the important role for cytokines in the ongoing inflammation in the joint. Interestingly, also the ability to convert $CD4^+CD25^-$ T cells to a "suppressor" phenotype [85, 86] was lost in $CD4^+CD25^+$ T cells from active RA patients before treatment. These deficiencies in suppressor-T-cell function were reversed in patients that responded clinically to the anti-TNF-α treatment. Furthermore it was found that upon treatment with anti-TNF-α the $CD4^+CD25^+$ T-cell population was significantly increased in the responding patients. Even though in our own studies on the effect of anti-TNF-α therapy we did not find significant changes in the percentage of $CD4^+CD25^+$ T cells (bright and total), whether patients responded to treatment or not, we did observe interesting differences in patients treated with MTX: patients that did not respond to treatment had a slight decrease in the percentage of $CD4^+CD25^+$ T cells (J.M. van Amelsfort, unpublished observations). Together these data suggest that clinical response to treatment is correlated with a maintenance/increase in the $CD4^+CD25^+$ T-cell population. However, it will be difficult to reach a consensus between the different studies investigating the influence of disease activity or treatment on percentage/function of $CD4^+CD25^+$ T cells due to different patient groups, different treatment strategies and different definitions of clinical response.

$CD4^+CD25^+$ T_R cells as an approach for immunotherapy in RA

$CD4^+CD25^+$ T_R cells play an important role in the regulatory mechanisms of the immune system. Enhancing these regulatory mechanisms in autoimmune diseases will restore the imbalance in the immune response and cease the chronic inflammation. $CD4^+CD25^+$ T_R cells are therefore an interesting target for immunotherapy in RA. Several issues have to be considered before such a therapy can be fully developed. The first consideration is how a more functional and/or larger $CD4^+CD25^+$ T_R-cell population can be obtained in patients with RA. One of the options is to expand or generate $CD4^+CD25^+$ T cells *ex vivo* and give them back to the patient. Several studies have succeeded in expanding or generating functional $CD4^+CD25^+$ T_R cells *in vitro*. Bluestone and co-workers [87] demonstrated recently that an expansion of up to 200-fold was reached in purified murine $CD4^+CD25^+$ T cells using a combination of anti-CD3, anti-CD28 and IL-2. Yamazaki et al. [88] showed that stimulation with antigen-loaded DCs led to an expansion of $CD4^+CD25^+$ T cells, which was partially dependent on IL-2 and required cell contact. Several other studies described a distinctive role for TGF-β in the *ex vivo* generation of $CD4^+CD25^+$ T_R cells. Naïve human $CD4^+$ T cells stimulated with alloantigens in the presence of TGF-β differentiated into $CD4^+CD25^+$ T cells with a phenotype and functional properties similar to naturally occurring $CD4^+CD25^+$ T cells: they expressed CTLA-4 and were able to suppress $CD8^+$ T-cell proliferation. Depletion of $CD4^+CD25^+$ T cells before priming with TGF-β markedly decreased the generation of suppressive activity, suggesting that $CD4^+CD25^+$ T cells generated *ex vivo* were the progeny of naturally occurring $CD4^+CD25^+$ T cells [89]. It was also shown that $CD4^+CD25^+$ T_R cells could be generated from $CD4^+CD25^-$ T cells. Zheng et al. [67] stimulated $CD4^+$ T cells with TGF-β, staphylococcal enterotoxin B (SEB) and irradiated B cells. The $CD4^+CD25^+$ T cells were able to inhibit B-cell IgG production and this was mediated by TGF-β and also cell contact. Furthermore, Foxp3 levels were induced when naïve mouse and human $CD4^+$ T cells were cultured in the presence of TGF-β [66, 90]. These TGF-β-converted $CD4^+CD25^+$ T cells were able to inhibit antigen-specific $CD4^+$ T-cell responses and were shown to prevent antigen-induced allergic pathogenesis of the lung in a murine model for allergic asthma [90].

Besides expanding $CD4^+CD25^+$ T_R cells *in vitro* and giving them back to the patient, another approach is to expand the cells or enhance their function *in vivo*. This has the advantage that no cells from patients have to be isolated and cultured under sterile circumstances and no intravenously administration of $CD4^+CD25^+$ T_R cells is necessary. Lin and Hünig found that $CD4^+CD25^+$ T_R cells can be generated *in vitro* and *in vivo* with a superagonistic CD28-specific mAb, leading to an increase in percentage of $CD4^+CD25^+$ T cells *in vivo* from 5 to 18% [91]. This increase was due to proliferation of the $CD4^+CD25^+$ T_R cells and not to the induction of CD25 on $CD4^+CD25^-$ T cells. These *in vivo*-generated $CD4^+CD25^+$ T_R cells were able to suppress $CD4^+CD25^-$ T cells *in vitro*.

Interestingly a recent study investigated the effects on $CD4^+CD25^{+bright}$ cells upon treatment of RA patients with dnaJP1 [92]. dnaJP1 is a peptide derived from the bacterial heat-shock protein dnaJ, which shares sequence homology with the shared epitope, a five-amino-acid stretch in common among RA-associated HLA alleles. Culturing of cells from patients treated with dnaJp1 showed a significant increase in Foxp3 expression in $CD4^+CD25^{+bright}$ cells upon treatment, suggesting that these cells had gained suppressive function *in vivo*.

Although the development of new protocols to expand $CD4^+CD25^+$ T_R cells *in vivo* or *ex vivo* suggests that treatment with these cells might be started in the near future, several shortcomings of such therapies have to be considered. Most importantly, so far only limited evidence has been provided that $CD4^+CD25^+$ T_R cells can actually control established disease. Using a $CD4^+CD45RB^{high}$ T-cell-transfer model of inflammatory bowel disease, Mottet et al. [55] showed that transfer of $CD4^+CD25^+$ T cells in mice with colitis resulted in $CD4^+CD25^+$ T cell proliferation in the mesenteric lymph nodes and inflamed colon and a reversal of clinical and pathophysiological characteristics of disease. A different study showed that transfer of $CD4^+CD25^+$ T cells into early stage diabetic NOD (non-obese diabetes) mice reversed diabetes. Transfer of $CD4^+CD25^+$ T cells in combination with islet transplantation in advanced diabetic mice, i.e. with complete islet cell destruction, resulted in tolerance to the transplant and to normoglycaemia in these mice [87]. In murine arthritis models, as described previously, although it is possible to prevent disease with $CD4^+CD25^+$ T_R cells, no data are as yet available to show that arthritis can be reversed or cured by improving $CD4^+CD25^+$ T_R function.

The fact that RA is a systemic disease raises another issue. Although RA is mainly located in the joint, all organs can be involved in the disease. This often makes systemic therapy necessary, and local administration of $CD4^+CD25^+$ T_R cells into the joints might therefore not completely control disease. However, local injections with corticosteroids are commonly used in the treatment of persistent monoarthritis and can lead to a sustained reduction of arthritis. Moreover, $CD4^+CD25^+$ T_R cells injected in the joint could possibly expand *in vivo* and migrate to the PB and other organs.

Systemic administration of $CD4^+CD25^+$ T_R cells in the PB leads to different considerations, since the cells will have to survive in the PB and migrate to the site of inflammation. The latter can be achieved by expression of pro-inflammatory chemokines and chemokine receptors, such as CCR5 and CXCR3. Survival *in vivo* could turn out to be a major issue, since $CD4^+CD25^+$ T_R cells are prone to apoptosis and *ex vivo*-generated $CD4^+CD25^+$ T cells have been suggested to have limited regenerative potential *in vivo* [51, 93].

Another possible RA-specific drawback is the unknown source of the disease-causing autoantigen. $CD4^+CD25^+$ T_R cells require contact with an antigen to become activated and subsequently suppressive. At the current stage, it is impossible to make antigen-specific $CD4^+CD25^+$ T_R cells that can specifically halt the response to the antigen that causes the disease. To circumvent this, one could use known joint-specific antigens, such as collagen type II, to activate $CD4^+CD25^+$ T_R cells. Once activated, $CD4^+CD25^+$ T_R cells might halt inflammation by means of bystander immunosuppression. Bystander suppression however could cause another problem in $CD4^+CD25^+$ T_R-mediated immunotherapy, namely the phenomenon of infectious tolerance [94]. With regard to $CD4^+CD25^+$ T_R cells, infectious tolerance describes the finding that $CD4^+CD25^+$ T cells induce long-lasting anergy and production of IL-10 and/or TGF-β in $CD4^+CD25^-$ T cells. This $CD4^+CD25^-$ T-cell population in turn inhibits the proliferation of freshly isolated, naïve $CD4^+$ T cells [85, 86]. This suppression is not cell-contact-dependent and could therefore have a wide range of effects which are hard to predict. This could be beneficial because a small dose of $CD4^+CD25^+$ T_R cells might induce profound immunosuppression. But excessive numbers of regulatory cells could impair host defence to infectious and tumour antigens.

Besides these practical problems in developing a suitable therapy with $CD4^+CD25^+$ T_R cells, the most important question is whether this therapy will be effective in blocking the chronic inflammation. The observation that in the SF an increased population of highly suppressive $CD4^+CD25^+$ T cells is present suggests that the immunoregulatory systems are activated to stop the inflammation, but not successfully. Would further enhancement of these cells be able to halt the suppression or has immunoregulation by $CD4^+CD25^+$ T cells somehow reached its biological limit? The data on the *in vivo* generation of $CD4^+CD25^+$ T_R cells suggests that it is possible to increase the presence and/or function of these cells. Further research will have to determine whether this increase in $CD4^+CD25^+$ T cells leads to a downregulation of the inflammation in RA. The enhancement of one's own immunoregulatory mechanisms might be a significant step in the treatment of RA. Until now, most treatment strategies have focused on inhibiting the pro-inflammatory mechanism, which has not led to a sufficient therapeutic strategy in RA. The combination of inhibiting the pro-inflammatory mechanisms and enhancing the anti-inflammatory mechanisms could lead to a more complete restoration of the balance and stop the ongoing inflammation.

References

1 Appelboom T, Halberg B (2004) Rheumatoid arthritis and other synovial disorders. In: MC Hochberg, AJ Silman, JS Smolen, ME Weinblatt, MH Weisman (eds): *Rheumatology*, 3rd ed., Mosby, London, 753–937

2 Hochberg MC, Spector TD (1990) Epidemiology of rheumatoid arthritis: update. *Epidemiol Rev* 12: 247–252

3 Hochberg MC (1983) Epidemiology of rheumatoid arthritis in developed countries. *J Rheumatol* (Suppl) 10: 7–10

4 Silman AJ, Pearson JE (2002) Epidemiology and genetics of rheumatoid arthritis. *Arthritis Res* 4 (Suppl 3): S265–S272

5 Jarvinen P, Aho K (1994) Twin studies in rheumatic diseases. *Semin Arthritis Rheum* 1: 19–28

6 Durie FH, Fava RA, Foy TM, Aruffo A, Ledbetter JA, Noelle RJ (1993) Prevention of collagen-induced arthritis with an antibody to gp39, the ligand for CD40. *Science* 261: 1328–1330

7 Kim WU, Cho ML, Jung YO, Min SY, Park SW, Min DJ, Yoon JH, Kim HY (2004) Type II collagen autoimmunity in rheumatoid arthritis. *Am J Med Sci* 4: 202–211

8 Cook AD, Rowley MJ, Mackay IR, Gough A, Emery P (1996) Antibodies to type II collagen in early rheumatoid arthritis. Correlation with disease progression. *Arthritis Rheum* 10: 1720–1727

9 van Eden W (1991) Heat-shock proteins as immunogenic bacterial antigens with the potential to induce and regulate autoimmune arthritis. *Immunol Rev* 121: 5–28

10 McDevitt H (2000) A new model for rheumatoid arthritis? *Arthritis Res* 2: 85–89

11 Verheijden GF, Rijnders AW, Bos E, Coenen-de Roo CJ, van Staveren CJ, Miltenburg AM, Meijerink JH, Elewaut D, de Keyser F, Veys E et al (1997) Human cartilage glycoprotein-39 as a candidate autoantigen in rheumatoid arthritis. *Arthritis Rheum* 6: 1115–1125

12 Corrigall VM, Bodman-Smith MD, Fife MS, Canas B, Myers LK, Wooley P, Soh C, Staines NA, Pappin DJ, Berlo SE et al (2001) The human endoplasmic reticulum molecular chaperone BiP is an autoantigen for rheumatoid arthritis and prevents the induction of experimental arthritis. *J Immunol* 3: 1492–1498

13 Myers LK, Rosloniec EF, Cremer MA, Kang AH (1997) Collagen-induced arthritis, an animal model of autoimmunity. *Life Sci* 19: 1861–1878
14 Luross JA, Williams NA (2001) The genetic and immunopathological processes underlying collagen-induced arthritis. *Immunology* 4: 407–416
15 Holoshitz J, Matitiau A, Cohen IR (1985) Role of the thymus in induction and transfer of vaccination against adjuvant arthritis with a T lymphocyte line in rats. *J Clin Invest* 2: 472–477
16 Cush JJ, Lipsky PE (1988) Phenotypic analysis of synovial tissue and peripheral blood lymphocytes isolated from patients with rheumatoid arthritis. *Arthritis Rheum* 10: 1230–1238
17 Kidd P (2003) Th1/Th2 balance: the hypothesis, its limitations, and implications for health and disease. *Altern Med Rev* 3: 223–246
18 van Roon JA, Bijlsma JW, Lafeber FP (2002) Suppression of inflammation and joint destruction in rheumatoid arthritis may require a concerted action of Th2 cytokines. *Curr Opin Invest Drugs* 7: 1011–1016
19 Salmon M, Scheel-Toellner D, Huissoon AP, Pilling D, Shamsadeen N, Hyde H, D'Angeac AD, Bacon PA, Emery P, Akbar AN (1997) Inhibition of T cell apoptosis in the rheumatoid synovium. *J Clin Invest* 3: 439–446
20 Buckley CD, Amft N, Bradfield PF, Pilling D, Ross E, Arenzana-Seisdedos F, Amara A, Curnow SJ, Lord JM, Scheel-Toellner D et al (2000) Persistent induction of the chemokine receptor CXCR4 by TGF-beta 1 on synovial T cells contributes to their accumulation within the rheumatoid synovium. *J Immunol* 6: 3423–3429
21 Iannone F, Corrigal VM, Panayi GS (1996) CD69 on synovial T cells in rheumatoid arthritis correlates with disease activity. *Br J Rheumatol* 35: 397
22 Liu MF, Chao SC, Wang CR, Lei HY (2001) Expression of CD40 and CD40 ligand among cell populations within rheumatoid synovial compartment. *Autoimmunity* 2: 107–113
23 Pacholczyk R, Kraj P, Ignatowicz L (2002) Peptide specificity of thymic selection of CD4+CD25+ T cells. *J Immunol* 2: 613–620
24 MacDonald KP, Nishioka Y, Lipsky PE, Thomas R (1997) Functional CD40 ligand is expressed by T cells in rheumatoid arthritis. *J Clin Invest* 9: 2404–2414
25 Ranges GE, Sriram S, Cooper SM (1985) Prevention of type II collagen-induced arthritis by *in vivo* treatment with anti-L3T4. *J Exp Med* 3: 1105–1110
26 Levitt NG, Fernandez-Madrid F, Wooley PH (1992) Pristane induced arthritis in mice. IV. Immunotherapy with monoclonal antibodies directed against lymphocyte subsets. *J Rheumatol* 9: 1342–1347
27 Van den Broek MF, Van de Langerijt LG, Van Bruggen MC, Billingham ME, Van den Berg WB (1992) Treatment of rats with monoclonal anti-CD4 induces long-term resistance to streptococcal cell wall-induced arthritis. *Eur J Immunol* 1: 57–61
28 Pohlers D, Nissler K, Frey O, Simon J, Petrow PK, Kinne RW, Brauer R (2004) Anti-CD4 monoclonal antibody treatment in acute and early chronic antigen-induced arthritis: influence on T helper cell activation. *Clin Exp Immunol* 3: 409–415

29 Moreland LW, Pratt PW, Mayes MD, Postlethwaite A, Weisman MH, Schnitzer T, Lightfoot R, Calabrese L, Zelinger DJ, Woody JN (1995) Double-blind, placebo-controlled multicenter trial using chimeric monoclonal anti-CD4 antibody, cM-T412, in rheumatoid arthritis patients receiving concomitant methotrexate. *Arthritis Rheum* 11: 1581–1588

30 van der Lubbe PA, Dijkmans BA, Markusse HM, Nassander U, Breedveld FC (1995) A randomized, double-blind, placebo-controlled study of CD4 monoclonal antibody therapy in early rheumatoid arthritis. *Arthritis Rheum* 8: 1097–1106

31 Choy EH, Panayi GS, Emery P, Madden S, Breedveld FC, Kraan MC, Kalden JR, Rascu A, Brown JC, Rapson N, Johnston JM (2002) Repeat-cycle study of high-dose intravenous 4162W94 anti-CD4 humanized monoclonal antibody in rheumatoid arthritis. A randomized placebo-controlled trial. *Rheumatology (Oxford)* 10: 1142–1148

32 Kremer JM, Westhovens R, Leon M, Di Giorgio E, Alten R, Steinfeld S, Russell A, Dougados M, Emery P, Nuamah IF et al (2003) Treatment of rheumatoid arthritis by selective inhibition of T-cell activation with fusion protein CTLA4Ig. *N Engl J Med* 20: 1907–1915

33 Moreland LW, Alten R, Van den Bosch F, Appelboom T, Leon M, Emery P, Cohen S, Luggen M, Shergy W, Nuamah I et al (2002) Costimulatory blockade in patients with rheumatoid arthritis: a pilot, dose-finding, double-blind, placebo-controlled clinical trial evaluating CTLA-4Ig and LEA29Y eighty-five days after the first infusion. *Arthritis Rheum* 6: 1470–1479

34 Kim HJ, Berek C (2000) B cells in rheumatoid arthritis. *Arthritis Res* 2: 126–131

35 Edwards JC, Szczepanski L, Szechinski J, Filipowicz-Sosnowska A, Emery P, Close DR, Stevens RM, Shaw T (2004) Efficacy of B-cell-targeted therapy with rituximab in patients with rheumatoid arthritis. *N Engl J Med* 25: 2572–2581

36 Kinne RW, Brauer R, Stuhlmuller B, Palombo-Kinne E, Burmester GR (2000) Macrophages in rheumatoid arthritis. *Arthritis Res* 3: 189–202

37 Mulherin D, Fitzgerald O, Bresnihan B (1996) Synovial tissue macrophage populations and articular damage in rheumatoid arthritis. *Arthritis Rheum* 1: 115–124

38 van der Heide A, Jacobs JW, Bijlsma JW, Heurkens AH, Booma-Frankfort C, van der Veen MJ, Haanen HC, Hofman DM, Albada-Kuipers GA, ter Borg EJ et al. (1996) The effectiveness of early treatment with "second-line" antirheumatic drugs. A randomized, controlled trial. *Ann Intern Med* 8: 699–707

39 Emery P, Zeidler H, Kvien TK, Guslandi M, Naudin R, Stead H, Verburg KM, Isakson PC, Hubbard RC, Geis GS (1999) Celecoxib *versus* diclofenac in long-term management of rheumatoid arthritis: randomised double-blind comparison. *Lancet* 354: 2106–2111

40 Bijlsma JW, Van Everdingen AA, Huisman M, De Nijs RN, Jacobs JW (2002) Glucocorticoids in rheumatoid arthritis: effects on erosions and bone. *Ann NY Acad Sci* 966: 82–90

41 Lundberg IE, Grundtman C, Larsson E, Klareskog L (2004) Corticosteroids – from an idea to clinical use. *Best Pract Res Clin Rheumatol* 1: 7–19

42 Pincus T, Ferraccioli G, Sokka T, Larsen A, Rau R, Kushner I, Wolfe F (2002) Evidence

from clinical trials and long-term observational studies that disease-modifying antirheumatic drugs slow radiographic progression in rheumatoid arthritis: updating a 1983 review. *Rheumatology* 12: 1346–1356

43 Breedveld FC, Kalden JR (2004) Appropriate and effective management of rheumatoid arthritis. *Ann Rheum Dis* 6: 627–633

44 Chan ES, Cronstein BN (2002) Molecular action of methotrexate in inflammatory diseases. *Arthritis Res* 4: 266–273

45 Cronstein BN (1995) The antirheumatic agents sulphasalazine and methotrexate share an anti-inflammatory mechanism. *Br J Rheumatol* 34 (Suppl 2): 30–32

46 Feldmann M, Maini RN (2001) Anti-TNF alpha therapy of rheumatoid arthritis: what have we learned? *Annu Rev Immunol* 19: 163–196

47 Feldmann M, Brennan FM, Williams RO, Woody JN, Maini RN (2004) The transfer of a laboratory based hypothesis to a clinically useful therapy: the development of anti-TNF therapy of rheumatoid arthritis. *Best Pract Res Clin Rheumatol* 1: 59–80

48 Nishimoto N, Yoshizaki K, Miyasaka N, Yamamoto K, Kawai S, Takeuchi T, Hashimoto J, Azuma J, Kishimoto T (2004) Treatment of rheumatoid arthritis with humanized anti-interleukin-6 receptor antibody: a multicenter, double-blind, placebo-controlled trial. *Arthritis Rheum* 6: 1761–1769

49 Taylor PC (2003) Antibody therapy for rheumatoid arthritis. *Curr Opin Pharmacol* 3: 323–328

50 Sakaguchi S, Sakaguchi N, Asano M, Itoh M, Toda M (1995) Immunologic self-tolerance maintained by activated T cells expressing IL-2 receptor alpha-chains (CD25). Breakdown of a single mechanism of self-tolerance causes various autoimmune diseases. *J Immunol* 3: 1151–1164

51 Taams LS, Smith J, Rustin MH, Salmon M, Poulter LW, Akbar AN (2001) Human anergic/suppressive CD4(+)CD25(+) T cells: a highly differentiated and apoptosis-prone population. *Eur J Immunol* 4: 1122–1131

52 Itoh M, Takahashi T, Sakaguchi N, Kuniyasu Y, Shimizu J, Otsuka F, Sakaguchi S (1999) Thymus and autoimmunity: production of CD25+CD4+ naturally anergic and suppressive T cells as a key function of the thymus in maintaining immunologic self-tolerance. *J Immunol* 9: 5317–5326

53 Stephens LA, Mottet C, Mason D, Powrie F (2001) Human CD4(+)CD25(+) thymocytes and peripheral T cells have immune suppressive activity *in vitro*. *Eur J Immunol* 4: 1247–1254

54 Woo EY, Yeh H, Chu CS, Schlienger K, Carroll RG, Riley JL, Kaiser LR, June CH (2002) Cutting edge: Regulatory T cells from lung cancer patients directly inhibit autologous T cell proliferation. *J Immunol* 9: 4272–4276

55 Mottet C, Uhlig HH, Powrie F (2003) Cutting edge: cure of colitis by CD4(+)CD25(+) regulatory T cells. *J Immunol* 8: 3939–3943

56 Cavani A, Nasorri F, Ottaviani C, Sebastiani S, De Pita O, Girolomoni G (2003) Human CD25+ regulatory T cells maintain immune tolerance to nickel in healthy, nonallergic individuals. *J Immunol* 11: 5760–5768

57 Suri-Payer E, Amar AZ, Thornton AM, Shevach EM (1998) $CD4^+CD25^+$ T cells inhibit both the induction and effector function of autoreactive T cells and represent a unique lineage of immunoregulatory cells. *J Immunol* 3: 1212–1218
58 Shevach EM (2001) Certified professionals: CD4(+)CD25(+) suppressor T cells. *J Exp Med* 11: F41–F46
59 Piccirillo CA, Shevach EM (2001) Cutting edge: control of $CD8^+$ T cell activation by $CD4^+CD25^+$ immunoregulatory cells. *J Immunol* 3: 1137–1140
60 Takahashi T, Kuniyasu Y, Toda M, Sakaguchi N, Itoh M, Iwata M, Shimizu J, Sakaguchi S (1998) Immunologic self-tolerance maintained by $CD25^+CD4^+$ naturally anergic and suppressive T cells: induction of autoimmune disease by breaking their anergic/suppressive state. *Int Immunol* 12: 1969–1980
61 Read S, Malmstrom V, Powrie F (2000) Cytotoxic T lymphocyte-associated antigen 4 plays an essential role in the function of CD25(+)CD4(+) regulatory cells that control intestinal inflammation. *J Exp Med* 2: 295–302
62 Shimizu J, Yamazaki S, Takahashi T, Ishida Y, Sakaguchi S (2002) Stimulation of CD25(+)CD4(+) regulatory T cells through GITR breaks immunological self-tolerance. *Nat Immunol* 2: 135–142
63 McHugh RS, Whitters MJ, Piccirillo CA, Young DA, Shevach EM, Collins M, Byrne MC (2002) CD4(+)CD25(+) immunoregulatory T cells: gene expression analysis reveals a functional role for the glucocorticoid-induced TNF receptor. *Immunity* 2: 311–323
64 Hara M, Kingsley CI, Niimi M, Read S, Turvey SE, Bushell AR, Morris PJ, Powrie F, Wood KJ (2001) IL-10 is required for regulatory T cells to mediate tolerance to alloantigens *in vivo*. *J Immunol* 6: 3789–3796
65 Nakamura K, Kitani A, Fuss I, Pedersen A, Harada N, Nawata H, Strober W (2004) TGF-beta1 plays an important role in the mechanism of $CD4^+CD25^+$ regulatory T cell activity in both humans and mice. *J Immunol* 2: 834–842
66 Chen W, Jin W, Hardegen N, Lei KJ, Li L, Marinos N, McGrady G, Wahl SM (2003) Conversion of peripheral $CD4^+$. *J Exp Med* 12: 1875–1886
67 Zheng SG, Gray JD, Ohtsuka K, Yamagiwa S, Horwitz DA (2002) Generation *ex vivo* of TGF-beta-producing regulatory T cells from $CD4^+CD25$- precursors. *J Immunol* 8: 4183–4189
68 Hori S, Nomura T, Sakaguchi S (2003) Control of regulatory T cell development by the transcription factor Foxp3. *Science* 299: 1057–1061
69 Cederbom L, Hall H, Ivars F (2000) $CD4^+CD25^+$ regulatory T cells down-regulate co-stimulatory molecules on antigen-presenting cells. *Eur J Immunol* 6: 1538–1543
70 Misra N, Bayry J, Lacroix-Desmazes S, Kazatchkine MD, Kaveri SV (2004) Cutting Edge: Human $CD4^+CD25^+$ T cells restrain the maturation and antigen-presenting function of dendritic cells. *J Immunol* 8: 4676–4680
71 Bardos T, Czipri M, Vermes C, Finnegan A, Mikecz K, Zhang J (2003) $CD4^+CD25^+$ immunoregulatory T cells may not be involved in controlling autoimmune arthritis. *Arthritis Res Ther* 2: R106–R113
72 Morgan ME, Sutmuller RP, Witteveen HJ, Van Duivenvoorde LM, Zanelli E, Melief CJ,

Snijders A, Offringa R, De Vries RR, Toes RE (2003) CD25$^+$ cell depletion hastens the onset of severe disease in collagen- induced arthritis. *Arthritis Rheum* 5: 1452–1460

73 Cao D, Malmstrom V, Baecher-Allan C, Hafler D, Klareskog L, Trollmo C (2003) Isolation and functional characterization of regulatory CD25^{bright}CD4$^+$ T cells from the target organ of patients with rheumatoid arthritis. *Eur J Immunol* 1: 215–223

74 van Amelsfort J, Jacobs K, Bijlsma JW, Lafeber FP, Taams L (2004) CD4$^+$CD25$^+$ regulatory T cells in rheumatoid arthritis: differences in presence, phenotype and function between peripheral blood and synovial fluid. *Arthritis Rheum* 50: 2775–2785

75 Baecher-Allan C, Brown JA, Freeman GJ, Hafler DA (2001) CD4$^+$CD25high regulatory cells in human peripheral blood. *J Immunol* 3: 1245–1253

76 de Kleer IM, Wedderburn LR, Taams LS, Patel A, Varsani H, Klein M, de Jager W, Pugayung G, Giannoni F, Rijkers G et al (2004) CD4$^+$CD25(bright) regulatory T cells actively regulate inflammation in the joints of patients with the remitting form of juvenile idiopathic arthritis. *J Immunol* 10: 6435–6443

77 Takahashi T, Tagami T, Yamazaki S, Uede T, Shimizu J, Sakaguchi N, Mak TW, Sakaguchi S (2000) Immunologic self-tolerance maintained by CD25(+)CD4(+) regulatory T cells constitutively expressing cytotoxic T lymphocyte-associated antigen 4. *J Exp Med* 2: 303–310

78 Baecher-Allan C, Viglietta V, Hafler DA (2002) Inhibition of human CD4(+) CD25(+high) regulatory T cell function. *J Immunol* 11: 6210–6217

79 Dieckmann D, Plottner H, Berchtold S, Berger T, Schuler G (2001) Ex vivo isolation and characterization of CD4(+)CD25(+) T cells with regulatory properties from human blood. *J Exp Med* 11: 1303–1310

80 Jonuleit H, Schmitt E, Stassen M, Tuettenberg A, Knop J, Enk AH (2001) Identification and functional characterization of human CD4(+)CD25(+) T cells with regulatory properties isolated from peripheral blood. *J Exp Med* 11: 1285–1294

81 van Roon JA, Glaudemans KA, Bijlsma JW, Lafeber FP (2003) Interleukin 7 stimulates tumour necrosis factor alpha and Th1 cytokine production in joints of patients with rheumatoid arthritis. *Ann Rheum Dis* 2: 113–119

82 Dolhain RJ, Ter Haar NT, De Kuiper R, Nieuwenhuis IG, Zwinderman AH, Breedveld FC, Miltenburg AM (1998) Distribution of T cells and signs of T-cell activation in the rheumatoid joint: implications for semiquantitative comparative histology. *Br J Rheumatol* 3: 324–330

83 Cao D, Vollenhoven RR, Klareskog L, Trollmo C, Malmstrom V (2004) CD25^{bright}CD4$^+$ regulatory T cells are enriched in inflamed joints of patients with chronic rheumatic disease. *Arthritis Res Ther* 4: R335–R346

84 Ehrenstein MR, Evans JG, Singh A, Moore S, Warnes G, Isenberg DA, Mauri C (2004) Compromised function of regulatory T cells in rheumatoid arthritis and reversal by anti-TNF{alpha} therapy. *J Exp Med* 3: 277–285

85 Dieckmann D, Bruett CH, Ploettner H, Lutz MB, Schuler G (2002) Human CD4(+)CD25(+) regulatory, contact-dependent T cells induce interleukin 10-producing, contact-independent type 1-like regulatory T cells [corrected]. *J Exp Med* 2: 247–253

86 Jonuleit H, Schmitt E, Kakirman H, Stassen M, Knop J, Enk AH (2002) Infectious tolerance: human CD25(+) regulatory T cells convey suppressor activity to conventional CD4(+) T helper cells. *J Exp Med* 2: 255–260
87 Tang Q, Henriksen KJ, Bi M, Finger EB, Szot G, Ye J, Masteller EL, McDevitt H, Bonyhadi M, Bluestone JA (2004) *In vitro*-expanded antigen-specific regulatory T cells suppress autoimmune diabetes. *J Exp Med* 11: 1455–1465
88 Yamazaki S, Iyoda T, Tarbell K, Olson K, Velinzon K, Inaba K, Steinman RM (2003) Direct expansion of functional $CD25^+$ $CD4^+$ regulatory T cells by antigen-processing dendritic cells. *J Exp Med* 2: 235–247
89 Yamagiwa S, Gray JD, Hashimoto S, Horwitz DA (2001) A role for TGF-beta in the generation and expansion of $CD4^+CD25^+$ regulatory T cells from human peripheral blood. *J Immunol* 12: 7282–7289
90 Fantini MC, Becker C, Monteleone G, Pallone F, Galle PR, Neurath MF (2004) Cutting edge: TGF-beta induces a regulatory phenotype in $CD4^+$. *J Immunol* 9: 5149–5153
91 Lin CH, Hünig T (2003) Efficient expansion of regulatory T cells *in vitro* and *in vivo* with a CD28 superagonist. *Eur J Immunol* 3: 626–638
92 Prakken BJ, Samodal R, Le TD, Giannoni F, Yung GP, Scavulli J, Amox D, Roord S, de Kleer I, Bonnin D et al. (2004) Epitope-specific immunotherapy induces immune deviation of proinflammatory T cells in rheumatoid arthritis. *Proc Natl Acad Sci USA* 12: 4228–4233
93 Horwitz DA, Zheng SG, Gray JD, Wang JH, Ohtsuka K, Yamagiwa S (2004) Regulatory T cells generated *ex vivo* as an approach for the therapy of autoimmune disease. *Semin Immunol* 2: 135–143
94 Qin S, Cobbold SP, Pope H, Elliott J, Kioussis D, Davies J, Waldmann H (1993) "Infectious" transplantation tolerance. *Science* 259: 974–977

Potential for manipulation of regulatory T cells in treatment or prevention of allergic disease

Douglas S. Robinson and Eleanor M. Ling

Leukocyte Biology Section, Division of Biomedical Sciences and Allergy and Clinical Immunology, National Heart and Lung Institute, Sir Alexander Fleming Building, Imperial College London, Exhibition Road, London SW7 2AZ, UK

The T-cell basis of allergic disease

Allergic disease

Allergic diseases such as rhinitis, asthma and atopic eczema are increasing in prevalence in developed countries [1]. They are among the most common causes of chronic ill health and contribute greatly to national health-care costs [2, 3]. Allergy is often defined by the presence of immunoglobulin E (IgE)-mediated type I hypersensitivity and will be used in this sense in this chapter. The majority of humans mount significant IgE responses only as a defence against parasitic infections. Atopic individuals, however, mount an IgE response to certain ubiquitous antigens (allergens) after inhalation, ingestion or skin contact, and re-exposure can then lead to symptoms [4]. Atopy is common, affecting up to 50% of populations as defined by a positive skin-prick test detecting IgE to common allergens, and has a genetic basis. It is of note that many more people are sensitized to allergens than develop clinical symptoms of allergic disease: the reason for lack of disease in some individuals with IgE to allergens is unknown.

Th2-type T cells and allergic disease

The cytokine secretion pattern of Th1- and Th2-cell subsets defines their differential functions [5–7]. Interferon (IFN)-γ production by Th1 cells activates macrophages for killing of intracellular pathogens, and is associated with delayed-type hypersensitivity reactions. Th1 cells also promote complement fixing and opsonizing antibody production from B cells [8].

Th2 cells produce interleukin (IL)-4, IL-5 and IL-13 and they also express CCR3, CCR4 [5–11] and the orphan receptor T1/ST2 [12]. Th2 cells provide optimal B cell

Regulatory T Cells in Inflammation, edited by Leonie S. Taams, Arne N. Akbar and Marca H.M. Wauben

help and induce immunoglobulin class switching to IgG1, IgA and IgE antibody production [13]. Th2 responses are therefore potentially important in allergen and parasitic responses, through IgE-induced mast cell degranulation and in mucosal defence through induction of IgA secretion. Th2-associated cytokines have multiple effects on cells such as eosinophils, mast cells, B cells, T cells and antigen-presenting cells (APCs).

The high levels of Th2 cytokines that are found at sites of allergic inflammation [14–18] are consistent with a Th2-based aetiology of atopic disease. The cytokine-secretion profile of peripheral blood mononuclear cells (PBMCs) in an *in vitro* allergen-driven cell-culture system also differs between atopic and non-atopic donors. PBMCs from atopic donors produce relatively higher levels of the Th2 cytokines IL-5 and IL-13 compared to the Th1 cytokine IFN-γ than those from non-atopic donors [19, 20]. When PBMCs are stimulated with an antigen that tends to induce a Th1 cytokine profile, such as *Mycobacterium tuberculosis*-derived purified protein derivative (PPD), there is no difference in cytokine production between atopic and non-atopic donors [20]. The reason for the difference in allergen-stimulated T-cell cytokine profiles between atopic and non-atopic donors is unknown. One possibility is that Th2-type responses are actively suppressed in non-atopic individuals. Understanding the mechanism of this suppression may provide a clue to the cause of allergic disease and assist in the development of novel therapeutic strategies.

In mouse models of allergic airway disease, inflammation and airway hyper-reactivity (AHR) is dependent on $CD4^+$ T cells, and more specifically is seen when T-cell-receptor (TCR)-transgenic Th2, but not Th1, cells are adoptively transferred [21–23]. The main role of IL-4 is during the initial priming of Th2 effector cells [24–26], whereas IL-5 and IL-13 are more directly involved in characteristic allergic pathology such as eosinophilia and mucus hypersecretion [27–30].

The cellular basis of atopic allergic reactions

When an atopic individual is exposed to an allergen to which they have previously been sensitized and to which they have specific IgE, they will immediately mount an immune reaction which, depending on the site of exposure, will result in either a cutaneous wheal-and-flare reaction, sneezing, runny nose or wheezing. This is known as the early-phase reaction, and may be followed some 6–9 h later by an oedematous, red swelling of the skin, sustained blockage of the nose or further airway obstruction in the lung, which is dependent on the concentration of allergen encountered and is known as the late-phase reaction [4].

Molecules such as histamine, tryptase, leukotrienes, prostaglandins and platelet-activating factor are released by mast cells and basophils when an allergen interacts with membrane-bound IgE, causing immediate hypersensitivity. This is the basis of acute allergic reactions such as anaphylaxis, rhinoconjunctovitis and urticaria. The

IgE is bound to mast cells and basophils by the α chain of the high-affinity IgE receptor (FcεR1-α) [4].

Late-phase reactions can be provoked by the activation of mast cells or T cells. In the cutaneous late-phase reaction, eosinophils and neutrophils accumulate, followed by an infiltration of $CD4^+$ T cells and basophils [31]. APCs, such as dendritic cells (DCs) and macrophages, are critical in both initiating and controlling allergic inflammation. They present antigen to $CD4^+$ Th2 cells in an MHC II-restricted fashion. When alveolar macrophages from atopic asthmatic subjects present allergen to $CD4^+$ T cells they stimulate secretion of Th2-type cytokines, whereas alveolar macrophages from non-atopic subjects do not [32]. IL-4, IL-5, IL-9 and IL-13 are the classic Th2-type cytokines produced by allergen-stimulated cells from atopic subjects [4].

Infiltration of eosinophils into the bronchial wall and lumen is a prominent pathological feature of asthma. Skin diseases such as atopic dermatitis, chronic urticaria and episodic angiodema are also associated with eosinophilia. IL-5 causes release of both mature and immature eosinophils from the bone marrow [33]. The eosinophil, which is a tissue inflammatory cell that matures in the bone marrow, is transported in the bloodstream to sites of inflammation. There it releases mediators which may play an important role in the later inflammatory stages of asthma, rhinitis, conjuctovitis and cutaneous allergies [4].

Regulatory T cells and allergic disease

The $CD4^+CD25^+$ regulatory population isolated from human peripheral blood and thymus has now been extensively characterized *in vitro* [34–40]. It is anergic to polyclonal stimulation, for example by cross-linking CD3 and CD28, and it can suppress T-cell responses to polyclonal stimulation and to antigens such as cows' milk antigen (CMA), tetanus toxoid (TT) and human heat-shock protein 60 (hHSP60) [41]. The $CD4^+CD25^+$ T-cell population also suppresses $CD8^+$ T-cell responses [42–44] and mixed lymphocyte reactions (MLRs) [45]. $CD4^+CD25^+$ T cells constitutively express CTLA-4 [46, 47], they also express high levels of the glucocorticoid-induced tumour necrosis factor (TNF) receptor family-related gene (GITR) [48–50] and the transcription factor Foxp3 [51–53].

As well as naturally occurring $CD4^+CD25^+$ T cells, a spectrum of other types of regulatory T cells have been induced *in vivo* and *in vitro*. Many are induced in response to antigen presented in a "tolerogenic form", for example as excess soluble antigen or as an altered peptide ligand with concomitant co-stimulatory blockade or with IL-4 and/or IL-10, or by a "tolerogenic route", for example mucosal administration [54, 55]. At least some subsets of $CD4^+CD25^+$ regulatory T cells can be induced in the periphery [55, 56], so the exact distinction between the "naturally occurring" and "experimentally induced" subsets is at present unclear and some overlap between groups is likely.

At present, regulatory T cells are defined primarily by their function, as no definitive marker has been identified. Regulatory T cells are therefore defined operationally as T cells that suppress active T-cell responses by naïve or effector memory T cells. Two properties that are shared by many known regulatory-T-cell populations are an impaired capacity to respond to proliferative signals *in vitro* and the ability to inhibit immune cell functions either directly through cell contact or through secretion of anti-inflammatory mediators.

In vitro, the hyporesponsive and suppressive qualities of regulatory T cells are shared by many anergic T cells [57, 58], and the relationship between anergic and regulatory-T-cell populations is also as yet unknown. *In vivo*, the suppressive nature of regulatory T cells is demonstrated most convincingly by their ability to suppress autoimmune disease and transplant rejection in animal models. An important issue to consider with these experiments, however, is that in most transfer experiments the hosts are lymphopenic and therefore regulatory functions cannot easily be distinguished from homeostatic competition for proliferation and clonal expansion of the transferred cells [59, 60].

In the context of allergic disease regulatory T cells could be considered as T cells that suppress Th2 responses to allergen. This will encompass "innate" regulatory cells such as $CD4^+CD25^+$ T cells, "adaptive" or induced regulatory cells such as IL-10-producing T cells, but also possibly other T-cell subsets such as Th1 cells that are functionally suppressive in this context [61]. We will discuss these different regulatory T cells in the context of allergic sensitization and disease and their potential manipulation for prevention and treatment of allergic diseases in the context of current drug treatments, allergen injection immunotherapy and possible future therapeutic approaches.

$CD4^+CD25^+$ T cells may prevent allergic sensitization in non-atopic donors

The fact that PBMCs from non-atopic individuals proliferate on *in vitro* stimulation with an allergen to which they are regularly exposed is not unexpected. The differences seen between non-atopic and atopic donors are usually in the pattern of cytokines produced. We hypothesized that one factor preventing Th2 responses to allergens in non-atopic individuals was suppression of such responses by $CD4^+CD25^+$ regulatory T cells [62]. When we isolated $CD4^+CD25^-$ and $CD4^+CD25^+$ T cells from peripheral blood of non-atopic donors several observations suggested that there is normally a state of active suppression of Th2 responses to allergen. Firstly, the depletion, by immunomagnetic beads, of $CD4^+CD25^+$ T cells from the $CD4^+$ T-cell population resulted in significantly increased proliferation and IL-5 production to *in vitro* allergen stimulation (Fig. 1). The IL-5/IFN-γ ratio in such cultures of $CD4^+CD25^-$ T cells from non-atopics was similar to that seen with cells from atopic subjects, suggesting that non-atopics do have T cells

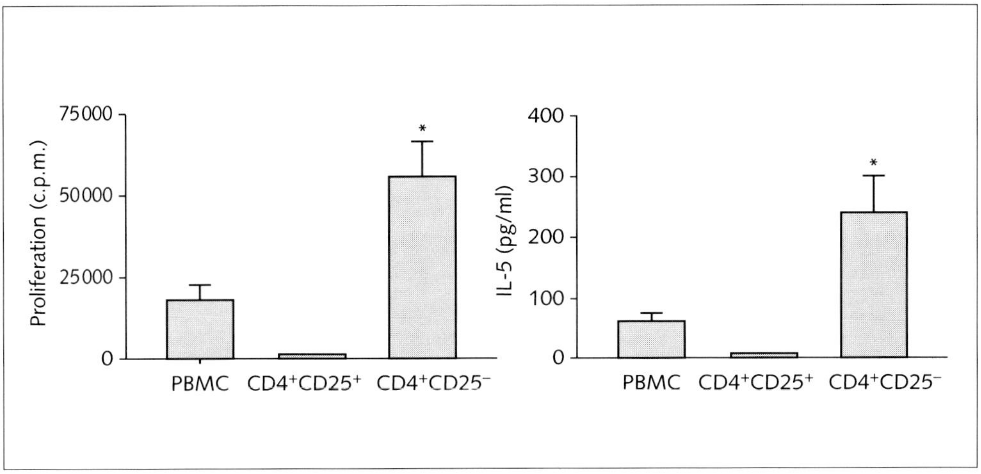

Figure 1
Human peripheral blood $CD4^+CD25^+$ T cells show reduced proliferation and cytokine production to allergen stimulation compared to PBMCs, whereas responses are increased in $CD4^+CD25^-$ T cells
*$CD4^+CD25^+$ T cells did not proliferate or produce IL-5 after 6 days of culture with cat or grass-pollen extracts. In contrast $CD4^+CD25^-$ T cells proliferated significantly more and made more IL-5 upon stimulation with allergen than did unseparated PBMC. Data are means S.E. from nine separate non-allergic donors. *$P<0.001$ compared to PBMC [62].*
Reprinted with permission from Elsevier [62].

capable of making a Th2 response. Secondly, $CD4^+CD25^+$ T cells did not themselves proliferate or produce cytokines upon allergen stimulation, but were able to suppress responses of $CD4^+CD25^-$ T cells in co-cultures. Suppression was nearly 100% at a 1:2 ($CD25^+/CD25^-$) ratio (Fig. 2). These findings are in accord with other reports for human $CD4^+CD25^+$ T cells from non-atopic donors and suppression of allergen responses [63, 64]. In addition, $CD4^+CD25^+$ T cells had also been implicated in preventing contact dermatitis (an allergic Th1 type of delayed-type hypersensitivity) in elegant studies from Cavani et al. [65]: these authors showed infiltration of skin-patch tests to nickel by $CD25^+$ T cells in non-allergic subjects.

Taken together, these data imply that naturally occurring $CD4^+CD25^+$ T cells in non-atopic individuals suppress Th2 responses to allergen, therefore affording protection from atopic allergic reactions following allergen exposure. This raises a number of important questions about allergy: firstly, is there a deficiency of such regulatory responses in atopic allergic individuals, and secondly how do these protective regulatory T cells develop, and can this process be manipulated for therapy?

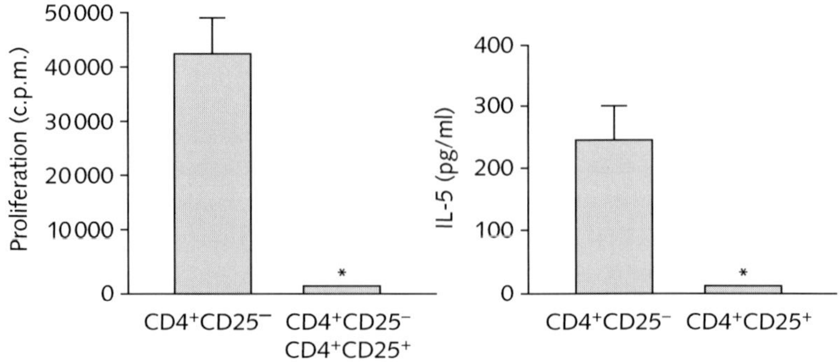

Figure 2
Human peripheral blood CD4⁺CD25⁺ T cells suppress allergen-induced proliferation and cytokine secretion by CD4⁺CD25⁻ T cells.
Addition of CD4⁺CD25⁺ to CD4⁺CD25⁻ cells in a ratio of 1: 2 resulted in suppression of proliferation and IL-5 production upon allergen stimulation, compared to addition of CD25⁻ T cells ($P < 0.0001$ for both proliferation and IL-5 production). Bars are means ± S.E. from nine non-allergic donors.*

Deficiency in suppression of Th2 responses to allergen by $CD4^+CD25^+$ T cells from atopic donors

We next went on to address the question of whether Th2 responses and IgE sensitization in atopic subjects results from a failure of $CD4^+CD25^+$ T-cell suppression. $CD4^+CD25^+$ regulatory T cells from atopic donors were less effective at suppressing *in vitro* allergen-stimulated T-cell proliferation and cytokine production than those from non-atopic individuals [62]. Results from this study also indicated that after natural exposure to allergen, $CD4^+CD25^+$ T cells isolated from donors with hay fever were less efficient at suppressing both the allergen-specific $CD4^+CD25^-$ T-cell proliferation and IL-5 production than those from atopic donors who were not exposed to allergen (Fig. 3).

We have subsequently confirmed these differences in $CD4^+CD25^+$ T-cell suppression of allergen-stimulated T cells between atopic and non-atopic in a separate study [66]. However, using a different *in vitro* protocol, Bellinghausen et al. [63] suggested that most atopic individuals had $CD4^+CD25^+$ T cells that were as suppressive to allergen-stimulated T-cell responses as those from non-atopic donors, although it is of note that a subset of patients had high levels of IL-4 and IL-10 production from $CD4^+CD25^+$ T cells and these did not suppress allergen-stimulated cultures. These studies raise two intriguing possibilities. The Bellinghausen study

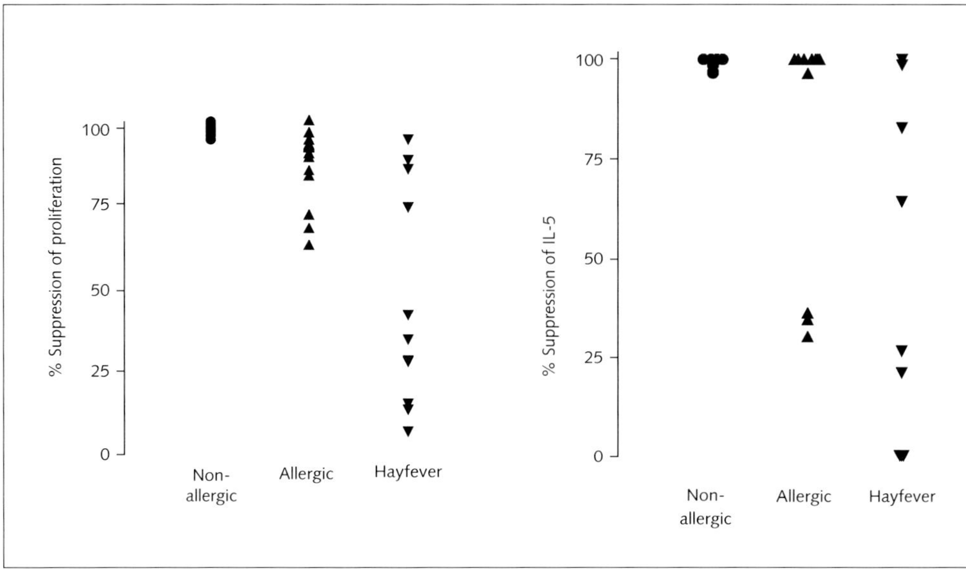

Figure 3
Suppression of allergen-induced T-cell proliferation and IL-5 production by $CD4^+CD25^+$ T cells is deficient in allergic donors
*Percentage suppression by $CD4^+CD25^+$ cells is shown for proliferation and IL-5 production from non-allergic (•), allergic (▲) and hay fever patients during the pollen season (▼) (reduction by addition of $CD4^+CD25^+$ to $CD4^+CD25^-$ as a percentage of $CD4^+CD25^-$ alone). *$P < 0.001$ for both non-allergic* versus *allergic donors and for hay fever patients* versus *non-allergic donors for suppression of proliferation; **$P < 0.01$ for hay fever patients* versus *non-allergic donors for suppression of IL-5 production ($n = 9$ for non-allergic volunteers, 12 for allergic volunteers and 11 for hay fever patients during the season).*
Reprinted with permission from Elsevier [62].

used *in vitro*-derived DCs as APCs: it is therefore possible that differences seen between atopic and non-atopic groups in our system reflected differential *in vivo* activation of APCs (used straight from the patient in our study), which could overcome regulation [62, 63]. The effect of seasonal exposure might thus be explained by effects of *in vivo* allergen exposure on APC function, for example by ligation of Toll-like receptors and IL-6 production [67]. We did not see any difference between atopic and non-atopic donors in suppression of anti-CD3 stimulated cultures by $CD4^+CD25^+$ T cells [62], a finding which might also suggest a role for the APCs in the differences observed for allergen. Another observation in both studies is an apparently wide spread of suppressive activity between patients without obvious clinical correlates. In our studies the degree of suppression of allergen-stimulated

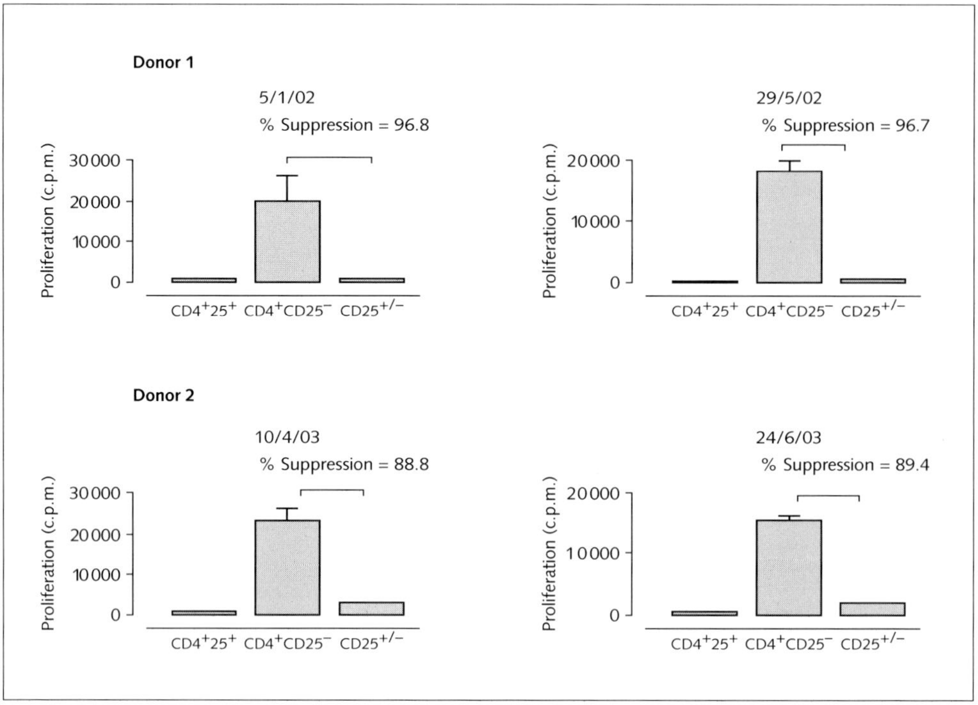

Figure 4
Reproducibility of the assay. CD4⁺CD25⁺ T cells at 2 x 10^6 cells/ml or CD4⁺CD25⁻ T cells at 3 x 10^6 cells/ml were stimulated with grass-pollen allergen extract at 100 μg/ml or media alone. CD4⁺CD25⁺ regulatory T cells were added to CD4⁺CD25⁻ T cells at a ratio of 1: 2 (marked CD25⁺/⁻). All wells contained irradiated autologous PBMCs at 2 × 10^6 c/ml. Allergen-specific proliferation of cells was measured by tritiated thymidine incorporation at day 7 and in the two representative donors shown the experiment was repeated at a later date. Mean values + S.E. from triplicate cultures are shown. Donor 1 was non-atopic and donor 2 was atopic.

cultures by $CD4^+CD25^+$ T cells from a single individual is remarkably stable over time (except for seasonal effects; Fig. 4). This raises a possible genetic influence on suppression and we suggest that the relative importance of failure of $CD4^+CD25^+$ T-cell suppression in development of allergic sensitization may differ between different atopic individuals. These questions will require further studies of the mechanism of suppression.

Without a specific marker for regulatory cells, the possibility that there were more effector cells in the $CD4^+CD25^+$ T-cell population isolated from symptomatic atopic donors cannot be completely ruled out. Similarly, there was no evidence of activated $CD25^-$ blood T cells in atopic individuals, since these cells did not express CD69 and their IL-5 production to allergen stimulation *in vitro* was not different to those from non-atopic donors.

Out-of-season suppression of allergen-stimulated $CD4^+CD25^-$ T cells from hay fever patients by $CD4^+CD25^+$ T cells was similar to that seen in the asymptomatic atopic donors, but still significantly less than that seen for non-atopic donors. This may result from a defect in allergen-specific $CD4^+CD25^+$ T cells in atopic donors or from expansion of allergen-reactive $CD25^-$ T cells, making these relatively resistant to suppression. $CD4^+CD25^+$ T cells are suggested to arise in the thymus [68], but may also develop in the periphery [69]. Whether $CD4^+CD25^+$ T cells that regulate allergic responses are selected by allergen peptides during the neonatal period, represent cross-reactive T cells directed against self-antigens or arise in the periphery requires investigation if these processes are to be manipulated for therapy.

$CD4^+CD25^+$ T cells in regulation of animal models of allergic sensitization

Animal models of allergic-airway sensitization have been useful in defining potential immunological mechanisms in asthma and allergic disease [70]. However, data on $CD4^+CD25^+$ T-cell regulation of mouse Th2 airway inflammation and airway hyper-responsiveness is sparse. When $CD4^+CD25^+$ T cells were depleted in one model airway inflammation actually decreased [71]. This may be because CD25 is also a marker of recently activated T cells or memory effector cells, so both regulators and effectors had been removed. However, co-transfer of *ex vivo*-expanded $CD4^+CD25^+$ ovalbumin-specific T cells together with Th2 cells had no effect on subsequent inhaled OVA challenge in another mouse model [72], and OVA-specific airway $CD4^+CD25^+$ T cells in another complex double transgenic model reduced airway inflammation but not airway hyper-responsiveness in response to inhaled challenge [73].

Mechanism of regulation of allergen-stimulated T cells by human $CD4^+CD25^+$ T cells

The mechanism of suppression by $CD4^+CD25^+$ T cells remains unknown. These cells suppress IL-2 transcription by non-$CD25^+$ T cells. Most *in vitro* studies of murine and human $CD4^+CD25^+$ T cells show that suppression can still occur in the absence of signals *via* IL-10, transforming growth factor (TGF)-β, or CTLA-4, and this was also the case with blocking antibody experiments in our system [34–40, 74]. We also confirmed that human $CD4^+CD25^+$ T cells preferentially express mRNA for the transcription factor Foxp3 when compared to $CD4^+CD25^-$ T cells. In a limited number of experiments suppression of allergen-stimulated $CD4^+CD25^-$

T cells by $CD4^+CD25^+$ T cells was still seen when these were separated across a transwell system (Fig. 5). This finding is at variance with reports from other systems [34–40, 74] and requires further investigation, although it could reflect the 6-day culture system, as induction of regulatory T cells from $CD25^-$ T cells has been described in co-culture systems, and these induced regulatory cells suppressed by soluble factors including IL-10 [75].

Non-$CD25^+$ regulatory T cells in allergic disease

Although rodent systems established an important role for $CD4^+CD25^+$ T cells in immune tolerance, it is of note that suppression by $CD4^+CD25^-$ T cells is also observed in a number of systems. For example, Lafaille and colleagues [76] demonstrated that the development of experimental autoimmune encephalomyelitis (EAE) in myelin basic protein (MBP)-specific TCR transgenic mice can be effectively controlled by $CD25^-$ T cells. Furthermore, $CD25^-$ cells have also been shown to be protective in both transgenic and non-transgenic models of diabetes [77, 78], and both $CD25^+$ and $CD25^-$ populations were able to confer protection from inflammatory bowel disease in a lymphopenic setting [79]. This suggests a further population of naturally occurring $CD25^-$ T cells. Are such cells active in prevention of allergic sensitization, and how do they relate to both $CD4^+CD25^+$ T cells and induced (or adaptive) regulatory T cells such as the IL-10 regulatory T cells induced by peptide treatment *in vivo* [80], dexamethasone, vitamin D_3 [81] or cytokines *in vitro* [82], or Th3 cells induced by oral tolerance [83]?

Human IL-10-producing regulatory T cells and suppression of allergic responses

IL-10 is a cytokine produced by both Th1 and Th2 cells, and other cell types, with immunosuppressive activity on T-cell responses through its ability to reduce APC expression of MHC class II and co-stimulatory molecules [84]. Neutralization of IL-10 or TGF-β uncovered proliferative and cytokine responses to the house-dust-mite allergen, Der p1, and the major birch-pollen allergen, Bet v1 [85]. These data suggested that such responses are normally suppressed but did not identify the cellular source of these cytokines. More recently the same investigators went on to isolate T cells making IL-10 upon short-term (12-h) stimulation with allergen *in vitro*, using an immunomagnetic cell-capture system [86]. They demonstrated an increased frequency of IL-10-producing T cells to allergen stimulation in non-atopic individuals (with more IL-4-producing cells in atopics). Furthermore, the IL-10-producing T cells were suppressive in co-cultures with IL-4-producing T cells and this suppression was specific for the allergen used in their isolation. Interestingly these cells were $CD4^+CD25^+$,

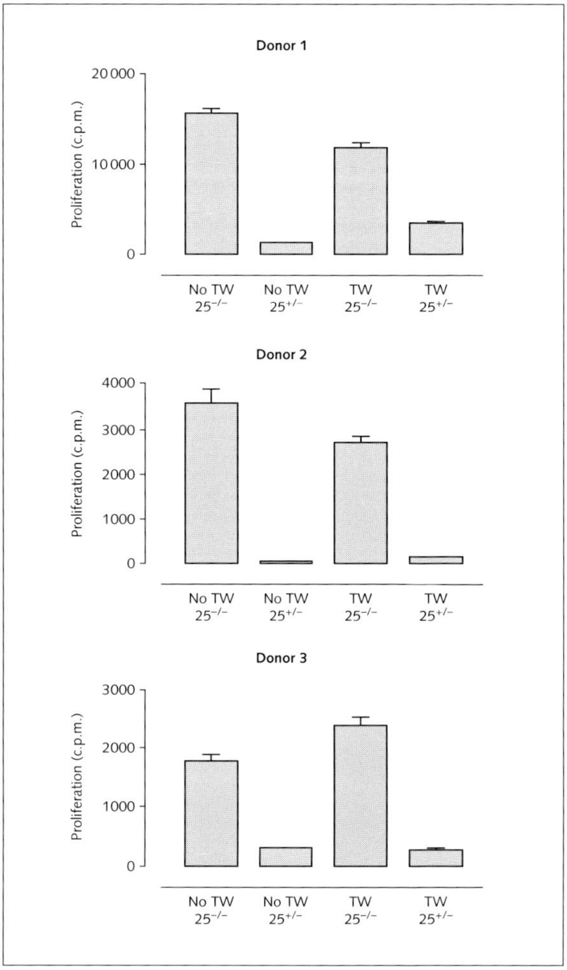

Figure 5

The effect of separating $CD4^+CD25^+$ regulatory T cells from $CD4^+CD25^-$ responder T cells using a transwell system, on the ability of the $CD4^+CD25^+$ subset to suppress allergen-driven T-cell responses

$CD4^+CD25^-$ T cells at 3×10^6 cells/ml were stimulated with grass-pollen allergen extract at 100 μg/ml or media alone (marked No TW $25^{-/-}$). $CD4^+CD25^+$ regulatory T cells were added to $CD4^+CD25^-$ T cells at a ratio of 1: 2 (marked No TW $25^{+/-}$). In some cultures, a transwell insert separated responder $CD4^+CD25^-$ cells from regulatory $CD4^+CD25^+$ cells (marked TW $25^{+/-}$) or, in the control, from added $CD4^+CD25^-$ cells (marked TW $25^{-/-}$). All wells contained irradiated autologous PBMCs at 2×10^6 c/ml. Allergen-specific proliferation of cells from three representative non-atopic donors was measured by tritiated thymidine incorporation at day 7. Mean values + S.E. from triplicate cultures are shown.

although whether expression of $CD25^+$ was present before culture or resulted from *in vitro* stimulation in unclear. These data suggest the presence of suppressive IL-10-producing T cells which contribute to prevention of atopic sensitization in non-atopics. We have recently isolated IL-10-producing T cells from the CD4$^+$CD25- fraction of peripheral blood from normal individuals [87]. Such T cells could be isolated in the absence of *in vitro* stimulation and were suppressive in co-cultures with allergen-stimulated $CD4^+CD25^-IL\text{-}10^-$ T cells. These data raise the possibility of suppression of Th2 responses in non-topic individuals by at least two populations of "naturally occurring" T cells: $CD4^+CD25^+$ T cells and $CD4^+CD25^-$ $IL\text{-}10^+$ T cells (see Fig. 6). As with $CD4^+CD25^+$ T cells we found diminished suppressive activity in allergen-stimulated cultures for $CD4^+CD25^-IL\text{-}10^+$ T cells isolated from atopic patients. These findings suggest that atopic sensitization and/or allergic disease may result when two immunological "brakes" fail: whether this differs between those with IgE sensitization but no disease and those with active disease will require further study. The derivation of both $CD4^+CD25^+$ T cells and $CD4^+CD25^-IL\text{-}10^+$ T cells *in vivo* in humans remains uncertain: they may arise from thymic selection, but may also differentiate *in vivo* in the face of natural allergen exposure. Understanding the mechanisms for derivation and persistence of suppressive T cells will be important in designing strategies for enhancing their activity in atopic subjects.

Induced regulatory T cells: potential for control of allergic disease

If a failure of regulation contributes to development of allergic sensitization, can regulatory T cells be induced in attempt to control disease?

Induced regulatory T cells

Numerous mouse and human model systems have been developed to induce T cells with *in vitro* or *in vivo* suppressive potential. Human $CD4^+CD25^+$ T cells were reported to induce a regulatory phenotype in $CD4^+CD25^-$ T cells upon co-culture, such cells became $CD25^+$, $Foxp3^+$ and suppressed through IL-10 and TGF-β [75, 88]. Development of $CD4^+CD25^+$ T cells from $CD4^+CD25^-$ T cells is described upon stimulation with TGF-β, inducing up-regulation of Foxp3 [89]. Whether this occurs in *in vitro* cultures with allergen in human cells is uncertain, but it is of note that a recent study reported development of $Foxp3^+$ cells that suppressed through surface TGF-β after repeated low-dose allergen challenge in mice [90].

In addition to $CD4^+CD25^+$ T cells, many other regulatory subtypes are described. These include T_R1 cells initially described for mouse and human as clones derived in the presence of IL-10, but also expanded in the presence of IFN-α and IL-10 [82, 91]. Mouse T_R1 clones could suppress colitis and prevented IgE sensitiza-

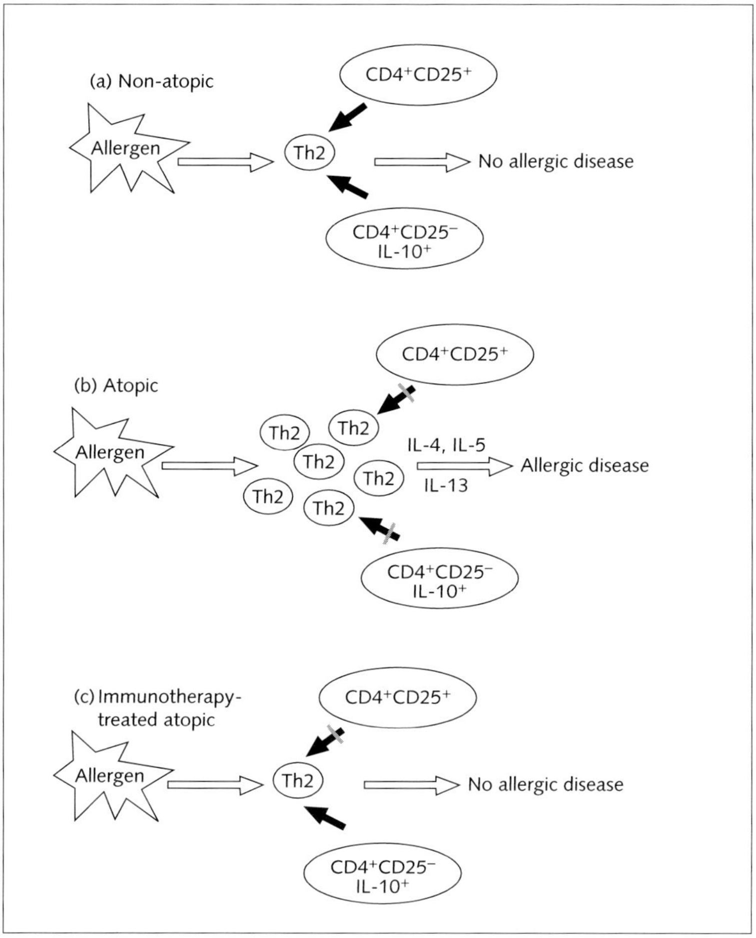

Figure 6
Schematic illustrating potential mechanisms of the suppression of Th2-cell responses to allergen in (a) a non-atopic donor, (b) an atopic donor and (c) an allergen-immunotherapy-treated atopic donor.

tion in mouse models [82, 92]. These T_R1 cells produce IL-10 but also IFN-γ and IL-5. IL-10-producing regulatory T cells were also produced from mouse and human naïve $CD4^+$ T cells by *in vitro* stimulation in the presence of the immunosuppressive drugs dexamethasone and vitamin D_3, which act to prevent expression of Th2 and Th1 cytokines respectively [81].

Using a murine model of asthma, it has been shown that, on exposure to respiratory allergen, pulmonary DCs present in the bronchial lymph nodes can induce regulatory T cells which block the development of airway hyper-responsiveness [93]. These cells developed through an intermediate stage that they share with developing Th2 cells, during which they produce IL-4 and IL-10. It is therefore possible that under normal circumstances exposure to respiratory allergen results in the development of regulatory T cells and subsequent tolerance, whereas Th2 cells may develop in atopic individuals as an aberration of regulatory-T-cell development [94]. *In vivo* administration of *Mycobacterium vaccae* prior to antigenic sensitization *via* inhaled allergen challenge also reduces the cellular inflammation and airway hyper-responsiveness after further allergen challenge, an effect attributed to induction of an IL-10-producing regulatory-T-cell population [95].

The role of regulatory T cells in current therapies for the treatment of allergic disease

Many of the current treatments for allergic disease function by suppressing immune responses in a general manner. Several recent studies suggest that some of these treatments, including corticosteroids and specific allergen immunotherapy, may work by inducing regulatory T cells or by enhancing their suppressive activity.

Corticosteroids

To suppress allergic inflammation, topical glucocorticoids are efficacious and the recommended therapy in current guidelines for asthma, rhninitis and allergic skin disease [96–98]. These drugs act by complexing to intracytoplasmic receptors, and directly increase or repress gene transcription, or interact with transcription factors, inhibiting T-cell activation and production of Th2 cytokines *in vitro* [99], and have been shown to reduce expression of Th2 cytokines in the airway mucosa of asthmatics *in vivo* [100, 101]. Glucocorticoids increase IL-10-production both *in vivo* and *in vitro* by a process that may involve regulatory T cells, as IL-10-producing regulatory T cells can be induced by a combination of corticosteroids and vitamin D_3 [81]. Furthermore, there is a defect in corticosteroid-induced IL-10 production by T cells from corticosteroid resistant asthmatic patients [102], suggesting that regulatory T cells in these individuals may not be functional. We recently showed that pretreatment of $CD4^+CD25^+$ regulatory T cells with the corticosteroid, fluticasone propionate, increased the ability of $CD4^+CD25^+$ regulatory T cells from atopic donors to suppress allergen-driven T-cell responses *in vitro* [66]. Interestingly this increase in suppressive activity was associated with increased IL-10 production and the increase could be blocked by antibody to IL-10 (though this did not prevent "baseline" suppression). Thus corticosteroids either induced additional (and distinct) suppressive activity in existing regulatory T cells by enhancing IL-10 production, or induced separate IL-10-producing T cells *in vitro* amongst the $CD25^+$ population isolated from peripheral blood (which in humans is a mixture of regulatory, effector and memory T cells). Whether corticosteroids (and other immunosuppressive agents) can enhance suppression *in vivo*, or could be co-adminstered with allergen to induce a regulatory-T-cell population, remain to be determined.

Allergen immunotherapy treatment

Specific allergen immunotherapy involves administering increasing concentrations of allergen extracts subcutaneously over a long period of time. This is an extremely effective treatment for patients suffering from seasonal allergic rhinitis and to a lesser extent those with perennial rhinitis. The effects are long lasting, especially if the treatment is continued for several years [103]. Allergen immunotherapy is limited by the risk of anaphylaxis, particularly during the "up-dosing" phase of treatment [4], and in the UK such treatment is not recommended for asthma, although it is widely used elsewhere.

The mechanism by which allergen-specific immunotherapy exerts its beneficial effects is extremely complex. The influence of immunotherapy on T cells is thought to be central to its inhibition of the late-phase response. Successful allergen immunotherapy treatment is associated with a decreased *ex vivo* allergen-specific T-cell proliferative response [104–106] accompanied by decreased production of the Th2 cytokines IL-4, IL-5 and IL-13 relative to the production of IFN-γ and increased concentration of IL-10 [107, 108].

Allergen-induced recruitment of inflammatory cells such as mast cells, basophils and eosinophils to sites of allergic disease is decreased after treatment [109–111]. This is thought to be due in part to the reduced levels of the inflammatory cytokines IL-4, IL-5 and IL-13, together with increased levels of both allergen-specific and total IgG4 [104]. IgG4 competes with IgE for allergen, reducing IgE-receptor cross-linking and subsequent release of inflammatory mediators from mast cells and basophils [104].

$CD4^+CD25^+$ T cells with regulatory-like features can be generated by oral or intranasal administration of whole protein or peptides in mice [56, 80] and several groups have suggested that regulatory T cells are involved in the mechanism through which specific allergen immunotherapy protects from allergic disease [85, 112]. Increased numbers of $CD4^+CD25^+$ T cells were present in cultures of PBMCs stimulated with the major grass-pollen allergen *Phleum pratense V*, from donors who had received allergen immunotherapy treatment than from untreated control subjects, and a higher proportion of these $CD4^+CD25^+$ cells from immunotherapy-treated donors were IL-10^+ by intracellular cytokine staining [112]. From this study, however, it was not clear whether the cells expressing the CD25 marker were truly regulatory T cells, or T cells activated during the course of the *in vitro* allergen-driven culture. A further study investigated the effect of allergen immunotherapy treatment on the ability of regulatory T cells to suppress allergen-specific T-cell responses [85]. The results showed that after immunotherapy treatment for allergy to house dust mite, $CD4^+CD25^+$ T cells were more efficient at suppressing the proliferation of T cells stimulated with this allergen *in vitro*. This increase in suppression was dependent on IL-10 and TGF-β.

We recently examined the suppression of allergen-stimulated $CD4^+CD25^-$ T-cell proliferation and cytokine production by $CD4^+CD25^+$ and $CD4^+CD25^-IL\text{-}10^+$ T cells from patients treated with grass pollen immunotherapy compared with patients with hay fever who had not been treated with immunotherapy [87]. Allergen-immunotherapy-treated patients showed increased suppression by $CD4^+CD25^-IL\text{-}10^+$ T cells when compared to an untreated control group, whereas there was no difference between the two for suppression by $CD4^+CD25^+$ T cells. It is possible that allergen immunotherapy may induce IL-10-producing T cells in both the $CD25^+$ and $CD25^-$ T-cell compartment, although our findings suggest that the IL-10-producing regulatory T cells induced by immunotherapy are distinct from $CD4^+CD25^+$ T cells.

Several approaches have been developed to avoid the risk of IgE cross-linking and thus anaphylaxis, which limits conventional allergen immunotherapy. One such approach is to administer peptides derived from allergen which preserve T-cell epitopes but do not cross-link IgE [113]. Such treatment using peptides derived from the major cat allergen, Fel d1, was shown to reduce skin early- and late-phase response to allergen challenge and to reduce airway hyperresponsiveness (AHR) [114]. In addition, peptide treatment resulted in reduced PBMC proliferative and cytokine responses to *in vitro* culture with cat allergen, and increase IL-10 production [114, 115]. We recently showed that these changes in the $CD4^+$ T-cell response to allergen after cat peptide treatment were not associated with any changes in suppressive activity of $CD4^+CD25^+$ T cells in cat-allergen-stimulated cultures [116]. However, $CD4^+$ T cells isolated after peptide tratment could suppress the response of $CD4^+$ T cells separated before therapy from the same patients (A. Verhoef et al., personal communication). Taken together these data also support a role for non-$CD25^+$ IL-10-producing regulatory T cells in allergen peptide immunotherapy.

CpG oligonucleotides act through toll-like receptor (TLR)9 to induce IFN-α production from plasmacytoid DCs and have been shown to inhibit Th2 responses *in vitro* and *in vivo* in mouse models. This approach has been refined by coupling the oligonucleoide to allergen molecules, thus masking IgE epitopes and ensuring co-delivery of CpG and allergen to APCs [117]. Whether this approach acts through induction of a regulatory-T-cell response in addition to promoting a Th1 response is unclear. It is of note that increased IFN-γ production is reported after allergen immunotherapy and corticosteroid therapy *in vivo* [100, 104]. Such a switch to a Th1 response (or immune deviation) could also reduce IgE switching and inhibit Th2 responses to allergen, and Th1 cells may act as "regulatory" T cells in the setting of allergic inflammation. It is of note however that transfer of Th1 cells in mouse models of Th2-induced inflammation had mixed effects: in most reports this did not prevent inflammation or IgE sensitization [118, 119].

Potential for induction or manipulation of regulatory cells for treatment of allergic disease

It may be that regulatory T cells induced by allergen immunotherapy treatment are similar to IL-10-producing regulatory-T cells induced in animal models after nasal or oral delivery of protein or peptide [80, 120]. If so, it may be possible to develop more efficient ways of inducing this regulatory-T-cell population during immunotherapy treatment of human allergic disease. Perhaps immunotherapy treatment could be used in conjunction with IL-10 or TGF-β treatment, or with glucocorticoids or other immunosuppressive drugs, which as mentioned above have been shown to induce regulatory-T-cell development. Other approaches could include combining immunotherapy treatment with adjuvants such mycobacterial products which have been shown to induce the development of regulatory T cells. Increased co-stimulation through inducible co-stimulator (ICOS) or CTLA-4 could also increase regulatory T-cell numbers, as could increasing the expression of the transcription factor Foxp3.

Notch signalling is an evoluntionarily conserved system regulating cell-fate decisions in organogenesis in the embryo and has been shown to have a role on T-cell development in the thymus. Recent evidence also raises the possibility that this system regulates peripheral T-cell function [121]. Over-expression of the Notch ligand Jagged-1 on DCs leads these cells to induce a regulatory phenotype in naïve murine $CD4^+$ T cells which could then inhibit house-dust-mite-induced airway changes [122]. In addition it was suggested that mRNA for Notch receptors and ligands were over-expressed by human $CD4^+CD25^+$ regulatory T cells [39], although our recent data did not confirm this, nor did we find any effect of blocking the cleavage of Notch (one important element in Notch signalling) upon regulatory function of human $CD4^+CD25^+$ T cells in allergen or anti-CD3 stimulated cultures (Pridgeon, personal communication). Over-expression of Notch could enhance graft survival in a murine transplant model [123] and thus manipulation of Notch signalling may also hold therapeutic potential in enhancing suppression of allergic disease. Recent work has also suggested a role for this system in Th1-/Th2-cell development [124] and the balance of ligands engaged and strength of signal is likely important in determining T-cell response to Notch ligation.

Utilizing regulatory T cells: potential new strategies for the treatment of atopic allergic disease?

In the previous section, *in vivo* induction of regulatory T cells was suggested as a method that could be used in conjunction with other therapies, to increase the efficacy of such treatment. It is also possible that such cells could be generated *ex vivo* from peripheral blood, using such mechanisms. The precise phenotype of regulatory T cells could then be determined, and it may be feasible to transfer these autologous cells back to the patient. Although regulatory T cells were defined as refractory to *in vitro* proliferation, recent reports have described successful expansion of both $CD4^+CD25^+$ T cells and IL-10-producing regulatory T cells [125–127]. The antigen specificity of such cells would need to be determined in advance, but even if the cells were not specific to the relevant antigen inflammation may be inhibited if the regulatory cells were activated at the site of disease. This could however provoke safety concerns, as the transferred or induced regulatory T cells may also suppress T-cell responses to pathogens or cancer antigens, which may cause the patient to become susceptible to such diseases. The effects of this treatment should therefore be investigated extensively in animal models of allergic disease. If successful, this treatment may also be applicable for the suppression of other inflammatory diseases such as autoimmune disease and transplant rejection.

Studies that have shown increases in the suppressive ability of regulatory T cells, such as steroid treatment [66], should now focus on how this treatment has modified the cells. Microarray analysis could be carried out pre- and post-treatment of the cells to ascertain which genes are up- or down-regulated after treatment. These genes should then be analysed further as they may be the key to the elusive mechanism of suppression by regulatory T cells.

The role of regulatory T cells in preventing the development of allergic sensitization

The "hygiene hypothesis" suggests that improved hygiene in industrialized societies, together with improved public-health measures and the use of vaccines and antibiotics, have reduced the incidence of infections that stimulate the immune system and protect against the development of allergic disease [128]. It is suggested that Mycobacterium tuberculosis [129], hepatitis A [130] and measles [131] infection during childhood prevent the development of allergic disease in later life. Originally, the proposed mechanism for this hypothesis was that infections would cause the weakly Th2-biased immune system of newborns to develop into more Th1-biased immune responses [132]. However, Th2-dominated immune responses such as helminth infections are associated with protection against the development of atopic disease [133] and the incidence of Th1-mediated autoimmune diseases has increased in parallel with atopic allergic disease [134], indicating that a lack of Th1-stimulating immune responses is not an adequate explanation for the observed epidemiological trends.

The hygiene hypothesis has recently been modified, since microbial exposure has been found to have a huge impact on regulatory-T-cell development [135]. Germ-free rodents, for example, are unable to elicit tolerance to allergen [136] and are highly sensitive to arthritis induction compared with specific pathogen-free

(SPF) or conventional animals [137]. The hygiene hypothesis may therefore be explained by the fact that a lack of exposure to microbial antigen hampers the development of regulatory T cells and therefore regulation of both Th1 and Th2 responses.

The mechanism by which microbial exposure induces regulatory-T-cell development may include signalling through TLRs. Mycobacteria, which are present in soil, water and dust in a natural environment signal through the pattern-recognition receptors TLRs present on host phagocytes and immune cells [138]. Activation of these receptors can aid the development of adaptive immunity, including generation of regulatory T cells. Indeed, exposure to endotoxin has also been found to promote regulatory-T-cell survival and proliferation, and to enhance their suppressive functions [139].

Results from animal models support a role for regulatory T cells in the protection from atopic disease afforded by mycobacteria. For example, treatment of mice with heat-killed *M. vaccae* during allergen immunization reduced airway hyperresponsiveness, eosinophilia, IgE and production of Th2 cytokines. This down-regulation was dependent on production of IL-10 and TGF-β by regulatory T cells [95]. The relationship between mycobacterial infection and the development of allergy and asthma in humans is controversial. Although *M. tuberculosis* infection rates are significantly inversely correlated with the prevalence of atopic disease [140], whether bacillus Calmette-Gurin (BCG) vaccination is associated with protection against development of atopy is still a matter of debate [141–143].

Another major environmental factor that affects the development of allergic sensitization is exposure to allergen. It has been proposed that avoidance measures can be used to prevent sensitization of individuals with negative skin-test results or even from birth (i.e., primary avoidance) [144–146]. It is not always possible to avoid exposure to allergen however. The major cat allergen Fel d 1, for example, which is carried on dander particles, spreads throughout the community. Thus not having a cat in the home is not an effective measure to prevent sensitization and indeed might have the opposite effect. The fact that families who chose not to own cats because of a family history of allergy had a higher prevalence of cat sensitization than comparable families who had a cat [147] may also be related to regulatory-T-cell development. This observation may be explained by the fact that, although increasing exposure to allergen is associated with increased prevalence of sensitization, exposure to high concentrations of cat dander is associated with decreased sensitization and a higher prevalence of IgG antibody to Fel d 1. A population of regulatory T cells may drive this "modified Th2-cell" response through increased production of IL-10 [148]. High-dose antigen exposure has previously been shown in many models to induce regulatory T cells or to increase the suppressive ability of such a population, although it is of note that similar protective effects of high-dose allergen exposure were not observed for house dust mite [148].

Prevention of allergic disease: vaccination strategies

If development of atopy results from a failure of regulation then a logical approach would be to enhance regulatory T-cell activity through vaccination. Such an approach will require further work to identify infants at risk and to define which are likely to progress from atopic sensitization to allergic disease. Further information on the duration of the regulatory T-cell response and suppressor memory will also be required: current information on natural desensitization in bee keepers suggests that protection from high-dose allergen exposure is relatively short-lived [107]. In contrast the effects of allergen immunotherapy persisted from several years [103]. If high-dose allergen exposure is to be used to prevent animal or food sensitization this will present the difficulty of preventing disease exacerbation in those already sensitized. Allergen immunotherapy has shown to reduce development of new allergen sensitization and development of asthma when used to treat children with pollen sensitization and rhinitis only: whether immunotherapy regimens can be modified to small numbers of injections suitable for mass vaccination of infants before they develop IgE sensitization remains to be seen [149]. It may be that use of adjuvants such as CpG oligonucleotides, probiotics or mycobacterial products will be useful in designing allergen vaccines for preventive treatment.

Conclusions

New therapeutic strategies to control Th2-mediated responses are required to address the increasing incidence of atopic disease. Recent data have illustrated the ability of human regulatory T cells to suppress allergen-driven Th2-type T-cell responses *in vitro*. The fact that $CD4^+CD25^+$ regulatory T cells isolated from atopic individuals are less able to suppress allergen-driven T-cell responses than those isolated from non-atopic individuals illustrates the potential therapeutic importance of this population of cells. Furthermore, induction of regulatory T cells after allergen immunotherapy treatment may provide a clue to the mechanism by which such treatment exerts its clinical effects. Therapeutic manipulation of the balance between regulatory T cells and activated effector T cells during the allergic response may therefore provide hope for better prevention, treatment or even cure of human allergic disease.

Acknowledgements
The work described in this chapter was supported by a Wellcome Trust Research Leave Award for Clinical Academics, the National Kidney Research Fund, Glaxo SmithKline and the Clinical Research Committee of the Royal Brompton Hospital. E.M.L. is a recipient of the Allergopharma Prize.

References

1 Holgate ST (2004) The epidemic of asthma and allergy. *J R Soc Med* 97: 103–110
2 Law AW, Reed SD, Sundy JS, Schulman KA (2003) Direct costs of allergic rhinitis in the United States: estimates from the 1996 Medical Expenditure Panel Survey. *J Allergy Clin Immunol* 111: 296–300
3 Smith DH, Malone DC, Lawson KA, Okamoto LJ, Battista C, Saunders WB (1997) A national estimate of the economic costs of asthma. *Am J Respir Crit Care Med* 156: 787–793
4 Kay AB (2001) Allergy and allergic diseases. First of two parts. *N Engl J Med* 344: 30–37
5 Mosmann TR, Cherwinski H, Bond MW, Giedlin MA, Coffman RL (1986) Two types of murine helper T cell clone. I. Definition according to profiles of lymphokine activities and secreted proteins. *J Immunol* 136: 2348–2357
6 Wierenga EA, Snoek M, de Groot C, Chretien I, Bos JD, Jansen HM, Kapsenberg ML (1990) Evidence for compartmentalization of functional subsets of $CD2^+$ T lymphocytes in atopic patients. *J Immunol* 144: 4651–4656
7 Parronchi P, Macchia D, Piccinni MP, Biswas P, Simonelli C, Maggi E, Ricci M, Ansari AA, Romagnani S (1991) Allergen- and bacterial antigen-specific T-cell clones established from atopic donors show a different profile of cytokine production. *Proc Natl Acad Sci USA* 88: 4538–4542
8 Mosmann TR, Coffman RL (1989) TH1 and TH2 cells: different patterns of lymphokine secretion lead to different functional properties. *Annu Rev Immunol* 7: 145–173
9 Bonecchi R, Bianchi G, Bordignon PP, D'Ambrosio D, Lang R, Borsatti A, Sozzani S, Allavena P, Gray PA, Mantovani A, Sinigaglia F (1998) Differential expression of chemokine receptors and chemotactic responsiveness of type 1 T helper cells (Th1s) and Th2s. *J Exp Med* 187: 129–134
10 Sallusto F, Lenig D, Mackay CR, Lanzavecchia A (1998) Flexible programs of chemokine receptor expression on human polarized T helper 1 and 2 lymphocytes. *J Exp Med* 187: 875–883
11 Sallusto F, Mackay CR, Lanzavecchia A (1997) Selective expression of the eotaxin receptor CCR3 by human T helper 2 cells. *Science* 277: 2005–2007
12 Lohning M, Stroehmann A, Coyle AJ, Grogan JL, Lin S, Gutierrez-Ramos JC, Levinson D, Radbruch A, Kamradt T (1998) T1/ST2 is preferentially expressed on murine Th2 cells, independent of interleukin 4, interleukin 5, and interleukin 10, and important for Th2 effector function. *Proc Natl Acad Sci USA* 95: 6930–6935
13 DeKruyff RH, Turner T, Abrams JS, Palladino MA, Umetsu DT (1989) Induction of human IgE synthesis by $CD4^+$ T cell clones. Requirement for interleukin 4 and low molecular weight B cell growth factor. *J Exp Med* 170: 1477–1493
14 Robinson DS, Hamid Q, Ying S, Tsicopoulos A, Barkans J, Bentley AM, Corrigan C,

Durham SR, Kay AB (1992) Predominant TH2-like bronchoalveolar T-lymphocyte population in atopic asthma. *N Engl J Med* 326: 298–304
15 van Reijsen FC, Bruijnzeel-Koomen CA, Kalthoff FS, Maggi E, Romagnani S, Westland JK, Mudde GC (1992) Skin-derived aeroallergen-specific T-cell clones of Th2 phenotype in patients with atopic dermatitis. *J Allergy Clin Immunol* 90: 184–193
16 Walker C, Bode E, Boer L, Hansel TT, Blaser K, Virchow JC Jr (1992) Allergic and non-allergic asthmatics have distinct patterns of T-cell activation and cytokine production in peripheral blood and bronchoalveolar lavage. *Am Rev Respir Dis* 146: 109–115
17 Robinson D S, Ying S, Bentley AM, Meng Q, North J, Durham SR, Kay AB, Hamid Q (1993) Relationships among numbers of bronchoalveolar lavage cells expressing messenger ribonucleic acid for cytokines, asthma symptoms, and airway methacholine responsiveness in atopic asthma. *J Allergy Clin Immunol* 92: 397–403
18 Huang SK, Xiao HQ, Kleine-Tebbe J, Paciotti G, Marsh DG, Lichtenstein LM, Liu MC (1995) IL-13 expression at the sites of allergen challenge in patients with asthma. *J Immunol* 155: 2688–2694
19 Till S, Durham S, Dickason R, Huston D, Bungre J, Walker S, Robinson D, Kay AB, Corrigan C (1997) IL-13 production by allergen-stimulated T cells is increased in allergic disease and associated with IL-5 but not IFN-gamma expression. *Immunology* 91: 53–57
20 Till S, Dickason R, Huston D, Humbert M, Robinson D, Larche M, Durham S, Kay AB, Corrigan C (1997) IL-5 secretion by allergen-stimulated $CD4^+$ T cells in primary culture: relationship to expression of allergic disease. *J Allergy Clin Immunol* 99: 563–569
21 Nakajima H, Iwamoto I, Tomoe S, Matsumura R, Tomioka H, Takatsu K, Yoshida S (1992) $CD4^+$ T-lymphocytes and interleukin-5 mediate antigen-induced eosinophil infiltration into the mouse trachea. *Am Rev Respir Dis* 146: 374–377
22 Gavett SH, Chen X, Finkelman F, Wills-Karp M (1994) Depletion of murine $CD4^+$ T lymphocytes prevents antigen-induced airway hyperreactivity and pulmonary eosinophilia. *Am J Respir Cell Mol Biol* 10: 587–593
23 Cohn L, Tepper JS, Bottomly K (1998) IL-4-independent induction of airway hyperresponsiveness by Th2, but not Th1, cells. *J Immunol* 161: 3813–3816
24 Corry DB, Folkesson HG, Warnock ML, Erle DJ, Matthay MA, Wiener-Kronish JM, Locksley RM (1996) Interleukin 4, but not interleukin 5 or eosinophils, is required in a murine model of acute airway hyperreactivity. *J Exp Med* 183: 109–117
25 Coyle AJ, Le Gros G, Bertrand C, Tsuyuki S, Heusser CH, Kopf M, Anderson GP (1995) Interleukin-4 is required for the induction of lung Th2 mucosal immunity. *Am J Respir Cell Mol Biol* 13: 54–59
26 Herrick CA, MacLeod H, Glusac E, Tigelaar RE, Bottomly K (2000) Th2 responses induced by epicutaneous or inhalational protein exposure are differentially dependent on IL-4. *J Clin Invest* 105: 765–775
27 Foster PS, Hogan SP, Ramsay AJ, Matthaei KI, Young IG (1996) Interleukin 5 deficiency abolishes eosinophilia, airways hyperreactivity, and lung damage in a mouse asthma model. *J Exp Med* 183: 195–201

28 Lee JJ, McGarry MP, Farmer SC, Denzler KL, Larson KA, Carrigan PE, Brenneise IE, Horton MA, Haczku A, Gelfand EW et al (1997). Interleukin-5 expression in the lung epithelium of transgenic mice leads to pulmonary changes pathognomonic of asthma. *J Exp Med* 185: 2143–2156

29 Grunig G, Warnock M, Wakil AE, Venkayya R, Brombacher F, Rennick DM, Sheppard D, Mohrs M, Donaldson DD, Locksley RM, Corry DB (1998) Requirement for IL-13 independently of IL-4 in experimental asthma. *Science* 282: 2261–2263

30 Wills-Karp M, Luyimbazi J, Xu X, Schofield B, Neben TY, Karp CL, Donaldson DD (1998) Interleukin-13: central mediator of allergic asthma. *Science* 282: 2258–2261

31 Ying S, Robinson DS, Meng Q, Barata LT, McEuen AR, Buckley MG, Walls AF, Askenase PW, Kay AB (1999) C-C chemokines in allergen-induced late-phase cutaneous responses in atopic subjects: association of eotaxin with early 6-hour eosinophils, and of eotaxin-2 and monocyte chemoattractant protein-4 with the later 24-hour tissue eosinophilia, and relationship to basophils and other C-C chemokines (monocyte chemoattractant protein-3 and RANTES). *J Immunol* 163: 3976–3984

32 Larche M, Till SJ, Haselden BM, North J, Barkans J, Corrigan CJ, Kay AB, Robinson DS (1998) Costimulation through CD86 is involved in airway antigen-presenting cell and T cell responses to allergen in atopic asthmatics. *J Immunol* 161: 6375–6382

33 Palframan RT, Collins PD, Severs NJ, Rothery S, Williams TJ, Rankin SM (1998) Mechanisms of acute eosinophil mobilization from the bone marrow stimulated by interleukin 5: the role of specific adhesion molecules and phosphatidylinositol 3-kinase. *J Exp Med* 188: 1621–1632

34 Levings MK, Sangregorio R, Roncarolo MG (2001) Human CD25(+)CD4(+) T regulatory cells suppress naive and memory T cell proliferation and can be expanded *in vitro* without loss of function. *J Exp Med* 193: 1295–1302

35 Jonuleit H, Schmitt E, Stassen M, Tuettenberg A, Knop J, Enk AH (2001) Identification and functional characterization of human CD4(+)CD25(+) T cells with regulatory properties isolated from peripheral blood. *J Exp Med* 193: 1285–1294

36 Taams LS, Smith J, Rustin MH, Salmon M, Poulter LW, Akbar AN (2001) Human anergic/suppressive CD4(+)CD25(+) T cells: a highly differentiated and apoptosis-prone population. *Eur J Immunol* 31: 1122–1131

37 Dieckmann D, Plottner H, Berchtold S, Berger T, Schuler G (2001) Ex vivo isolation and characterization of CD4(+)CD25(+) T cells with regulatory properties from human blood. *J Exp Med* 193: 1303–1310

38 Stephens LA, Mottet C, Mason D, Powrie F (2001) Human CD4(+)CD25(+) thymocytes and peripheral T cells have immune suppressive activity *in vitro*. *Eur J Immunol* 31: 1247–1254

39 Ng WF, Duggan PJ, Ponchel F, Matarese G, Lombardi G, Edwards AD, Isaacs JD, Lechler RI (2001) Human CD4(+)CD25(+) cells: a naturally occurring population of regulatory T cells. *Blood* 98: 2736–2744

40 Baecher-Allan C, Brown JA, Freeman GJ, Hafler DA (2001) $CD4^+$CD25high regulatory cells in human peripheral blood. *J Immunol* 167: 1245–1253

41 Taams LS, Vukmanovic-Stejic M, Smith J, Dunne PJ, Fletcher JM, Plunkett FJ, Ebeling SB, Lombardi G, Rustin MH, Bijlsma JW et al (2002) Antigen-specific T cell suppression by human $CD4^+CD25^+$ regulatory T cells. *Eur J Immunol* 32: 1621–1630
42 van Maurik A, Herber M, Wood KJ, Jones ND (2002) Cutting edge: $CD4^+CD25^+$ alloantigen-specific immunoregulatory cells that can prevent $CD8^+$ T cell-mediated graft rejection: implications for anti-CD154 immunotherapy. *J Immunol* 169: 5401–5404
43 Lin CY, Graca L, Cobbold SP, Waldmann H (2002) Dominant transplantation tolerance impairs $CD8^+$ T cell function but not expansion. *Nat Immunol* 3: 1208–1213
44 Piccirillo CA, Shevach EM (2001) Cutting edge: control of $CD8^+$ T cell activation by $CD4^+CD25^+$ immunoregulatory cells. *J Immunol* 167: 1137–1140
45 Chen J, Huoam C, Plain K, He XY, Hodgkinson SJ, Hall BM (2001) CD4(+), CD25(+) T cells as regulators of alloimmune responses. *Transplant Proc* 33: 163–164
46 Takahashi T, Tagami T, Yamazaki S, Uede T, Shimizu J, Sakaguchi N, Mak TW, Sakaguchi S (2000) Immunologic self-tolerance maintained by CD25(+)CD4(+) regulatory T cells constitutively expressing cytotoxic T lymphocyte-associated antigen 4. *J Exp Med* 192: 303–310
47 Read S, Malmstrom V, Powrie F (2000) Cytotoxic T lymphocyte-associated antigen 4 plays an essential role in the function of CD25(+)CD4(+) regulatory cells that control intestinal inflammation. *J Exp Med* 192: 295–302
48 McHugh RS, Whitters MJ, Piccirillo CA, Young DA, Shevach EM, Collins M, Byrne MC (2002) CD4(+)CD25(+) immunoregulatory T cells: gene expression analysis reveals a functional role for the glucocorticoid-induced TNF receptor. *Immunity* 16: 311–323
49 Zelenika D, Adams E, Humm S, Graca L, Thompson S, Cobbold SP, Waldmann H (2002) Regulatory T cells overexpress a subset of Th2 gene transcripts. *J Immunol* 168: 1069–1079
50 Shimizu J, Yamazaki S, Takahashi T, Ishida Y, Sakaguchi S (2002) Stimulation of CD25(+)CD4(+) regulatory T cells through GITR breaks immunological self-tolerance. *Nat Immunol* 3: 135–142
51 Hori S, Nomura T, Sakaguchi S (2003) Control of regulatory T cell development by the transcription factor Foxp3. *Science* 299: 1057–1061
52 Fontenot JD, Gavin MA, Rudensky AY (2003) Foxp3 programs the development and function of $CD4^+CD25^+$ regulatory T cells. *Nat Immunol* 4: 330–336
53 Khattri R, Cox T, Yasayko SA, Ramsdell F (2003) An essential role for Scurfin in $CD4^+CD25^+$ T regulatory cells. *Nat Immunol* 4: 337–342
54 von Herrath MG, Harrison LC (2003) Antigen-induced regulatory T cells in autoimmunity. *Nat Rev Immunol* 3: 223–232
55 Zhang X, Izikson L, Liu L, Weiner HL (2001) Activation of CD25(+)CD4(+) regulatory T cells by oral antigen administration. *J Immunol* 167: 4245–4253
56 Thorstenson KM, Khoruts A (2001) Generation of anergic and potentially immunoregulatory $CD25^+CD4$ T cells *in vivo* after induction of peripheral tolerance with intravenous or oral antigen. *J Immunol* 167: 188–195
57 Takahashi T, Kuniyasu Y, Toda M, Sakaguchi N, Itoh M, Iwata M, Shimizu J, Sakaguchi

S (1998) Immunologic self-tolerance maintained by $CD25^+CD4^+$ naturally anergic and suppressive T cells: induction of autoimmune disease by breaking their anergic/suppressive state. *Int Immunol* 10: 1969–1980
58 Gavin MA, Clarke SR, Negrou E, Gallegos A, Rudensky A (2002) Homeostasis and anergy of CD4(+)CD25(+) suppressor T cells *in vivo*. *Nat Immunol* 3: 33–41
59 Bach JF, Francois BJ (2003) Regulatory T cells under scrutiny. *Nat Rev Immunol* 3: 189–198
60 Barthlott T, Kassiotis G, Stockinger B (2003) T cell regulation as a side effect of homeostasis and competition. *J Exp Med* 197: 451–460
61 Bluestone JA, Abbas AK (2003) Natural *versus* adaptive regulatory T cells. *Nat Rev Immunol* 3: 253–257
62 Ling EM, Smith TRF, Nguyen XD, Pridgeon CP, Dallman MJ, Arbery J, Carr VA, Robinson DS (2004) Relation of $CD4^+CD25^+$ regulatory T-cell suppression of allergen-driven T-cell activation to atopic status and expression of allergic disease. *Lancet* 363: 608–615
63 Bellinghausen I, Klostermann B, Knop J, Saloga J (2003) Human $CD4^+CD25^+$ T cells derived from the majority of atopic donors are able to suppress TH1 and TH2 cytokine production. *J Allergy Clin Immunol* 111: 862–868
64 Tiemessen MM, Van Hoffen E, Knulst AC, Van Der Zee JA, Knol EF, Taams LS (2002) CD4 CD25 regulatory T cells are not functionally impaired in adult patients with IgE-mediated cow's milk allergy. *J Allergy Clin Immunol* 110: 934–936
65 Cavani A, Nasorri F, Ottaviani C, Sebastiani S, De Pita O, Girolomoni G (2003) Human $CD25^+$ regulatory T cells maintain immune tolerance to nickel in healthy, nonallergic individuals. *J Immunol* 171: 5760–5768
66 Nguyen XD, Robinson DS (2004) Fluticasone propionate increases $CD4^+CD25^+$ T cell suppression of allergen-stimulated $CD4^+CD25^-$ T cells by an IL-10-dependent mechanism. *J Allergy Clin Immunol* 114: 296–301
67 Pasare C, MedzhitovR (2003) Toll pathway-dependent blockade of $CD4^+CD25^+$ T cell-mediated suppression by dendritic cells. *Science* 299: 1033–1036
68 Jordan MS, Boesteanu A, Reed AJ, Petrone AL, Holenbeck AE, Lerman MA, Naji A, Caton AJ (2001) Thymic selection of $CD4^+CD25^+$ regulatory T cells induced by an agonist self-peptide. *Nat Immunol* 2: 301–306
69 Horwitz DA, Zheng SG, Gray JD (2003) The role of the combination of IL-2 and TGF-beta or IL-10 in the generation and function of $CD4^+$ $CD25^+$ and $CD8^+$ regulatory T cell subsets. *J Leukoc Biol* 74: 471–478
70 Lloyd CM, Gonzalo JA, Coyle AJ, Gutierrez-Ramos JC (2001) Mouse models of allergic airway disease. *Adv Immunol* 77: 263–295
71 Suto A, Nakajima H, Kagami SI, Suzuki K, Saito Y, Iwamoto I (2001) Role of CD4(+) CD25(+) regulatory T cells in T helper 2 cell-mediated allergic inflammation in the airways. *Am J Respir Crit Care Med* 164: 680–687
72 Jaffar Z, Sivakuru T, Roberts K (2004) $CD4^+CD25^+$ T cells regulate airway eosinophilic inflammation by modulating the Th2 cell phenotype. *J Immunol* 172: 3842–3849

73 Hadeiba H, Locksley RM (2003) Lung CD25 CD4 regulatory T cells suppress type 2 immune responses but not bronchial hyperreactivity. *J Immunol* 170: 5502–5510
74 Thornton AM, Shevach EM (1998) $CD4^+CD25^+$ immunoregulatory T cells suppress polyclonal T cell activation *in vitro* by inhibiting interleukin 2 production. *J Exp Med* 188: 287–296
75. Jonuleit H, Schmitt E, Kakirman H, Stassen M, Knop J, Enk AH (2002) Infectious tolerance: human CD25(+) regulatory T cells convey suppressor activity to conventional CD4(+) T helper cells. *J Exp Med* 196: 255–260
76 Olivares-Villagomez D, Wensky AK, Wang Y, Lafaille JJ (2000) Repertoire requirements of $CD4^+$ T cells that prevent spontaneous autoimmune encephalomyelitis. *J Immunol* 164: 5499–5507
77 Gonzalez A, Andre-Schmutz I, Carnaud C, Mathis D, Benoist C (2001) Damage control, rather than unresponsiveness, effected by protective $DX5^+$ T cells in autoimmune diabetes. *Nat Immunol* 2: 1117–1125
78 Stephens LA, Mason D (2000) CD25 is a marker for $CD4^+$ thymocytes that prevent autoimmune diabetes in rats, but peripheral T cells with this function are found in both $CD25^+$ and $CD25^-$ subpopulations. *J Immunol* 165: 3105–3110
79 Annacker O, Pimenta-Araujo R, Burlen-Defranoux O, Barbosa TC, Cumano A, Bandeira A (2001) $CD25^+$ $CD4^+$ T cells regulate the expansion of peripheral CD4 T cells through the production of IL-10. *J Immunol* 166: 3008–3018
80 Sundstedt A, O'Neill EJ, Nicolson KS, Wraith DC (2003) Role for IL-10 in suppression mediated by peptide-induced regulatory T cells *in vivo*. *J Immunol* 170: 1240–1248
81 Barrat FJ, Cua DJ, Boonstra A, Richards DF, Crain C, Savelkoul HF, Waal-Malefyt R, Coffman RL, Hawrylowicz CM, O'Garra A (2002) *In vitro* generation of interleukin 10-producing regulatory CD4(+) T cells is induced by immunosuppressive drugs and inhibited by T helper type 1 (Th1)- and Th2-inducing cytokines. *J Exp Med* 195: 603–616
82 Groux H, O'Garra A, Bigler M, Rouleau M, Antonenko S, de Vries JE, Roncarolo MG (1997) A $CD4^+$ T-cell subset inhibits antigen-specific T-cell responses and prevents colitis. *Nature* 389: 737–742
83 Chen Y, Kuchroo VK, Inobe J, Hafler DA, Weiner HL (1994) Regulatory T cell clones induced by oral tolerance: suppression of autoimmune encephalomyelitis. *Science* 265: 1237–1240
84 Moore KW, de Waal Malefyt R, Coffman RL, O'Garra A (2001) Interleukin-10 and the interleukin-10 receptor. *Annu Rev Immunol* 19: 683–765
85 Jutel M, Akdis M, Budak F, Aebischer-Casaulta C, Wrzyszcz M, Blaser K, Akdis CA (2003) IL-10 and TGF-beta cooperate in the regulatory T cell response to mucosal allergens in normal immunity and specific immunotherapy. *Eur J Immunol* 33: 1205–1214
86 Akdis M, Verhagen J, Taylor A, Karamloo F, Karagiannidis C, Crameri R, Thunberg S, Deniz G, Valenta R, Fiebig H et al (2004) Immune responses in healthy and allergic Individuals are characterized by a fine balance between allergen-specific T regulatory 1 and T helper 2 cells. *J Exp Med* 199: 1567–1575

87 EM Ling, M Calderon, D Nguyen, K Powrie, SR Durham, DS Robinson (2004) Allergen immunotherapy increases suppressive activity by $CD4^+CD25^-$, IL-10 producing T cells, but does not affect suppression by $CD4^+CD25^+$ T cells. *J Allergy Clin Immunol* 113 (Suppl) (Abstract 333)
88 Walker MR, Kasprowicz DJ, Gersuk VH, Benard A, Van Landeghen M, Buckner JH, Ziegler SF (2003) Induction of FoxP3 and acquisition of T regulatory activity by stimulated human $CD4^+CD25^-$ T cells. *J Clin Invest* 112: 1437–1443
89 Chen W, Jin W, Hardegen N, Lei KJ, Li L, Marinos N, McGrady G, Wahl SM (2003) Conversion of peripheral $CD4^+CD25-$ naive T cells to $CD4^+CD25^+$ regulatory T cells by TGF-beta induction of transcription factor Foxp3. *J Exp Med* 198: 1875–1886
90 Ostroukhova M, Seguin-Devaux C, Oriss TB, Dixon-McCarthy B, Yang L, Ameredes BT, Corcoran TE, Ray A (2004) Tolerance induced by inhaled antigen involves CD4(+) T cells expressing membrane-bound TGF-beta and FOXP3. *J Clin Invest* 114: 28–38
91 Levings MK, Sangregorio R, Galbiati F, Squadrone S, de Waal Malefyt R, Roncarolo MG (2001) IFN-alpha and IL-10 induce the differentiation of human type 1 T regulatory cells. *J Immunol* 166: 5530–5539
92 Cottrez F, Hurst SD, Coffman RL, Groux H (2000) T regulatory cells 1 inhibit a Th2-specific response *in vivo*. *J Immunol* 165: 4848–4853
93 Akbari O, DeKruyff RH, Umetsu DT (2001) Pulmonary dendritic cells producing IL-10 mediate tolerance induced by respiratory exposure to antigen. *Nat Immunol* 2: 725–731
94 Akbari O, Stock P, DeKruyff RH, Umetsu DT (2003) Role of regulatory T cells in allergy and asthma. *Curr Opin Immunol* 15: 627–633
95 Zuany-Amorim C, Sawicka E, Manlius C, Le Moine A, Brunet LR, Kemeny DM, Bowen G, Rook G, Walker C (2002) Suppression of airway eosinophilia by killed Mycobacterium vaccae-induced allergen-specific regulatory T-cells. *Nat Med* 8: 625–629
96 British Thoracic Society; Scottish Intercollegiate Guidelines Network (2003) British guideline on the management of asthma. *Thorax* 58 (Suppl 1): i1–i94
97 Bousquet J, Van Cauwenberge P, Khaltaev N; Aria Workshop Group; World Health Organization (2001) Allergic rhinitis and its impact on asthma. *J Allergy Clin Immunol* 108: S147–S334
98 Ellis C, Luger T, Abeck D, Allen R, Graham-Brown RA, De Prost Y, Eichenfield LF, Ferrandiz C, Giannetti A, Hanifin J et al; ICCAD II Faculty (2003) International Consensus Conference on Atopic Dermatitis II (ICCAD II): clinical update and current treatment strategies. *Br J Dermatol* 148 (Suppl) 63: 3–10
99 Leung DY, Bloom JW (2003) Update on glucocorticoid action and resistance. *J Allergy Clin Immunol* 111: 3–22
100 Robinson D, Hamid Q, Ying S, Bentley A, Assoufi B, Durham S, Kay AB (1993) Prednisolone treatment in asthma is associated with modulation of bronchoalveolar lavage cell interleukin-4, interleukin-5, and interferon-gamma cytokine gene expression. *Am Rev Respir Dis* 148: 401–406
101 Bentley AM, Hamid Q, Robinson DS, Schotman E, Meng Q, Assoufi B, Kay AB, Durham SR (1996) Prednisolone treatment in asthma. Reduction in the numbers of

eosinophils, T cells, tryptase-only positive mast cells, and modulation of IL-4, IL-5, and interferon-gamma cytokine gene expression within the bronchial mucosa. *Am J Respir Crit Care Med* 153: 551–556

102 Hawrylowicz C, Richards D, Loke TK, Corrigan C, Lee T (2002) A defect in corticosteroid-induced IL-10 production in T lymphocytes from corticosteroid-resistant asthmatic patients. *J Allergy Clin Immunol* 109: 369–370

103 Durham SR, Walker SM, Varga EM, Jacobson MR, O'Brien F, Noble W, Till SJ, Hamid QA, Nouri-Aria KT (1999) Long-term clinical efficacy of grass-pollen immunotherapy. *N Engl J Med* 341: 468–475

104 Durham SR, Till SJ (1998) Immunologic changes associated with allergen immunotherapy. *J Allergy Clin Immunol* 102: 157–164

105 Jutel M, Pichler WJ, Skrbic D, Urwyler A, Dahinden C, Muller UR (1995) Bee venom immunotherapy results in decrease of IL-4 and IL-5 and increase of IFN-gamma secretion in specific allergen-stimulated T cell cultures. *J Immunol* 154: 4187–4194

106 Ebner C, Siemann U, Bohle B, Willheim M, Wiedermann U, Schenk S, Klotz F, Ebner H, Kraft D, Scheiner O (1997) Immunological changes during specific immunotherapy of grass pollen allergy: reduced lymphoproliferative responses to allergen and shift from TH2 to TH1 in T-cell clones specific for Phl p 1, a major grass pollen allergen. *Clin Exp Allergy* 27: 1007–1015

107 Akdis CA, Joss A, Akdis M, Blaser K (2001) Mechanism of IL-10-induced T cell inactivation in allergic inflammation and normal response to allergens. *Int Arch Allergy Immunol* 124: 180–182

108 Pierkes M, Bellinghausen I, Hultsch T, Metz G, Knop J, Saloga J (1999) Decreased release of histamine and sulfidoleukotrienes by human peripheral blood leukocytes after wasp venom immunotherapy is partially due to induction of IL-10 and IFN-gamma production of T cells. *J Allergy Clin Immunol* 103: 326–332

109 Wilson DR, Nouri-Aria KT, Walker SM, Pajno GB, O'Brien F, Jacobson MR, Mackay IS, Durham SR (2001) Grass pollen immunotherapy: symptomatic improvement correlates with reductions in eosinophils and IL-5 mRNA expression in the nasal mucosa during the pollen season. *J Allergy Clin Immunol* 107: 971–976

110 Durham SR, Varney VA, Gaga M, Jacobson MR, Varga EM, Frew AJ, Kay AB (1999) Grass pollen immunotherapy decreases the number of mast cells in the skin. *Clin Exp Allergy* 29: 1490–1496

111 Varney VA, Hamid QA, Gaga M, Ying S, Jacobson M, Frew AJ, Kay AB, Durham SR (1993) Influence of grass pollen immunotherapy on cellular infiltration and cytokine mRNA expression during allergen-induced late-phase cutaneous responses. *J Clin Invest* 92: 644–651

112 Francis JN, Till SJ, Durham SR (2003) Induction of IL-10$^+$CD4$^+$CD25$^+$ T cells by grass pollen immunotherapy. *J Allergy Clin Immunol* 111: 1255–1261

113. Larche M (2001) Inhibition of human T-cell responses by allergen peptides. *Immunology* 104: 377–382

114 Oldfield WL, Larche M, Kay AB (2002) Effect of T-cell peptides derived from Fel d 1

on allergic reactions and cytokine production in patients sensitive to cats: a randomised controlled trial. *Lancet* 360: 47–53

115 Oldfield WL, Kay AB, Larche M (2001) Allergen-derived T cell peptide-induced late asthmatic reactions precede the induction of antigen-specific hyporesponsiveness in atopic allergic asthmatic subjects. *J Immunol* 167: 1734–1739

116 Smith TRF, Alexander C, Kay AB, Larché M, Robinson DS (2004) Cat allergen peptide immunotherapy reduces $CD4^+$ T cell responses to cat allergen but does not alter suppression by $CD4^+CD25^+$ T cells. *Allergy* 59: 1097–1101

117 Tighe H, Takabayashi K, Schwartz D, Van Nest G, Tuck S, Eiden JJ, Kagey-Sobotka A, Creticos PS, Lichtenstein LM, Spiegelberg HL, Raz E (2000) Conjugation of immunostimulatory DNA to the short ragweed allergen amb a 1 enhances its immunogenicity and reduces its allergenicity. *J Allergy Clin Immunol* 106: 124–134

118 Cohn L, Homer RJ, Niu N, Bottomly K (1999) T helper 1 cells and interferon gamma regulate allergic airway inflammation and mucus production. *J Exp Med* 190: 1309–1318

119 Hansen G, Berry G, DeKruyff RH, Umetsu DT (1999) Allergen-specific Th1 cells fail to counterbalance Th2 cell-induced airway hyperreactivity but cause severe airway inflammation. *J Clin Invest* 103: 175–183

120 Wraith DC (2003) Role of interleukin-10 in the induction and function of natural and antigen-induced regulatory T cells. *J Autoimmun* 20: 273–275

121 Hoyne GF, Dallman MJ, Champion BR, Lamb JR (2001) Notch signalling in the regulation of peripheral immunity. *Immunol Rev* 182: 215–227

122 Hoyne GF, Le Roux I, Corsin-Jimenez M, Tan K, Dunne J, Forsyth LM, Dallman MJ, Owen MJ, Ish-Horowicz D, Lamb JR (2000) Serrate1-induced notch signalling regulates the decision between immunity and tolerance made by peripheral CD4(+) T cells. *Int Immunol* 12: 177–185

123 Wong KK, Carpenter MJ, Young LL, Walker SJ, McKenzie G, Rust AJ, Ward G, Packwood L, Wahl K, Delriviere L et al (2003) Notch ligation by Delta1 inhibits peripheral immune responses to transplantation antigens by a $CD8^+$ cell-dependent mechanism. *J Clin Invest* 112: 1741–1750

124 Amsen D, Blander JM, Lee GR, Tanigaki K, Honjo T, Flavell RA (2004) Instruction of distinct CD4 T helper cell fates by different notch ligands on antigen-presenting cells. *Cell* 117: 515–526

125 Yamazaki S, Iyoda T, Tarbell K, Olson K, Velinzon K, Inaba K, Steinman RM (2003) Direct expansion of functional $CD25^+$ $CD4^+$ regulatory T cells by antigen-processing dendritic cells. *J Exp Med* 198: 235–247

126 Jiang S, Camara N, Lombardi G, Lechler RI (2003) Induction of allopeptide-specific human $CD4^+CD25^+$ regulatory T cells *ex vivo*. *Blood* 102: 2180–2186

127 Bacchetta R, Sartirana C, Levings MK, Bordignon C, Narula S, Roncarolo MG (2002) Growth and expansion of human T regulatory type 1 cells are independent from TCR activation but require exogenous cytokines. *Eur J Immunol* 32: 2237–2245

128 Strachan DP (2000) Family size, infection and atopy: the first decade of the "hygiene hypothesis". *Thorax* 55 (Suppl 1): S2–S10
129 von Mutius E, Pearce N, Beasley R, Cheng S, von Ehrenstein O, Bjorksten B, Weiland S (2000) International patterns of tuberculosis and the prevalence of symptoms of asthma, rhinitis, and eczema. *Thorax* 55: 449–453
130 Matricardi PM, Rosmini F, Riondino S, Fortini M, Ferrigno L, Rapicetta M, Bonini S (2000) Exposure to foodborne and orofecal microbes *versus* airborne viruses in relation to atopy and allergic asthma: epidemiological study. *BMJ* 320: 412–417
131 Shaheen SO, Aaby P, Hall AJ, Barker DJ, Heyes CB, Shiell AW, Goudiaby A (1996) Measles and atopy in Guinea-Bissau. *Lancet* 347: 1792–1796
132 Holt PG, Macaubas C, Stumbles PA, Sly PD (1999) The role of allergy in the development of asthma. *Nature* 402: B12–B17
133 van den Biggelaar AH, van Ree R, Rodrigues LC, Lell B, Deelder AM, Kremsner PG, Yazdanbakhsh M (2000) Decreased atopy in children infected with *Schistosoma haematobium*: a role for parasite-induced interleukin-10. *Lancet* 356: 1723–1727
134 Black P (2001) Why is the prevalence of allergy and autoimmunity increasing? *Trends Immunol* 22: 354–355
135 Wills-Karp M, Santeliz J, Karp CL (2001) The germless theory of allergic disease: revisiting the hygiene hypothesis. *Nat Rev Immunol* 1: 69–75
136 Sudo N, Sawamura S, Tanaka K, Aiba Y, Kubo C, Koga Y (1997) The requirement of intestinal bacterial flora for the development of an IgE production system fully susceptible to oral tolerance induction. *J Immunol* 159: 1739–1745
137 Kohashi O, Kuwata J, Umehara K, Uemura F, Takahashi T, Ozawa A (1979) Susceptibility to adjuvant-induced arthritis among germfree, specific-pathogen-free, and conventional rats. *Infect Immun* 26: 791–794
138 Strohmeier GR Fenton MJ (1999) Roles of lipoarabinomannan in the pathogenesis of tuberculosis. *Microbes Infect* 1: 709–717
139 Caramalho I, Lopes-Carvalho T, Ostler D, Zelenay S, Haury M, Demengeot J (2003) Regulatory T cells selectively express toll-like receptors and are activated by lipopolysaccharide. *J Exp Med* 197: 403–411
140 Ohrui T, Zayasu K, Sato E, Matsui T, Sekizawa K, Sasaki H (2000) Pulmonary tuberculosis and serum IgE. *Clin Exp Immunol* 122: 13–15
141 Shirakawa T, Enomoto T, Shimazu S, Hopkin JM (1997) The inverse association between tuberculin responses and atopic disorder. *Science* 275: 77–79
142 Aaby P, Shaheen SO, Heyes CB, Goudiaby A, Hall AJ, Shiell AW, Jensen H, Marchant A (2000) Early BCG vaccination and reduction in atopy in Guinea-Bissau. *Clin Exp Allergy* 30: 644–650
143 Matricardi PM, Yazdanbakhsh M (2003) Mycobacteria and atopy, 6 years later: a fascinating, still unfinished, business. *Clin Exp Allergy* 33: 717–720
144 Arshad SH, Bateman B, Matthews SM (2003) Primary prevention of asthma and atopy during childhood by allergen avoidance in infancy: a randomised controlled study. *Thorax* 58: 489–493

145 Chan-Yeung M, Manfreda J, Dimich-Ward H, Ferguson A, Watson W, Becker A (2000) A randomized controlled study on the effectiveness of a multifaceted intervention program in the primary prevention of asthma in high-risk infants. *Arch Pediatr Adolesc Med* 154: 657–663
146 Custovic A, Simpson BM, Simpson A, Kissen P, Woodcock A (2001) Effect of environmental manipulation in pregnancy and early life on respiratory symptoms and atopy during first year of life: a randomised trial. *Lancet* 358: 188–193
147 Platts-Mills TA (2004) Allergen avoidance. *J Allergy Clin Immunol* 113: 388–391
148 Platts-Mills T, Vaughan J, Squillace S, Woodfolk J, Sporik R (2001) Sensitisation, asthma, and a modified Th2 response in children exposed to cat allergen: a population-based cross-sectional study. *Lancet* 357: 752–756
149 Moller C, Dreborg S, Ferdousi HA, Halken S, Host A, Jacobsen L, Koivikko A, Koller DY, Niggemann B, Norberg et al (2002) Pollen immunotherapy reduces the development of asthma in children with seasonal rhinoconjunctivitis (the PAT-study). *J Allergy Clin Immunol* 109: 251–256

The role of regulatory T cells in cutaneous disorders

Katie E. Birch[1,2], Milica Vukmanovic-Stejic[1], John R. Reed[1], Malcolm H.A. Rustin[2] and Arne N. Akbar[1]

[1]Department of Immunology and Molecular Pathology, Division of Infection and Immunity, University College London, 46, Cleveland Street, London W1T 4JF, UK; [2]Department of Dermatology, Royal Free Hospital, Pond Street, London NW3 2QG, UK

Introduction

The skin represents the largest surface area that is in contact with the environment and is constantly exposed to microbial pathogens, physical and chemical insults and a diverse range of potential allergens. Whereas the skin provides a physical barrier between the host and the environment, there is also a versatile system of immune surveillance in order to provide adequate immune defence against potential pathogens. However, the regulation of immune defence in the skin is crucial in order to prevent inappropriate responses to either non-pathogenic environmental antigens or self-antigens.

There is accumulating evidence to suggest that regulatory T cells have a critical role in maintaining a balance between immunity and pathology through their suppression of T-cell responses [1, 2]. This subset of T cells has been implicated in a variety of inflammatory cutaneous disorders including atopic dermatitis [3], acute contact dermatitis [4], fixed drug eruptions [5], psoriasis and cutaneous infections [6]. In addition, several therapeutic agents that are widely used in dermatology, including vitamin D derivatives [7] and ultraviolet light [8], have been shown to influence regulatory T cells. Animal studies have also indicated that regulatory-T-cell depletion enhances cutaneous B16 melanoma rejection and response to tumour vaccines [9].

The extent to which regulatory T cells are able to regulate immune responses within the skin, however, remains unclear. The presence of circulating antigen-specific regulatory T cells to allergens such as house dust mite (Der P 1) [10] or nickel [4] would suggest that regulatory T cells do have a protective role in the prevention of allergic responses to environmental allergens encountered on the skin. It is therefore possible that dysfunctional regulatory T cells may result in a loss of cutaneous anergy resulting in the development of allergic responses to innocuous agents.

Regulatory T cells have also been demonstrated in the skin at sites of inflammation, including inflammation induced by delayed-type hypersensitivity reaction to

Regulatory T Cells in Inflammation, edited by Leonie S. Taams, Arne N. Akbar and Marca H.M. Wauben

nickel [4], in psoriatic lesions and at sites of chronic cutaneous infection [6]. *Ex vivo*, these cells appear to have suppressive ability on stimulated responder T cells. This would suggest that regulatory T cells are present at sites of inflammation but lack the ability to successfully overcome inflammation once it has become established within the skin.

The aim of this review is to assess the role that regulatory T cells have to play in both the prevention of the development of pathological skin diseases and the suppression of ongoing inflammation.

Types of regulatory T cell

A naturally occurring subset of thymically derived suppressor $CD4^+$ T cells that expresses CD25 were first described by Sakaguchi et al. [11] in mice in the 1990s. Subsequently this population of cells has been described in humans and, like mice, constitutes around 5–10% of circulating T cells [12]. These cells are present in the peripheral blood, thymus and secondary lymphoid organs [13, 14]. Over the last decade it has also become apparent that, in addition to thymically derived $CD4^+CD25^+$ regulatory T cells, different regulatory-T-cell populations that can be induced in the periphery exist, including T_R1 cells [15] and Th3 cells [16]. Whereas $CD4^+CD25^+$ regulatory T cells are thought to mediate their suppressive effects through cell-to-cell contact [17], T_R1 and Th3 cells appear to be cytokine-dependent, producing interleukin (IL)-10 and transforming growth factor (TGF)-β respectively [18, 19]. T_R1 cells have been shown to regulate antigen-specific immune responses and inhibit pathologic autoantigen-specific T cells *in vivo* [13]. They have been implicated in several cutaneous disorders as discussed in this chapter. TGF-β-dependent Th3 cells have been identified following oral tolerance/specific immunotherapy and are also thought to have a role in the prevention of the development of autoimmunity [20].

Migration of regulatory T cells into the skin

In order to determine the role of regulatory T cells in the cutaneous immune system it is important to establish where and when regulatory T cells might be acting. The identification of the expression of certain receptors, chemokines and integrins suggests that populations of regulatory T cells are uniquely capable of migrating into the skin at sites of inflammation. A population of circulating $CD4^+CD25^+$ regulatory T cells in the skin has been found to express the cutaneous lymphocyte antigen (CLA) [21], indicating that regulatory T cells may be able to migrate to sites of inflammation within the skin using similar pathways to those used by effector T cells. The chemokine CCR8 is markedly upregulated following the activation of

$CD4^+CD25^+$ regulatory CLA^+ T cells *in vitro*, suggesting that CCR8 may also be involved in the localization of T cells to the skin by virtue of CCL1 production within the skin [21]. Interestingly, CCR8 expression on T cells has also recently been implicated in the routine immune-surveillance mechanisms present in the skin although such cells did not appear to have a regulatory phenotype [22].

Integrin expression may also have a role to play in determining the trafficking of regulatory T cells to specific sites [23]. Populations of both murine $CD4^+CD25^+$ and $CD4^+CD25^-$ regulatory T cells have recently been identified that express the integrin $\alpha\epsilon\beta_7$ (CD103) that recognizes epithelial cadherin [24]. However significant levels of expression of this integrin in human regulatory T cells have yet to be identified. The integrin $\alpha_4\beta_7$ is known to act as a homing receptor for T lymphocytes to inflamed tissues and has been found to be expressed on IL-10-producing T_R1-like human regulatory T cells [25]. It is therefore possible that recruitment of regulatory T cells to sites of cutaneous inflammation may be controlled in part through the expression of $\alpha_4\beta_7$.

Role of regulatory T cells in chronic inflammatory skin disease: atopic dermatitis and psoriasis

Atopic dermatitis is characterized by a multifactorial, but predominantly genetically determined, abnormal response to non-infectious environmental allergens and a shift towards the production of Th2 cytokines including IL-4, IL-5 and IL-13. Th2 cytokine production results in the recruitment of eosinophils to the site of inflammation and elevated IgE levels. The absence of functional regulatory T cells in the immune dysregulation, polyendocrinopathy, enteropathy, X-linked syndrome (IPEX syndrome) clearly demonstrates that regulatory T cells play a critical role in preventing the development of atopic responses. The IPEX syndrome, thought to be caused by mutations in the Foxp3 gene, is characterized by severe eczema, elevated IgE levels, eosinophilia and food allergy [26]. Further evidence for the role of regulatory T cells in preventing atopic responses is provided by the finding that in experiments *in vitro*, where $CD4^+CD25^+$ regulatory T cells are depleted from the T-cell pool of non-atopic individuals, T effector cells produce the Th2 cytokines IL-5 and IL-13 [27] in response to aeroallergens. In addition, animal studies have demonstrated that the infusion of T_R1 cells results in the down-regulation of Th2 and hyper-IgE responses to ovalbumin in sensitized mice [28, 29].

In vitro experiments have demonstrated that $CD4^+CD25^+$ regulatory T cells isolated from the peripheral circulation of atopic individuals are able to suppress Th1 and Th2 cytokine production [30], suggesting that regulatory T cells are functional in these individuals. However, in this study a small subset of atopics had $CD4^+CD25^+$ regulatory T cells with reduced suppressive capacity, produced IL4 and IL10 and did not suppress Th2 responses *in vitro* [30]. These cells were phenotypically indistinguishable from functional regulatory T cells. Other studies have also indicated that $CD4^+CD25^+$ regulatory T cells from atopic individuals may be inherently dysfunctional [27].

The increased expression of E-selectin on inflamed dermal post-capillary venules and endothelial cells in skin affected by atopic dermatitis would favour increased recruitment of CLA^+ regulatory T cells into the skin [31, 32]. Thus, the presence of ongoing inflammation in spite of the isolation of functional regulatory T cells from atopic individuals may be explained by either changes in the local environment of the skin or an intrinsic inability to suppress different subsets of T cells involved in atopic responses *in vivo*. Local cytokine production may influence regulatory-T-cell function. In particular, the Th2 cytokines IL-4, IL-15, IL-7 and IL-9 that are raised within atopic dermatitis lesions have been shown to induce $CD4^+CD25^+$ regulatory T cells with reduced suppressive capacity [33]. $CD4^+CD25^+$ regulatory T cells from both atopic and non-atopic individuals are less effective at suppressing Th2 cells than Th1 cells [33], thus explaining a relative lack of efficacy in atopic dermatitis patients where Th2 cells predominate.

The suppressive ability of regulatory T cells may also be dependent on the strength of the antigenic stimulus and the degree of activation of effector T cells. In one study, the suppressive capability of $CD4^+CD25^+$ regulatory T cells was reduced in patients suffering with hay fever during the pollen season [27]. The reduced suppressive affect appeared to be antigen-specific in that global stimulation with anti-CD3 antibody resulted in similar suppression between regulatory T cells from atopics and non-atopics. This observation may be explained by the strength of T-cell receptor (TCR) stimulation by certain allergens. Experiments with co-culture of $CD4^+CD25^+$ regulatory T cells and responder T cells has indicated that strong TCR stimulation is able to abrogate suppression by increasing the resistance of T effector cells to suppression by regulatory T cells [34]. Inhibition of $CD4^+CD25^+$ regulatory T cells may also occur as a result of bacterial colonization of the skin by *Staphylococcus aureus*. One study has shown that although patients with atopic dermatitis have functional circulating regulatory T cells, staphylococcal superantigenic stimulation *in vitro* significantly reduced their suppressive capacity [3]. Thus, for certain antigens or in situations where there is a high antigenic load, regulatory T cells may be unable to overcome effector-T-cell stimulation.

Current data on the role of regulatory T cells in psoriasis is limited; however, $CD4^+CD25^+$ regulatory T cells have been isolated from psoriatic lesions. The isolated regulatory T cells from individuals with psoriasis were unable to suppress the hyper-reactive responder T cells found in skin affected by psoriasis. However, $CD4^+CD25^+$ regulatory T cells isolated from peripheral blood mononuclear cells of normal individuals without psoriasis were able to suppress the hyper-reactive psoriatic effector T cells *in vitro*. The investigators therefore concluded that the regulatory-T-cell population was dysfunctional in individuals with psoriasis (H. Sugiyama, K. Cooper et al., poster abstract, Society of Investigative Dermatology, 2004).

Infectious disease

In addition to the suppression of undesirable immune reaction to autoantigens and non-pathogenic environmental agents, regulatory T cells may also have an important role in modulating immune responses to pathogenic organisms. The failure to control overwhelming immunological responses to pathogenic organisms can result in significant tissue damage at the site of infection resulting in deleterious pathology. For example, severe pneumonitis due to *Pneumocystis carinii* infection was observed in SCID mice following immune reconstitution with $CD4^+$ T cells depleted of $CD25^+$ cells and could be prevented by the addition of regulatory T cells [35].

Conversely, it is also possible that regulatory T cells favour the long-term survival of pathogens by suppressing the immune response. The persistence of pathogens after clinical cure is the hallmark of certain viral, bacterial and parasitic infections including *Mycobacterium tuberculosis* [36] and leishmaniasis. In cutaneous leishmaniasis, low numbers of viable organisms can be found within lymphoid tissue and/or the skin after skin healing [37]. Infection of the skin of mice with *Leishmania major* has been shown to result in the accumulation of IL-10-producing $CD4^+CD25^+$ regulatory T cells at the site of infection in the dermis where they are able to suppress the ability of $CD4^+CD25^-$ effector cells to eliminate the parasite from the site [6, 38]. This suggests that regulatory T cells may be responsible for persistent latent infection with leishmaniasis. The containment and cure of active tuberculosis is dependent on an effective cell-mediated response. The absence of a cutaneous delayed-type hypersensitivity response to mycobacterial antigens, such as tuberculin purified protein derivative (PPD), has been shown to be associated with poor clinical outcome in humans infected with tuberculosis [39, 40]. Cutaneous anergy to PPD in spite of active tuberculosis infection is found in around 15% of cases and has been shown to be due to IL-10-producing cells with a T_R1 regulatory phenotype. Anergic individuals with active tuberculosis infection also have reduced $CD4^+$ T-cell responses to PPD *in vitro* in comparison with individuals with strong cutaneous PPD responses. Increased proliferative responses are seen with the addition of anti-IL-10 antibodies, suggesting IL-10-mediated suppression by regulatory T cells [41].

Latent infection by pathogenic organisms has potentially serious clinical consequences in that immunosuppression, for example following co-infection with HIV, can result in life-threatening reactivation disease such as visceral leishmaniasis or pulmonary tuberculosis. Treatment with anti-IL-10 antibody in mice infected with leishmaniasis has been shown to result in sterile cure, indicating a possible therapeutic approach to eliminate latency and the subsequent risk of reactivation disease. However, impaired IL-10 activity was also associated with a loss of immunity to re-infection, suggesting that incomplete clearance of the organism is advantageous in order to generate a memory response [42]. Thus, it seems that regulatory T cells are involved with both host and parasite survival strategies in chronic infection.

Delayed-type hypersensitivity reactions

The skin is constantly exposed to a large variety of chemicals within the environment yet only a limited number of individuals will develop detrimental inflammatory reactions to such agents. Allergic contact dermatitis is an inflammatory skin disease that occurs predominantly in adults at sites of contact with small lipophilic chemicals known as haptens. Sensitizing haptens are able to stimulate the immune response by binding in precise conformation to peptide–MHC complexes. Following sensitization to a hapten, subsequent exposure results in the recruitment of $CD8^+$ and $CD4^+$ T cells to the site of challenge. $CD8^+$ T cells are thought to be the primary mediators of inflammation in acute contact dermatitis through the release of type 1 cytokines and mediation of tissue damage although $CD4^+$ Th1 cells also augment the inflammatory response [43, 44].

In humans, $CD4^+CD25^+$ regulatory T cells [4] and T_R1 type regulatory T cells [45, 46] have been found at the site of challenge with haptens suggesting that regulatory T cells may have a role in preventing the initial development of acute contact dermatitis. In addition, circulating and cutaneous nickel-specific, skin-homing CLA^+ regulatory T_R1 cells have been shown to release high levels of IL-10 and inhibit the ability of dendritic cells to activate hapten-specific T cells. Functional T_R1 cells have been isolated from the peripheral blood of sensitized and non-sensitized individuals; however, the number of regulatory T cells in allergic individuals may be reduced. In individuals with negative patch tests to nickel, functional $CD4^+CD25^+$ regulatory T cells that are able to suppress the *in vitro* response of $CD25^-$ T cells to nickel have been isolated in the skin at the site of nickel application [4]. In non-sensitized individuals significant stimulation of effector T cells with nickel *in vitro* only occurred after depletion of $CD25^+CD4^+$ cells. However, $CD4^+CD25^+$ regulatory T cells from the blood of nickel-allergic individuals have been shown to have a reduced ability to inhibit metal specific CD4 and CD8 responses *in vitro* [4]. These observations suggest that regulatory T cells may be either dysfunctional or present in insufficient numbers in individuals with acute contact dermatitis reactions, resulting in failure to prevent the activation and proliferation of hapten-specific effector T cells upon antigenic stimulation.

It is interesting to note that certain haptens, such as dinitrochlorobenzene (DNCB), will invariably induce sensitization in most individuals that are exposed to them if they are applied in sufficient concentrations [47]. This may be due to strong antigenic stimulation of effector T cells so that they become resistant to suppression by regulatory T cells [34]. Repeated exposure to DNCB has been shown in a limited number of cases to result in the recruitment of increasing numbers of T_R1 regulatory T cells in to the skin [45]. This would further support the hypothesis that regulatory T cells can be generated in the periphery from highly differentiated T cells that have undergone chronic antigenic stimulation [48]. It is also possible that an individual's ability to mobilize sufficient numbers of regulatory T cells upon exposure to haptens may determine whether or not they become sensitized to that particular antigen. DNCB has been shown to down-regulate the expression of E-cadherin on dendritic cells [49, 50], which may affect the ability of CLA^+ regulatory T cells to migrate in to the skin [51].

Following the development of inflammation, regulatory T_R1 cells have been found within acute contact dermatitis lesions [46]. The upregulation of CCR8 expression on the cell surface of Th^{IL-10} T cells, with a phenotype similar to T_R1 cells, within such lesions but not surrounding skin, in addition to increased production of CCL1 (I309), would appear to favour the recruitment of regulatory T cells once inflammation has been established [52]. It is possible that the recruitment of regulatory T cells within the lesion may determine the duration of the inflammatory response to the hapten following removal of the agent from the skin; however, it is currently unknown as to whether regulatory T cells are functional in this context.

Therapeutics

The manipulation of regulatory T cells by either enhancing or reducing their suppressive effects may be useful clinically. In spite of the apparent anergic nature of regulatory T cells, large-scale *in vitro* expansion of polyclonal human $CD4^+CD25^{+high}$ T cells has recently been described [53], suggesting that *ex vivo* manipulation and subsequent allogeneic transfusion of regulatory T cells may be possible. In addition to *ex vivo* expansion it would be useful clinically to use protocols that would induce their development or enhance their activity *in vivo*. Several immunomodulatory agents currently in use for the treatment of skin disorders and induction of transplantation tolerance have been demonstrated to influence regulatory T cells which may in part explain the mechanism of action of such agents.

The therapeutic enhancement of the number and activity of regulatory T cells is a potential target for the treatment of both autoimmune and allergic skin disease. The active form of vitamin D_3, 1α,25-Dihydroxyvitamin D_3, is currently used topically to treat psoriasis [54]. 1α,25-Dihydroxyvitamin D_3, in combination with dexamethasone, has been shown both in human and mouse studies *in vitro* to modulate the phenotype and function of dendritic cells, favouring the induction of $CD4^+CD25^+$ regulatory rather than effector T cells [7]. Dexamethasone alone has also been shown to induce regulatory $CD4^+CD25^+$ T cells [55] and this may partly explain the beneficial immunosuppressive effects of both topical and systemic steroids. Mycophenolate mofetil (MMF) is also used to treat inflammatory skin disorders such as psoriasis. [56] In combination with vitamin D_3, MMF has been shown in mice to induce tolerance to islet allografts. Tolerance can be transferred to naïve mice by the adoptive transfer of $CD25^+$ regulatory T cells [57].

Certain immunomodulatory agents used to treat inflammation in the skin may work independently of regulatory T cells, however. We have demonstrated (M. Vukmanovic-Stejic, A. McQuaid, K.E. Birch, J.R. Reed, C. MacGregor, M.H.A. Rustin, A.N. Akbar; unpublished observations) that CD25$^+$CD4$^+$ regulatory T cells and tacrolimus, a calcineurin inhibitor, have an additive effect on the suppression of *in vitro* effector-T-cell responses to tuberculin PPD and house-dust-mite antigen (Der P 1) in cells isolated from atopic dermatitis patients. The effects of tacrolimus on suppression were not dependent on the presence of regulatory CD4$^+$CD25$^+$ T cells. Tacrolimus has recently been introduced as a topical treatment for atopic dermatitis [58]. The additive effect of tacrolimus and CD4$^+$CD25$^+$ regulatory T cells on the suppression of proliferation indicates that the integration of intrinsic immunoregulation (mediated by regulatory T cells) and therapeutically used immunosuppression is possible. This may allow for lower doses of potentially harmful immunosuppressants to be used in conjunction with regimes devised to enhance regulatory-T-cell function.

Phototherapy is commonly used to treat a range of dermatological disorders including inflammatory skin diseases such as atopic dermatitis and psoriasis. The immunosuppressive effects of ultraviolet irradiation are still incompletely understood; however, ultraviolet-induced immunosuppression using cutaneous irradiation has been shown to reduce the number of Langerhans cells and change their antigen-presenting function, favouring the development of suppressor T cells [8, 59]. Low-dose irradiation with ultraviolet light in mice can also prevent the development of contact-hypersensitivity reactions to haptens applied to the skin by inducing hapten-specific regulatory T_R1 cells [60, 61].

CD4$^+$CD25$^+$ regulatory T cells also play a role in transplantation tolerance by preventing graft rejection and delaying graft-*versus*-host disease [62, 63]. This may have implications for cutaneous graft-*versus*-host disease and also skin grafting. In mouse studies, CD25$^+$ and CD25$^-$ regulatory T cells have been found within tolerated skin allografts [64] that could transfer tolerance to naive recipients [65]. Transplant-tolerance-inducing protocols using monoclonal antibodies against CD4, CD8, CD3 or co-stimulatory molecules have been shown to induce regulatory T cells and this may explain their therapeutic effects [66, 67]. The anti-CD25 monoclonal antibodies basiliximab and dacluzimab have been used clinically in transplantation-tolerance regimes and more recently small studies have suggested a role for these agents in the treatment of psoriasis [68, 69]. The beneficial effects of anti-CD25 monoclonal antibodies observed in patients with psoriasis is perhaps surprising in that enhancement rather than depletion of the CD25$^+$ subset was found to be beneficial in this pro-inflammatory disorder. The observed benefit, however, is likely to be explained by the depletion of activated effector T cells that also express CD25. In fact, the non-specific depletion of the regulatory-T-cell subset by these agents may reduce their efficacy. The identification of a specific regulatory-T-cell marker would aid the future development of more specific antibodies in order to circumvent this problem with T-cell-depletion therapies.

Whereas increased generation of regulatory T cells may be beneficial in treating inflammatory skin conditions or in transplantation, the depletion of regulatory T cells may also be beneficial for certain diseases. It has been postulated that regulatory T cells may be detrimental in patients with tumours due to suppression of the host anti-tumour response. Indeed, increased numbers of tumour-specific $CD4^+CD25^+$ regulatory T cells have been demonstrated in patients with epithelial malignancies [70]. Cutaneous melanoma has a rapidly increasing incidence [71] and there is limited available treatment for metastatic disease. The development of both new treatments and a vaccine for melanoma is therefore highly desirable. Depletion of $CD4^+CD25^+$ T cells in mice has been shown to result in enhanced anti-tumour responses to syngeneic tumours, including B16 melanoma, resulting in tumour eradication [72, 73]. In addition, CD25 depletion and blocking of CTLA-4 has been shown to promote immunity to a B16 melanoma vaccine in mice [74].

In spite of the apparent advantages of modulation of regulatory-T-cell activity extreme caution will need to be employed in order to ensure that no significant side effects occur, such as carcinogenesis or autoimmunity, following enhancement or depletion therapy respectively. Selective targeting of antigen-specific regulatory T cells will be crucial in order to suppress pathogenic T cells without causing generalized immunosuppression of T cells directed against infectious pathogens or tumours and activation of autoreactive T cells [75].

Conclusions

Although the precise role of regulatory T cells in the skin has yet to be characterized, there is accumulating evidence to suggest that they play a key role in a multitude of immune-mediated responses within the skin as discussed in this chapter. The identification of functional regulatory T cells within inflamed skin suggests that they may have a beneficial role, although their ability to adequately suppress the response may be limited by numerous factors including local cytokine production, the influence of colonizing bacteria and the antigenic stimulation of effector T cells. The development of sensitization to innocuous environmental agents may, however, be representative of inherent regulatory-T-cell dysfunction. In the future, enhancement of the suppressive effects of the regulatory-T-cell subset may be used to treat autoimmune disorders and allergic responses by enhancing their suppressive function, whereas depletion may be employed in order to clear persistent infection, improve response to vaccines or enhance tumour immunity. Further understanding of the mechanisms by which currently used therapeutic agents modify the action of regulatory T cells will not only aid further therapeutic development but also provide an insight into the mechanisms that control regulatory T cells within the skin *in vivo*.

References

1 Shevach EM (2002) $CD4^+$ $CD25^+$ suppressor T cells: more questions than answers. *Nat Rev Immunol* 2: 389–400

2 Baecher-Allan C, Hafler DA (2004) Suppressor T cells in human diseases. *J Exp Med* 200: 273–276

3 Ou LS, Goleva E, Hall C, Leung DY (2004) T regulatory cells in atopic dermatitis and subversion of their activity by superantigens. *J Allergy Clin Immunol* 113: 756–763

4 Cavani A, Nasorri F, Ottaviani C, Sebastiani S, De Pita O, Girolomoni G (2003) Human $CD25^+$ regulatory T cells maintain immune tolerance to nickel in healthy, nonallergic individuals. *J Immunol* 171: 5760–5768

5 Teraki Y, Shiohara T (2003) IFN-gamma-producing effector $CD8^+$ T cells and IL-10-producing regulatory $CD4^+$ T cells in fixed drug eruption. *J Allergy Clin Immunol* 112: 609–615

6 Belkaid Y, Piccirillo CA, Mendez S, Shevach EM, Sacks DL (2002) $CD4^+CD25^+$ regulatory T cells control Leishmania major persistence and immunity. *Nature* 420: 502–507

7 Adorini L (2002) Immunomodulatory effects of vitamin D receptor ligands in autoimmune diseases. *Int Immunopharmacol* 2: 1017–1028

8 Elmets CA, Bergstresser PR, Tigelaar RE, Wood PJ, Streilein JW (1983) Analysis of the mechanism of unresponsiveness produced by haptens painted on skin exposed to low dose ultraviolet radiation. *J Exp Med* 158: 781–794

9 Steitz J, Bruck J, Lenz J, Knop J, Tuting T (2001) Depletion of CD25(+) CD4(+) T cells and treatment with tyrosinase-related protein 2-transduced dendritic cells enhance the interferon alpha-induced, CD8(+) T-cell-dependent immune defense of B16 melanoma. *Cancer Res* 61: 8643–8646

10 Taams LS, Vukmanovic-Stejic M, Smith J, Dunne PJ, Fletcher JM, Plunkett FJ, Ebeling SB, Lombardi G, Rustin MH, Bijlsma JW et al (2002) Antigen-specific T cell suppression by human $CD4^+CD25^+$ regulatory T cells. *Eur J Immunol* 32: 1621–1630

11 Sakaguchi S, Sakaguchi N, Asano M, Itoh M, Toda M (1995) Immunologic self-tolerance maintained by activated T cells expressing IL-2 receptor alpha-chains (CD25). Breakdown of a single mechanism of self-tolerance causes various autoimmune diseases. *J Immunol* 155: 1151–1164

12 Jonuleit H, Schmitt E, Stassen M, Tuettenberg A, Knop J, Enk AH (2001) Identification and functional characterization of human CD4(+)CD25(+) T cells with regulatory properties isolated from peripheral blood. *J Exp Med* 193: 1285–1294

13 Taams LS, Smith J, Rustin MH, Salmon M, Poulter LW, Akbar AN (2001) Human anergic/suppressive CD4(+)CD25(+) T cells: a highly differentiated and apoptosis-prone population. *Eur J Immunol* 31: 1122–1131

14 Stephens LA, Mottet C, Mason D, Powrie F (2001) Human CD4(+)CD25(+) thymocytes and peripheral T cells have immune suppressive activity *in vitro*. *Eur J Immunol* 31: 1247–1254

15 Groux H, O'Garra A, Bigler M, Rouleau M, Antonenko S, de Vries JE, Roncarolo MG

(1997) A CD4+ T-cell subset inhibits antigen-specific T-cell responses and prevents colitis. *Nature* 389: 737–742

16 Chen Y, Kuchroo VK, Inobe J, Hafler DA, Weiner HL (1994) Regulatory T cell clones induced by oral tolerance: suppression of autoimmune encephalomyelitis. *Science* 265: 1237–1240

17 Thornton AM, Shevach EM (1998) CD4+CD25+ immunoregulatory T cells suppress polyclonal T cell activation *in vitro* by inhibiting interleukin 2 production. *J Exp Med* 188: 287–296

18 Asseman C, Mauze S, Leach MW, Coffman RL, Powrie F (1999) An essential role for interleukin 10 in the function of regulatory T cells that inhibit intestinal inflammation. *J Exp Med* 190: 995–1004

19 Nakamura K, Kitani A, Fuss I, Pedersen A, Harada N, Nawata H, Strober W (2004) TGF-beta 1 plays an important role in the mechanism of CD4+CD25+ regulatory T cell activity in both humans and mice. *J Immunol* 172: 834–842

20 Jutel M, Akdis M, Budak F, Aebischer-Casaulta C, Wrzyszcz M, Blaser K, Akdis CA (2003) IL-10 and TGF-beta cooperate in the regulatory T cell response to mucosal allergens in normal immunity and specific immunotherapy. *Eur J Immunol* 33: 1205–1214

21 Colantonio L, Iellem A, Sinigaglia F, D'Ambrosio D (2002) Skin-homing CLA+ T cells and regulatory CD25+ T cells represent major subsets of human peripheral blood memory T cells migrating in response to CCL1/I-309. *Eur J Immunol* 32: 3506–3514

22 Schaerli P, Ebert L, Willimann K, Blaser A, Roos RS, Loetscher P, Moser B (2004) A Skin-selective homing mechanism for human immune surveillance T cells. *J Exp Med* 199: 1265–1275

23 Pribila JT, Quale AC, Mueller KL, Shimizu Y (2004) Integrins and T cell-mediated immunity. *Annu Rev Immunol* 22: 157–180

24 Lehmann J, Huehn J, de la Rosa M, Maszyna F, Kretschmer U, Krenn V, Brunner M, Scheffold A, Hamann A (2002) Expression of the integrin alpha E beta 7 identifies unique subsets of CD25+ as well as CD25− regulatory T cells. *Proc Natl Acad Sci USA* 99: 13031–13036

25 Stassen M, Fondel S, Bopp T, Richter C, Muller C, Kubach J, Becker C, Knop J, Enk AH, Schmitt S et al (2004) Human CD25+ regulatory T cells: two subsets defined by the integrins alpha4beta7 or alpha4beta1 confer distinct suppressive properties upon CD4+ T helper cells. *Eur J Immunol* 34: 1303–1311

26 Patel DD (2001) Escape from tolerance in the human X-linked autoimmunity-allergic disregulation syndrome and the Scurfy mouse. *J Clin Invest* 107: 155–157

27 Ling EM, Smith T, Nguyen XD, Pridgeon C, Dallman M, Arbery J, Carr VA, Robinson DS (2004) Relation of CD4+CD25+ regulatory T-cell suppression of allergen-driven T-cell activation to atopic status and expression of allergic disease. *Lancet* 363: 608–615

28 Cottrez F, Hurst SD, Coffman RL, Groux H (2000) T regulatory cells 1 inhibit a Th2-specific response *in vivo*. *J Immunol* 165: 4848–4853

29 Curotto de Lafaille MA, Muriglan S, Sunshine MJ, Lei Y, Kutchukhidze N, Furtado GC, Wensky AK, Olivares-Villagomez D, Lafaille JJ (2001) Hyper immunoglobulin E

response in mice with monoclonal populations of B and T lymphocytes. *J Exp Med* 194: 1349–1359

30 Bellinghausen I, Klostermann B, Knop J, Saloga J (2003) Human CD4+CD25+ T cells derived from the majority of atopic donors are able to suppress TH1 and TH2 cytokine production. *J Allergy Clin Immunol* 111: 862–868

31 Furue M, Koga T, Yamashita N (1999) Soluble E-selectin and eosinophil cationic protein are distinct serum markers that differentially represent clinical features of atopic dermatitis. *Br J Dermatol* 140: 67–72

32 Wakita H, Sakamoto T, Tokura Y, Takigawa M (1994) E-selectin and vascular cell adhesion molecule-1 as critical adhesion molecules for infiltration of T lymphocytes and eosinophils in atopic dermatitis. *J Cutan Pathol* 21: 33–39

33 Cosmi L, Liotta F, Angeli R, Mazzinghi B, Santarlasci V, Manetti R, Lasagni L, Vanini V, Romagnani P, Maggi E et al (2004) Th2 cells are less susceptible than Th1 cells to the suppressive activity of CD25+ regulatory thymocytes because of their responsiveness to different cytokines. *Blood* 103: 3117–3121

34 Baecher-Allan C, Viglietta V, Hafler DA (2002) Inhibition of human CD4(+) CD25(+high) regulatory T cell function. *J Immunol* 169: 6210–6217

35 Hori S, Carvalho TL, Demengeot J (2002) CD25+CD4+ regulatory T cells suppress CD4+ T cell-mediated pulmonary hyperinflammation driven by *Pneumocystis carinii* in immunodeficient mice. *Eur J Immunol* 32: 1282–1291

36 Stewart GR, Robertson BD, Young DB (2003) Tuberculosis: a problem with persistence. *Nat Rev Microbiol* 1: 97–105

37 Schubach A, Haddad F, Oliveira-Neto MP, Degrave W, Pirmez C, Grimaldi G Jr, Fernandes O (1998) Detection of Leishmania DNA by polymerase chain reaction in scars of treated human patients. *J Infect Dis* 178: 911–914

38 Xu D, Liu H, Komai-Koma M, Campbell C, McSharry C, Alexander J, Liew FY (2003) CD4+CD25+ regulatory T cells suppress differentiation and functions of Th1 and Th2 cells, *Leishmania major* infection, and colitis in mice. *J Immunol* 170: 394–399

39 Delgado JC, Tsai EY, Thim S, Baena A, Boussiotis VA, Reynes JM, Sath S, Grosjean P, Yunis EJ, Goldfeld AE (2002) Antigen-specific and persistent tuberculin anergy in a cohort of pulmonary tuberculosis patients from rural Cambodia. *Proc Natl Acad Sci USA* 99: 7576–7581

40 Sousa AO, Salem JI, Lee FK, Vercosa MC, Cruaud P, Bloom BR, Lagrange PH, David HL (1997) An epidemic of tuberculosis with a high rate of tuberculin anergy among a population previously unexposed to tuberculosis, the Yanomami Indians of the Brazilian Amazon. *Proc Natl Acad Sci USA* 94: 13227–13232

41 Boussiotis VA, Tsai EY, Yunis EJ, Thim S, Delgado JC, Dascher CC, Berezovskaya A, Rousset D, Reynes JM, Goldfeld AE (2000) IL-10-producing T cells suppress immune responses in anergic tuberculosis patients. *J Clin Invest* 105: 1317–1325

42 Belkaid Y, Hoffmann KF, Mendez S, Kamhawi S, Udey MC, Wynn TA, Sacks DL (2001) The role of interleukin (IL)-10 in the persistence of *Leishmania major* in the skin after

healing and the therapeutic potential of anti-IL-10 receptor antibody for sterile cure. *J Exp Med* 194: 1497–1506

43 Traidl C, Sebastiani S, Albanesi C, Merk HF, Puddu P, Girolomoni G, Cavani A (2000) Disparate cytotoxic activity of nickel-specific $CD8^+$ and $CD4^+$ T cell subsets against keratinocytes. *J Immunol* 165: 3058–3064

44 Wang B, Fujisawa H, Zhuang L, Freed I, Howell BG, Shahid S, Shivji GM, Mak TW, Sauder DN (2000) $CD4^+$ Th1 and $CD8^+$ type 1 cytotoxic T cells both play a crucial role in the full development of contact hypersensitivity. *J Immunol* 165: 6783–6790

45 Lecart S, Boulay V, Raison-Peyron N, Bousquet J, Meunier L, Yssel H, Pene J (2001) Phenotypic characterization of human $CD4^+$ regulatory T cells obtained from cutaneous dinitrochlorobenzene-induced delayed type hypersensitivity reactions. *J Invest Dermatol* 117: 318–325

46 Cavani A, Nasorri F, Prezzi C, Sebastiani S, Albanesi C, Girolomoni G (2000) Human $CD4^+$ T lymphocytes with remarkable regulatory functions on dendritic cells and nickel-specific Th1 immune responses. *J Invest Dermatol* 114: 295–302

47 Friedmann PS, Moss C, Shuster S, Simpson JM (1983) Quantitative relationships between sensitizing dose of DNCB and reactivity in normal subjects. *Clin Exp Immunol* 53: 709–715

48 Akbar AN, Taams LS, Salmon M, Vukmanovic-Stejic M (2003) The peripheral generation of $CD4^+$ $CD25^+$ regulatory T cells. *Immunology* 109: 319–325

49 Aiba S, Manome H, Yoshino Y, Tagami H (2000) *in vitro* treatment of human transforming growth factor-beta1-treated monocyte-derived dendritic cells with haptens can induce the phenotypic and functional changes similar to epidermal Langerhans cells in the initiation phase of allergic contact sensitivity reaction. *Immunology* 101: 68–75

50 Staquet MJ, Sportouch M, Jacquet C, Schmitt D, Guesnet J, Peguet-Navarro J (2004) Moderate skin sensitizers can induce phenotypic changes on *in vitro* generated dendritic cells. *Toxicol In Vitro* 18: 493–500

51 Hwang JM, Yamanouchi J, Santamaria P, Kubes P (2004) A critical temporal window for selectin-dependent $CD4^+$ lymphocyte homing and initiation of late-phase inflammation in contact sensitivity. *J Exp Med* 199: 1223–1234

52 Sebastiani S, Allavena P, Albanesi C, Nasorri F, Bianchi G, Traidl C, Sozzani S, Girolomoni G, Cavani A (2001) Chemokine receptor expression and function in $CD4^+$ T lymphocytes with regulatory activity. *J Immunol* 166: 996–1002

53 Hoffmann P, Eder R, Kunz-Schughart LA, Andreesen R, Edinger M (2004) Large-scale *in vitro* expansion of polyclonal human CD4(+)CD25high regulatory T cells. *Blood* 104: 895–903

54 Kira M, Kobayashi T, Yoshikawa K (2003) Vitamin D and the skin. *J Dermatol* 30: 429–437

55 Chen X, Murakami T, Oppenheim JJ, Howard OM (2004) Differential response of murine $CD4^+CD25^+$ and $CD4^+$. *Eur J Immunol* 34: 859–869

56 Lebwohl M, Ali S (2001) Treatment of psoriasis. Part 2. Systemic therapies. *J Am Acad Dermatol* 45: 649–661

57 Gregori S, Casorati M, Amuchastegui S, Smiroldo S, Davalli AM, Adorini L (2001) Regulatory T cells induced by 1 alpha,25-dihydroxyvitamin D3 and mycophenolate mofetil treatment mediate transplantation tolerance. *J Immunol* 167: 1945–1953
58 Reitamo S, Wollenberg A, Schopf E, Perrot JL, Marks R, Ruzicka T, Christophers E, Kapp A, Lahfa M, Rubins A et al (2000) Safety and efficacy of 1 year of tacrolimus ointment monotherapy in adults with atopic dermatitis. The European Tacrolimus Ointment Study Group. *Arch Dermatol* 136: 999–1006
59 Toews GB, Bergstresser PR, Streilein JW (1980) Epidermal Langerhans cell density determines whether contact hypersensitivity or unresponsiveness follows skin painting with DNFB. *J Immunol* 124: 445–453
60 Schwarz A, Maeda A, Wild MK, Kernebeck K, Gross N, Aragane Y, Beissert S, Vestweber D, Schwarz T (2004) Ultraviolet radiation-induced regulatory T cells not only inhibit the induction but can suppress the effector phase of contact hypersensitivity. *J Immunol* 172: 1036–1043
61 Shreedhar VK, Pride MW, Sun Y, Kripke ML, Strickland FM (1998) Origin and characteristics of ultraviolet-B radiation-induced suppressor T lymphocytes. *J Immunol* 161: 1327–1335
62 Trenado A, Charlotte F, Fisson S, Yagello M, Klatzmann D, Salomon BL, Cohen JL (2003) Recipient-type specific $CD4^{+}CD25^{+}$ regulatory T cells favor immune reconstitution and control graft-*versus*-host disease while maintaining graft-*versus*-leukemia. *J Clin Invest* 112: 1688–1696
63 Trenado A, Fisson S, Braunberger E, Klatzmann D, Salomon BL, Cohen JL (2004) *Ex vivo* selection of recipient-type alloantigen-specific CD4(+)CD25(+) immunoregulatory T cells for the control of graft-*versus*-host disease after allogeneic hematopoietic stem-cell transplantation. *Transplantation* 77: S32–S34
64 Graca L, Cobbold SP, Waldmann H (2002) Identification of regulatory T cells in tolerated allografts. *J Exp Med* 195: 1641–1646
65 Kingsley CI, Karim M, Bushell AR, Wood KJ (2002) $CD25^{+}CD4^{+}$ regulatory T cells prevent graft rejection: CTLA-4- and IL-10-dependent immunoregulation of alloresponses. *J Immunol* 168: 1080–1086
66 Morris ES, MacDonald KP, Rowe V, Johnson DH, Banovic T, Clouston AD, Hill GR (2004) Donor treatment with pegylated G-CSF augments the generation of IL-10-producing regulatory T cells and promotes transplantation tolerance. *Blood* 103: 3573–3581
67 Waldmann H, Graca L, Cobbold S, Adams E, Tone M, Tone Y (2004) Regulatory T cells and organ transplantation. *Semin Immunol* 16: 119–126
68 Krueger JG, Walters IB, Miyazawa M, Gilleaudeau P, Hakimi J, Light S, Sherr A, Gottlieb AB (2000) Successful *in vivo* blockade of CD25 (high-affinity interleukin 2 receptor) on T cells by administration of humanized anti-Tac antibody to patients with psoriasis. *J Am Acad Dermatol* 43: 448–458
69 Mrowietz U, Zhu K, Christophers E (2000) Treatment of severe psoriasis with anti-CD25 monoclonal antibodies. *Arch Dermatol* 136: 675–676

70 Wolf AM, Wolf D, Steurer M, Gastl G, Gunsilius E, Grubeck-Loebenstein B (2003) Increase of regulatory T cells in the peripheral blood of cancer patients. *Clin Cancer Res* 9: 606–612
71 Lens MB, Dawes M (2004) Global perspectives of contemporary epidemiological trends of cutaneous malignant melanoma. *Br J Dermatol* 150: 179–185
72 Onizuka S, Tawara I, Shimizu J, Sakaguchi S, Fujita T, Nakayama E (1999) Tumor rejection by *in vivo* administration of anti-CD25 (interleukin-2 receptor alpha) monoclonal antibody. *Cancer Res* 59: 3128–3133
73 Steitz J, Bruck J, Lenz J, Knop J, Tuting T (2001) Depletion of CD25(+) CD4(+) T cells and treatment with tyrosinase-related protein 2-transduced dendritic cells enhance the interferon alpha-induced, CD8(+) T-cell-dependent immune defense of B16 melanoma. *Cancer Res* 61: 8643–8646
74 Sutmuller RP, van Duivenvoorde LM, van Elsas A, Schumacher TN, Wildenberg ME, Allison JP, Toes RE, Offringa R, Melief CJ (2001) Synergism of cytotoxic T lymphocyte-associated antigen 4 blockade and depletion of CD25(+) regulatory T cells in antitumor therapy reveals alternative pathways for suppression of autoreactive cytotoxic T lymphocyte responses. *J Exp Med* 194: 823–832
75 O'Garra A, Vieira P (2004) Regulatory T cells and mechanisms of immune system control. *Nat Med* 10: 801–805

The potential role of CD25⁺CD4⁺ regulatory T cells in the induction and maintenance of transplantation tolerance in humans

Kathryn J. Wood and Ahmed Akl

Nuffield Department of Surgery, University of Oxford, John Radcliffe Hospital, Oxford OX3 9DU, UK

Introduction

Active control, in other words suppression or regulation, of immune responses is a relatively old concept in immunology. Lymphocytes that could suppress immune responses were first described more than 30 years ago [1]. Kilshaw, Brent and Pinto [2] soon followed up this observation by looking for the existence of such cells in a transplantation model. They were able to confirm that these so-called suppressor cells existed and were operational in transplantation tolerance. Suppressor cells, now known more commonly as regulatory T cells (T_R cells), were and still are a potentially important tool for controlling the immune response to a transplant. However, despite very active investigation of suppressor T cells through the 1980s, the cellular and molecular characteristics of the cells were never defined with any precision even though the phenomenon of suppression was and remains highly reproducible in transplant models *in vivo*.

Fortunately, the challenge to characterize T_R cells responsible for unresponsiveness to alloantigens at a molecular level has been taken up again and over the past few years a consensus has emerged in the literature that immunoregulation has the potential to play a key role in both the induction and maintenance of transplantation tolerance [3]. It has also emerged that immunoregulation does not have to operate as the only mechanism responsible for the development of unresponsiveness after transplantation, in other words in isolation, but more likely works alongside other mechanisms, particularly deletion of donor alloantigen reactive T cells to enable effective control of the recipient's response to the mismatched, donor alloantigens to be induced and maintained throughout the post-transplant course [4]. Defining the molecular mechanisms responsible for T-cell-mediated suppression/immunoregulation of immune responses to alloantigens will play a key role in current efforts to develop and translate protocols for the induction of transplantation tolerance in the clinic.

Regulatory T Cells in Inflammation, edited by Leonie S. Taams, Arne N. Akbar and Marca H.M. Wauben

Evidence for the existence of T_R cells that can control graft rejection

In both rodent and human systems, T cells that have immunoregulatory activity responsive to alloantigen stimulation have been found to be enriched within $CD4^+$ T cells. However, it is important to note that regulatory activity is not exclusive to $CD4^+$ T cells, as $CD8^+$ [5–7], $CD8^+CD28^-$ [8], $TCR^+CD4^-CD8^-$ ("double negative"; where TCR is T-cell receptor) [9] and natural killer T (NKT) [10, 11] cells may also have regulatory properties following transplantation. Indeed, it would appear that the development of T_R cells is contextual and that depending on the precise conditions that are created *in vivo*, different subsets of T cells may acquire regulatory functions and contribute to the unresponsiveness state.

The precise combination of T_R cells present may be a consequence of cross-talk between the innate and adaptive immune systems. Several examples where different populations of leukocytes with regulatory properties interact with one another have been reported in the literature. For example, after allogeneic bone marrow transplantation, $CD25^+CD4^+$ T_R cells along with NKT cells have been shown to prevent acute graft-*versus*-host disease [10, 12–14]. Thus regulatory mechanisms in both the innate and adaptive immune systems will most likely contribute to the overall outcome after transplantation with $CD4^+$ regulatory cells being only one of the populations involved. Having said that, for the purposes of this chapter, the discussion will focus on $CD25^+CD4^+$ T_R cells.

In solid organ transplantation, Hall and colleagues [15, 16] were the first to show that $CD25^+CD4^+$ T cells were involved in the long-term survival of cardiac allografts in rats. Subsequently, recipient-derived $CD25^+CD4^+$ T cells have been found to have potent regulatory properties in both the induction and maintenance phases of tolerance to alloantigens *in vivo* in mice (e.g. [17–20]) and later studies in the rat have confirmed Hall et al.'s data [21]. In bone marrow transplantation, donor $CD25^+CD4^+$ T cells present in the bone marrow inoculum can protect the host from graft-*versus*-host disease [12–14].

In humans, $CD25^+CD4^+$ T_R cells have been found to be present in the peripheral blood and in the thymus [22–27]. Their role in promoting allograft survival in humans is under active investigation. Results published to date suggest that $CD25^+CD4^+$ T_R cells are present in some recipients with long-term-surviving allografts [28, 29].

Identifying T_R cells that contribute to transplantation tolerance

CD25 is undoubtedly a very useful way of identifying potential $CD4^+$ T_R cells present in a mixture of other T cells with a range of functional properties [30]. Mouse and human $CD4^+$ T cells with the most potent regulatory properties have been shown to express high and sustained levels of CD25 [22, 31]. This is in contrast to recently activated $CD4^+$ T cells that only express CD25 transiently and at a lower level. Clearly CD25 expression, even at high levels, is not unique to T_R cells and therefore although useful is not sufficiently reliable or adequate for identifying and quantitating T cells with regulatory activity in settings such as clinical transplantation where a highly specific and reproducible test is required. Other phenotypic or molecular markers are therefore urgently needed.

Many new candidate markers for refining the identification of T_R cells have been reported in both mouse and human, including CD45RB, CTLA-4, glucocorticoid-induced tumour necrosis factor (TNF) receptor family-related gene (GITR), CD122, CD103 and Foxp3 [3]. We have analysed human $CD25^{high}CD4^+$ T cells for expression of these makers and found that expression levels are often heterogenous particularly when the cells are purified from renal transplant recipients (A. Akl and K.J. Wood, unpublished observations). For example, we have confirmed that CD122 is expressed on human $CD25^+CD4^+$ regulatory cells [23, 26, 32] and that in mice treatment with donor-specific blood transfusion leads to an increase in the proportion of $CD25^+CD4^+$ cells expressing CD122 [18]. However, in the mouse studies further analysis revealed that the majority of these $CD25^+CD122^+CD4^+$ cells were contained within the population of cells expressing intermediate rather than high levels of CD25, and when analysed alloantigen-specific regulatory activity was found within the $CD122^-$ fraction *in vitro* (M. Karim, A.R. Bushell and K.J. Wood, unpublished observations). These findings are consistent with those reported for self-antigen-specific regulatory cells and suggest that CD122 may also not be a reliable marker.

The ability of some markers to identify T cells with regulatory activity may also be contextual, as is the case for CD25. As mentioned above, this molecule can be expressed on both activated and regulatory T cells. One other example is GITR, a member of the TNF/nerve growth factor receptor gene superfamily that has been reported to be expressed by $CD25^+CD4^+$ alloantigen-specific regulatory cells in mice after exposure to alloantigens (manipulated hosts) [33, 34], as well as by naturally occurring regulatory cells [35, 36]. However, the ability of GITR to identify T_R cells may depend on the context in which the question is asked as, after activation through the TCR, GITR is up-regulated by naïve CD25-$CD4^+$ T cells and functions as a cell-survival signal for the activated cell [37].

Scurfin, the protein product of the *foxp3* gene, is a forkhead-family transcription factor that negatively regulates T-cell function. Foxp3 has been shown to be critical for the development and programming the functional activity of mouse $CD4^+$ T_R cells [38–40]. In human, Foxp3 has also been found to be associated with regulatory T cell in some settings function [41]. However, Foxp3 is not exclusive to $CD25^+CD4^+$ cells [42]. Further work is required to elucidate how useful monitoring Foxp3 expression will be as a diagnostic tool in clinical transplantation.

Unfortunately, to date, no one molecule or gene has been found to fulfill the criteria of exclusivity and stability for the identification of $CD25^+CD4^+$ T_R cells and, at present, it is probably easier to rule out rather than rule in molecules whose expression is exclusively associated with T_R cells. Nevertheless, the search continues and sooner or later a panel of markers that can be used to reliably identify T_R cells will be developed. This search for new markers is of the utmost importance for monitoring the efficacy and safety of novel therapeutic strategies, designed specifically for inducing or restoring tolerance, as they are introduced into clinical practice in the form of pilot studies or clinical trials. Several such protocols are under investigation as part of the portfolio of studies sponsored by the Immune Tolerance Network (www.immunetolerance.org) [43] – even those where immunoregulation is not thought to be the dominant mechanism in operation [44] – and the European Union under the RISET initiative [45].

Relationship between naturally occurring and alloantigen-responsive T_R cells

$CD4^+$ T cells with regulatory activity are generated and positively selected in the thymus (for reviews see [46, 47]) in a process mediated by MHC class II-positive cells [48, 49]. Naturally occurring T_R cells express high levels of CD25 and emerge into the periphery where they seed the peripheral lymphoid organs and circulate in the peripheral blood. Recent evidence shows that naturally occurring $CD25^+CD4^+$ regulatory T cells display a broad TCR Vβ repertoire in humans and mice, suggesting that $CD25^+CD4^+$ T_R cells can recognize a wide variety of antigens [49, 50] and that they may have more efficient interactions with MHC class II and self-peptides than $CD25^-CD4^+$ T cells [51].

Present in this naturally occurring population of T_R cells are likely to be cells that can recognize and respond to alloantigens, in the same way as for other subsets of T cells (for a review of allorecognition see [52]). Indeed, it has been shown that $CD25^+CD4^+$ purified from normal, unmanipulated adult mice – that is, mice that have not previously been exposed to alloantigens – can prevent skin-graft rejection when co-transferred to immunodeficient mice along with potential effector cells [20, 53]. However, in these studies the order 10 times more cells from naïve mice were required to protect grafts from rejection than when cells from mice previously exposed to alloantigen were used as the source of $CD25^+CD4^+$ cells. In other studies, where the capacity of $CD25^+CD4^+$ T cells from naïve mice to prevent rejection of fully allogeneic skin grafts has been tested, the naturally occurring T_R cells have been shown to be ineffective as compared to $CD25^+CD4^+$ from alloantigen-pretreated mice, when their functional activity has been examined under equivalent assay conditions [18]. These findings suggest that the number and/or potency of alloantigen-reactive T_R cells is low in unmanipulated mice and can be enhanced by exposing T cells to alloantigen *in vivo*.

The relationship between naturally occurring T_R cells and those that are generated following exposure to antigen, sometimes referred to as induced T_R cells, remains to be clarified. The thymus may not be necessary for the generation of T_R cells *in vivo* in adult mice [54]. This does not rule out the possibility that naturally occurring $CD25^+CD4^+$ cells already present in the periphery can expand when exposed to alloantigen under permissive conditions. Indeed it has been shown that this can occur [55]. However, there are also reports that T cells with regulatory activity can be generated in the periphery *in vivo* [54] or from peripheral T cells exposed to antigens *ex vivo* under appropriate conditions ([23, 50]; V. Oliveira, B. Sawitzki and K.J. Wood, unpulished observations) and that this can occur even when $CD25^+CD4^+$ cells are not present in the starting population. These latter findings are potentially of particular importance for transplantation since they suggest that specific manipulation for induction or expansion of these cells is achievable *in vivo* in older transplant recipients who may not have an active thymus [56] and that it may well be possible to supplement T_R cells generated *in vivo* with cells from the recipient manipulated *ex vivo* and stored for future use [57].

Mechanisms of immunoregulation by T_R cells

It is clear that the mechanisms by which T_R cells prevent allograft rejection *in vivo* and T cell activation/proliferation *in vitro* are also context-dependent. *In vitro*, T_R cells have been shown to function through a contact-dependent mechanism that does not involve cytokines (e.g. [58]). T_R cells prevent activation of the responding cells, either $CD4^+$ or $CD8^+$ T cells in mouse [18, 59] and human [60, 61], by inhibiting interleukin (IL)-2 production [62] and/or impacting on antigen-presenting-cell (APC) function [63].

Regulatory activity manifest *in vitro* does not appear to have specificity and is mediated through bystander activity [64]. In marked contrast, *in vivo*, T_R cells have been shown to require cytokines to be present in the microenviroment in which they operate to prevent allograft rejection in many studies, with IL-10 [17] and/or transforming growth factor (TGF)-β [65, 66] the most common cytokines that have been shown to be involved. Although the involvement of TGF-β was at first discounted, it has now been reported to play a role not only in their mechanism of action [67], but also in their generation [68] in some settings.

The effects of T_R cells on other T cells, either directly or indirectly through effects on APCs [69], can result in inhibition of cytokine production and secretion, down-regulation of co-stimulatory and/or adhesion-molecule expression, inhibition of proliferation, induction of anergy, elimination of the effector population by promoting cell death, or even conversion of the effector T cell to a regulatory phenotype, a process known as infectious tolerance [70, 71]. This latter effect of T_R cells clearly has very important implications for the stability of tolerance in recipients of long-term surviving grafts as any new T cells entering the repertoire after transplantation where immunoregulation is the dominant mechanism of tolerance could be converted to T_R cells, thereby propagating and reinforcing the tolerant state throughout the post-transplant course. The molecular mechanism responsible for this conversion requires further investigation. *In vivo* it has been suggested that Th2 cytokines may be involved [72], while *in vitro* a role for TGF-β has been proposed [71].

CTLA-4 is known to be an important negative regulator of T-cell function [73], but has also been specifically implicated in the mechanism of action of $CD25^+CD4^+$ T_R cells [74, 75]. Unlike other resting T-cell populations, $CD4^+CD25^+$ cells have been shown to express CTLA-4 constitutively in both mice [74, 75] and humans [24, 26]. The inhibition of human T-cell proliferation by CTLA-4 engagement operates through CD80 engagement and requires the presence of the $CD25^+CD4^+$ T_R cells, which also suggests that it plays a role in the mechanism of action of human T_R cells [76, 77]. Engagement of CTLA-4 plays a role in the induction of peripheral tolerance with signalling through CTLA-4 linked to the production of TGF-β [78].

The programmed death-1 (PD-1) pathway may also play a role in T_R-cell function affecting the level of suppression and thereby fine-tuning the response. However, since this reduction in suppression can be overcome by increasing the number of T_R cells in the co-culture assay, the PD-1/PD-L1 pathway is not essential for T_R-cell function [80].

Potential therapeutic applications of T_R cells

As mentioned above, there are several possible approaches for the clinical application of T_R cells in transplantation. The development of T_R cells could be promoted either (i) *in vivo* by the administration of immunosuppressive drugs, either alone or together with alloantigen, that will both protect the transplant from rejection while promoting the development of T_R cells or (ii) by isolating T cells from the recipient and generating T_R cells by manipulating the cells *in vitro* before reinfusing the characterized T_R cells into the transplant recipient. Such T_R cells may expand from naturally present $CD25^+CD4^+$ T_R cells upon exposure to alloantigens or could be generated from naïve T cells.

Recently there has been an explosion of protocols that have been shown to lead to the development of regulatory activity *in vitro* using either mouse or human T cells [14, 23, 79–82]. For example, culture of mouse $CD25^+CD4^+$ cells in the presence of IL-2 and allogeneic stimulators has been reported to result in a significant expansion of T_R cells, which retain their ability to prevent graft-*versus*-host disease in an alloantigen-specific manner when reinfused *in vivo* [14, 82]. For naive human peripheral-blood lymphocytes, culture in the presence of TGF-β leads to the generation of $CD4^+CD25^+$ cells that regulate T-cell responses *in vitro* [79, 83]. A combination of the active form of vitamin D_3 and dexamethasone has also been shown to lead to the generation of IL-10-producing $CD4^+$ T_R cells [80] and the presence of IL-10 has also been shown to result in expansion of T_R cells without any loss of function [23].

Homing and trafficking of T_R cells

Trafficking of T_R cells in the circulation and their ability to home to specific sites *in vivo* is of key importance to their ability to prevent rejection. If T_R cells generated either *in vivo* or *ex vivo* are unable to access either the draining lymphoid tissue and/or the graft, wherever it is placed in the body, they will not be effective in controlling effector cells and preventing rejection. This property is of major importance for the efficacy of T_R cells generated *ex vivo* where the culture conditions and particularly prolonged periods of culture might alter the ability of the cells produced to survive and/or migrate when they are introduced *in vivo*.

Human $CD25^+CD4^+$ T_R cells have been demonstrated to be particularly responsive to chemokines engaging the CCR4 and CCR8 receptors, potentially enabling the cells to home to sites of inflammation [84], while mouse $CD25^+CD4^+$ cells are potently attracted by the chemokine CCL4, possibly mediated *via* their expression of the receptor CCR5 [85]. *In vivo*, it has been noted in several studies that expression of CD62L, a member of the selectin family of adhesion molecules, endows the cells with more potent regulatory activity [86, 87]. Moreover, expression of CD103 has been linked to T_R cells activity [33, 88].

Impact of immunosuppressive drugs on T_R cells

At present, tolerance to donor alloantigens is sadly not a reality for the vast majority of transplant recipients. The reasons for this are complex, but one contributing factor may be the mechanisms of action of the immunosuppressive drugs that are currently used to prevent rejection.

Calcineurin inhibitors, such as cyclosporine and tacrolimus, are often used as part of an immunosuppressive-drug cocktail to block IL-2 gene transcription, thereby preventing graft rejection. As IL-2 has been shown to be necessary in experimental studies for long-term allograft survival [89], and signalling by IL-2 is required for the function of T_R cells [90], calcineurin inhibitors may have a negative impact on the development and function of T_R cells in transplant recipients. Experimentally, high doses of calcineurin inhibitors administered at the time of tolerance induction have been shown to prevent long-term graft survival [91, 92] and the development of $CD25^+CD4^+$ T_R cells (M. Dover, A.R. Bushell and K.J. Wood, unpublished observations). Interestingly, blocking IL-2 at a later stage, i.e. after the induction of tolerance, may not have such a deleterious effect. Delaying administration of calcineurin inhibitors may therefore be a strategy that could be used to ensure that T_R cells function is not affected.

In clinical transplantation, the administration of cyclosporine simultaneously with anti-CD3 therapy at the time of transplantation resulted in a poorer graft outcome than when cyclosporine therapy was delayed until later after transplantation when the transplanted kidney had achieved satisfactory function; typically a serum creatinine of < 3.0 mg/100 ml [93]. These findings suggest that, under some circumstances, high-dose calcineurin inhibition at the time of transplantation may prevent either the development or, possibly more likely, the function of T_R cells. To conclude from these data that calcineurin inhibitors are totally inhibitory to tolerance induction may be incorrect, as tolerance to donor alloantigens does develop in some recipients of living donor liver transplants even though these patients are treated with tacrolimus in the initial phase post-transplant [94]. Moreover, when the impact of cyclosporine on the immune response after transplantation was first being investigated in rodent models in the 1970s the possibility that it prevented rejection while sparing suppressor cells was highlighted [95]. Further studies in clinical transplantation are required to clarify these issues.

The impact of other immunosuppressive drugs currently in use in clinical transplantation on T_R-cell development and function also needs to be assessed. For example, mycophenolate mofetil, an anti-proliferative agent, has been shown in the mouse to promote an increase in the frequency of $CD25^+CD4^+$ T_R cells and induce transplantation tolerance when combined with the active form of vitamin D_3 [96]. Another factor that might also affect the development of tolerance in clinical transplantation, as mentioned above, is anti-CD25 therapy, which has been introduced into clinical practice. The impact of anti-CD25 therapy on alloantigen-specific regulatory T cells in the early phase of the response after transplantation needs to be clarified. It is possible that short-term treatment with anti-CD25, although effective at reducing the incidence of acute allograft rejection [97], may compromise the development of immunoregulation to donor alloantigens. A recent study examining the impact of anti-CD25 therapy on CD40/CD28 blockade of proliferative responses to alloantigen *in vitro* suggests that this may not be the case; anti-CD25 therapy was synergistic with co-stimulation blockade [98]. A recent study in our laboratory has shown that elimination of $CD25^+CD4^+$ cells from primary allograft recipients in whom specific unresponsiveness to donor alloantigens has been induced by pretreatment with alloantigen in combination with anti-CD4 monoclonal antibody therapy results in rejection of a cardiac allograft (A.R. Bushell and K.J. Wood, unpublished observations). These data provide direct evidence that $CD25^+CD4^+$ cells play a critical role in graft survival in this model.

Another biological agent currently under evaluation in a phase II clinical study is a mutated form of CTLA-4 Ig, designated LEA29Y. Given that CTLA-4 has been shown to be involved in the function of T_R cells *in vivo*, the impact of blocking engagement of CD80/CD86 by long-term administration of CTLA-4 Ig on tolerance induction needs to be established.

Obviously, as well as immunosuppressive drugs having the potential to inhibit T_R-cell generation and function, it should also be possible to identify drugs that promote the development and function of T_R cells. Some of the protocols for generating T_R cells *ex vivo*, referred to above, use small molecules to promote T_R-cell development *in vitro*, including cytokines, dexamethasone and vitamin D_3. Doubtless there will be other molecules identified as further information about the generation and function of T_R cells is obtained that will be used to promote the development of the potent regulatory mechanism *in vivo* or *ex vivo* in the future.

Acknowledgements
K.J. Wood holds a Royal Society Wolfson Research Merit Award. A. Akl is an International Society of Nephrology Research Fellow. Work in K.W.'s laboratory is funded by The Wellcome Trust, European Union, National Kidney Research Fund and Roche Organ Transplant Research Foundation.

References

1 Gershon R, Kondo K (1971) Infectious immunlogical tolerance. *Immunology* 21: 903–914
2 Kilshaw P, Brent L, Pinto M (1975) Suppressor T cells in mice made unresponsive to skin allografts. *Nature* 255: 489–491
3 Wood KJ, Sakaguchi S (2003) Regulatory T cells in transplantation tolerance. *Nat Immunol Rev* 3: 199–210
4 Lechler, R, Garden OA, Turka L (2003) The complementary roles of deletion and regulation in transplantation tolerance. *Nat Rev Immunol* 3: 147–158
5 Hutchinson IV (1986) Suppressor T cells in allogeneic models. *Transplantation* 41: 547–555
6 Gilliet M, Liu Y-J (2002) Generation of human CD8 T regulatory cells by CD40 ligand-activated plasmacytoid dendritic cells. *J Exp Med* 195(6): 695–704
7 Zhou J, Carr RI, Liwski RS, Stadnyk AW, Lee GTD (2001) Oral exposure to alloantigen generates intragraft CD8⁺ regulatory cells. *J Immunol* 167: 107–113
8 Ciubotariu R, Colovai AI, Pennesi G, Liu Z, Smith D, Berlocco P, Cortesini R, Suciu-Foca N (1998) Specific suppression of human CD4⁺ Th cell responses to pig MHC antigens by CD8⁺CD28- regulatory T cells. *J Immunol* 161: 5193–5202
9 Zhang Z, Yang L, Young K, DuTemple B, Zhang L (2000) Identification of a previously unknown antigen-specific regulatory T cell and its mechanism of suppression. *Nat Med* 6: 782–789
10 Zeng D, Lewis D, Dejbakhsh-Jones S, Lan F, Garcia-Ojeda M, Sibley R, Strober S (1999) Bone marrow NK1.1- and NK 1.1⁺ T cells reciprocally regulate acute graft-*versus*-host disease. *Blood* 99: 1449–1457
11 Seino K-I, Fukao K, Muramoto K, Yanagisawa K, Takada Y, Kakuta S, Iwakura Y, Van Kaer L, Takeda K, Nakayama T et al (2001) Requirement for natural killer T (NKT) cells in the induction of allograft tolerance. *Proc Natl Acad Sci* 98: 2577–2581
12 Taylor P, Noelle RJ, Blazar BR (2001) CD4⁺CD25⁺ immune regulatory cells are required

for induction of tolerance to alloantigen *via* costimulatory blockade. *J Exp Med* 193: 1311–1318

13 Hoffmann P, Ermann J, Edinger M, Fathman CG, Strober S (2002) Donor-type $CD4^+CD25^+$ regulatory T cells suppress lethal acute graft-*Versus*-host disease after allogeneic bone marrow transplantation. *J Exp Med* 196: 389–399

14 Cohen JL, Trenado A, Vasey D, Klatzmann D, Salomon BL (2002) $CD4^+CD25^+$ Immunoregulatory T cells: new therapeutics for graft-*Versus*-host disease. *J Exp Med* 196: 401–406

15 Hall B, Pearce N, Gurley K, Dorsch S (1990) Specific unresponsiveness in rats with prolonged cardiac allograft survival after treatment with cyclosporine. III. Further characterisation of the $CD4^+$ suppressor cell and its mechanisms of action. *J Exp Med* 171: 141–157

16 Pearce N, Berger M, Gurley K, Spinelli A, Hall B (1993) Specific unresponsiveness in rats with prolonged cardiac allograft survival after treatment with cyclosporine: vi. *in vitro* alloreactivity of T cell subsets from rats with long-surviving allografts. *Transplantation* 55: 380–389

17 Hara M, Kingsley C, Niimi M, Read S, Turvey S, Bushell A, Morris P, Powrie F, Wood KJ (2001) IL-10 is required for regulatory T cells to mediate tolerance to alloantigens *in vivo*. *J Immunol* 166: 3789–3796

18 Kingsley CI, Karim M, Bushell AR, Wood KJ (2002) $CD25^+CD4^+$ Regulatory T cells prevent graft rejection: CTLA-4^- and IL-10-dependent immunoregulation of alloresponses. *J Immunol* 168: 1080–1086

19 Sanchez-Fueyo A, Weber M, Domenig C, Strom T, Zheng X (2002) Tracking immunoregulatory mechanisms during allograft tolerance. *J Immunol* 168: 2274–2281

20 Graca L, Thompson S, Lin C-Y, Adams E, Cobbold SP, Waldmann H (2002) Both $CD4^+CD25^+$ and $CD4^+CD25^-$ regulatory cells mediate dominant transplantation tolerance. *J Immunol* 168: 5558–5565

21 Lehmann M, Graser E, Risch K, Hancock W, Muller A, Kuttler B, Hahn H, Kupiec-Weglinski J, Brock J, Volk H (1997) Anti-CD4 monoclonal antibody-induced allograft tolerance in rats despite persisitence of donor-reactive T cells. *Transplantation* 64: 1181–1187

22 Baecher-Allan C, Brown J, Freeman G, Hafler D (2001) $CD4^+CD25^+$ high regulatory cells in human peripheral blood. *J Immunol* 167: 1245–1253

23 Levings M, Sangregorio R, Roncarolo M-G (2001) Human $CD25^+CD4^+$ T regulatory cells suppress naive and memory T cell proliferation and can be expanded *in vitro* without loss of function. *J Exp Med* 193: 1295–1301

24 Jonuleit H, Stassen EM, Tuettenberg A, Knop J, Enk A (2001) Identification and functional characterisation of human $CD4^+CD25^+$ T cells with regulatory properties isolated from peripheral blood. *J Exp Med* 139: 1285–1294

25 Ng WF, Duggan PJ, Ponchel F, Matarese G, Lombardi G, Edwards AD, Isaacs JD, Lechler RI (2001) Human $CD4^+CD25^+$ cells: a naturally occurring population of regulatory T cells. *Blood* 98: 2736–2744

26 Stephens L, Mottet C, Mason D, Powrie F (2001) Human CD4+CD25+ thymocytes and peripheral T cells have immune suppressive activity *in vitro*. *Eur J Immunol* 31: 1247–1254

27 Taams L, Smith J, Rustin M, Salmon M, Poulter L, Akbar A (2001) Human anergic/suppresor CD4+CD25+ T cells: a highly differentiated and apoptosis-prone population. *Eur J Immunol* 31: 1122–1131

28 Meloni F, Vitulo P, Bianco AM, Paschetto E, Morosini M, Cascina A, Mazzucchelli I, Ciardelli L, Oggionni T, Fietta AM et al (2004) Regulatory CD4+CD25+ T cells in the peripheral blood of lung transplant recipients: correlation with transplant outcome. *Transplantation* 77: 762–766

29 Li Y, Koshiba T, Yoshizawa A, Yonekawa Y, Masuda K, Ito A, Ueda M, Mori T, Kawamoto H, Tanaka Y et al (2004) Analyses of peripheral blood mononuclear cells in operational tolerance after pediatric living donor liver transplantation. *Am J Transpl* 4: 2118–2125

30 Sakaguchi S, Sakaguchi N, Asano M, Itoh M, Toda M (1995) Immunologic self tolerance maintained by activated T cells expressing IL-2 receptor alpha chains (CD25) Breakdown of a single mechanism of self tolerance causes various autoimmune diseases. *J Immunol* 155: 1151–1164

31 Kuniyasu Y, Takahashi T, Itoh M, Shimizu J, Toda G, Sakaguchi S (2000) Naturally anergic and suppressive CD25+CD4+ T cells as a functionally and phenotypically distinct immunoregulatory T cell subpopulation. *Int Immunol* 12: 1145–1155

32 Dieckmann D, Plottner H, Berchtold S, Berger T, Schuler G (2001) *Ex vivo* isolation and characterisation of CD4+CD25+ T cells with regulatory properties from human blood. *J Exp Med* 139: 1303–1310

33 Zelenika D, Adams E, Humm S, Graca L, Thompson S, Cobbold SP, Waldmann H (2002) Regulatory T cells overexpress a subset of Th2 gene transpcripts. *J Immunol* 168: 1069–1079

34 Wood KJ, Ushigome H, Karim M, Bushell A, Hori S, Sakaguchi S (2003) Regulatory cells in transplantation. *Novartis Foundation Symp* 252: 177–188

35 Shimizu J, Yamazaki S, Takahashi T, Ishida Y, Sakaguchi S (2002) Stimulation of CD25+CD4+ regulatory T cells through GITR breaks immunological self tolerance. *Nat Immunol* 3: 135–142

36 McHugh RS, Whitters M, Piccirillo C, Young D, Shevach EM, Collins M, Byrne M (2002) CD4+CD25+ immunoregulatory T cells: gene expression analysis reveals a functional role for the glucocorticoid-induced TNF receptor. *Immunity* 16: 311–323

37 Nocentini G, Giunchi L, Ronchetti S, Krausz LT, Bartoli A, Moraca R, Migliorati G, Riccardi C (1997) A new member of the tumor necrosis factor/nerve growth factor receptor family inhibits T cell receptor-induced apoptosis. *Proc Natl Acad Sci* 94: 6216–6221

38 Hori S, Nomura T, Sakaguchi S (2003) Control of regulatory T cell development by transcription factor FOXP3. *Science* 299: 1507–1061

39 Fontenot J, Gavin M, Rudensky A (2003) Foxp3 programs the development and function of $CD4^+CD25^+$ regulatory T cells. *Nat Immunol* 4: 330–336

40 Khattri R, Cox T, Yasayko S, Ramsdell F (2003) An essential role for Scurfin in $CD4^+CD25^+$ T regulatory cells. *Nat Immunol* 4: 304–306

41 Stassen M, Schmitt E, Jonuleit H (2004) Human CD(4+)CD(25+) regulatory T cells and infectious tolerance. *Transplantation* 77: S23–S25

42 Bennett CL, Christie F, Ramsdell F, Brunkow PJ, Ferguson PJ, Whitesell L, Kelly TE, Saulsbury FT, Chance PF, Ochs HD (2001) The immune dysregulation, polyendocrinopathy, enteropathy, X-linked syndrome (IPEX) is caused by mutations of FOXP3. *Nat Genet* 27: 20–21

43 Herold K, Hagopian W, Auger J, Poumian-Ruiz E, Taylor L, Donaldson D, Gitelman S, Harlan D, Xu D, Zivin R et al (2002) Anti-CD3 monoclonal antibody in new onset type 1 diabetes mellitus. *N Engl J Med* 346: 1740–1742

44 Spitzer T, Delmonico F, Tolkoff-Rubin N, McAfee S, Sackstein R, Saidman S, Colby C, Sykes M, Sachs D, Cosimi A (1999) Combined histocompatibility leukocyte antigen-matched donor bone marrow and renal transplantation for multiple myeloma with end stage renal disease: the induction of allograft tolerance through mixed lymphohematopoietic chimersim. *Transplantation* 68: 480–484

45 Fandrich F, Zhou X, Schlemminger M, Lin X, Dresske B (2002) Future strategies for tolerance induction: a comparative study between hematopoietic stem cells adn macrophages. *Hum Immunol* 63: 805–812

46 Sakaguchi S, Sakaguchi N, Shimizu J, Yamazaki S, Sakihama T, Itoh M, Kuniyasu Y, Nomura T, Toda M, Takahashi T (2001) Immunologic tolerance maintained by $CD25^+CD4^+$ regulatory T cells: their common role in controlling autoimmunity, tumour immunity and transplantation tolerance. *Immunol Rev* 182: 18–32

47 Maloy K, Powrie F (2001) Regulatory T cells in the control of immune pathology. *Nat Immunol Rev* 2: 816–822

48 Kawahata K, Misaki Y, Yamauchi M, Tsunekawa S, Setoguchi K, Miyazaki J, Yamamoto K (2002) Generation of CD4(+)CD25(+) regulatory T cells from autoreactive T cells simultaneously with their negative selection in the thymus and from nonautoreactive T cells by endogenous TCR expression. *J Immunol* 168: 4399–4405

49 Sakaguchi S, Hori S, Fukui Y, Sasazuki T, Sakaguchi N, Takahashi T (2003) Thymic generation and selection of $CD25^+CD4^+$ regulatory T cells: Implications of their broad repertoire and high self-reactivity for the maintenance of immunologic self-tolerance. *Novartis Foundation Symp* 252: 6–16

50 Taams L, Vukmanovic-Stejic M, Smith J, Dunne P, Fletcher J, Plunkett F, Ebeling S, G Lombardi, Rustin M, Bijlsma J et al (2002) Antigen specific T cell suppression by human $CD4^+CD25^+$ regulatory T cells. *Eur J Immunol* 32: 1621–1630

51 Hsieh CS, Liang Y, Tyznik AJ, Self SG, Liggitt D, Rudensky AY (2004) Recognition of the peripheral self by naturally arising $CD25^+$ $CD4^+$ T cell receptors. *Immunity* 2: 267–277

52 Lechler RI, Lombardi G, Batchelor JR, Reinsmoen N, Bach FH (1990) The molecular basis of alloreactivity. *Immunol Today* 11: 83–88

53 Graca L, Le Moine A, Lin C-Y, Fairchild PJ, Cobbold SP, Waldmann H (2004) Donor-specific transplantation tolerance: The paradoxical behavior of $CD4^+CD25^+$ T cells. *Proc Natl Acad Sci* 101: 10122–10126

54 Karim M, Kingsley CI, Bushell A, Sawitzki B, Wood KJ (2004) Alloantigen-induced $CD25^+CD4^+$ regulatory T cells can develop *in vivo* from CD25-$CD4^+$ precursors in a thymus independent process. *J Immunol* 172: 923–928

55 Nishimura E, Sakihama T, Setoguchi R, Tanaka K, Sakaguchi S (2004) Induction of antigen-specific immunologic tolerance by *in vivo* and *in vitro* antigen-specific expansion of naturally arising $Foxp3^+CD25^+CD4^+$ regulatory T cells. *Int Immunol* 16: 1189–1202

56 Karim M, Bushell A, Wood KJ (2002) Regulatory T cells in transplantation. *Curr Opin Immunol* 14: 584–591

57 Taams L, Vukmanovic-Stejic M, Salama AD, Akbar A (2003) Immune regulation by $CD4^+CD25^+$ regulatory T cells: implications for transplantation tolerance. *Transpl Immunol* 11: 277–285

58 Piccirillo CA, Letterio JJ, Thornton AM, McHugh RS, Mamura M, Mizuhara H, Shevach EM (2002) $CD4^+CD25^+$ regulatory T cells can mediate suppressor function in the absence of transforming growth factor {beta}1 production and responsiveness. *J Exp Med* 196: 237–246

59 van Maurik A, Wood KJ, Jones N (2002) Cutting edge: CD4(+)CD25(+) alloantigen-specific immunoregulatory cells that can prevent CD8(+) T cell-mediated graft rejection: implications for anti-CD154 immunotherapy. *J Immunol* 169: 5401–5404

60 Dieckmann D, Bruett CH, Ploettner H, Lutz MB, Schuler G (2002) Human $CD4^+CD25^+$ regulatory, contact-dependent T cells induce interleukin 10-producing, contact-independent type 1-like regulatory T cells. *J Exp Med* 196: 247–253

61 Camara N, Sebille F, Lechler R (2003) Human $CD4^+CD25^+$ regulatory cells have marked and sustained effects on $CD8^+$ T cell activation. *Eur J Immunol* 33: 3473–3483

62 Thornton A, Shevach E (1998) $CD4^+CD25^+$ immunoregulatory T cells suppress polyclonal activation by inhibiting interleukin 2 production. *J Exp Med* 188: 287–296

63 Frasca L, Scotta C, Lombardi G, Piccolella E (2002) Human anergic $CD4^+$ T cells can act as suppressor cells by affecting autologous dendritic cell conditioning and survival. *J Immunol* 168: 1060–1068

64 Thornton A, Shevach EM (2000) Suppressor effector function of $CD4^+CD25^+$ immunoregulatory T cells is antigen non-specific. *J Immunol* 164: 183–190

65 Josien R, Douillard P, Guillot C, Muschen M, Anegon I, Chetritt J, Menoret S, Vigens C, Soulillou J, Cuturi M (1998) A critical role for transforming growth factor beta in donor transfusion induced allograft tolerance. *J Clin Invest* 102: 1920–1926

66 Cobbold SP, Castejon R, Adams E, Zelenika D, Graca L, Humm S, Waldmann H (2004) Induction of $foxP3^+$ regulatory T cells in the periphery of T cell receptor transgenic mice tolerized to transplants. *J Immunol* 172: 6003–6010

67 Chen W, Wahl SM (2003) TGF-beta: the missing link in $CD4^+CD25^+$ regulatory T cell-mediated immunosuppression. *Cytokine Growth Factor Rev* 2: 85–89

68 Zheng SG, Gray JD, Ohtsuka K, Yamagiwa S, Horwitz D (2002) Generation *ex vivo* of TGF-beta-producing regulatory T cells from $CD4^+CD25^-$ precursors. *J Immunol* 169: 4183–4189

69 Cederbom L, Hall H, Ivars F (2000) $CD4^+CD25^+$ regulatory T cells down-regulate costimulatory molecules on antigen presenting cells. *Eur J Immunol* 30: 1538–1543

70 Qin S, Cobbold SP, Pope H, Elliott J, Kioussis D, Davies J, Waldmann H (1993) "Infectious" transplantation tolerance. *Science* 259: 974–977

71 Jonuleit H, Schmitt E, Kakirman H, Stassen M, Knop J, Enk AH (2002) Infectious tolerance: human $CD25^+$ regulatory T cells convey suppressor activity to conventional $CD4^+$ T helper cells. *J Exp Med* 196(2): 255–260

72 Onodera K, Hancock WW, Graser E, Lehmann M, Sayegh MH, Strom TB, Volk HD, Kupiec-Weglinski JW (1997) Type 2 helper T cell-type cytokines and the development of "infectious" tolerance in rat cardiac allograft recipients. *J Immunol* 158: 1572–1581

73 Waterhouse P, Penninger J, Timms E, Wakeham A, Shahinian A, Lee K, Thompson C, Griesser H, Mak T (1995) Lymphoproliferative disorders with early lethality in mice deficient in CTLA-4. *Science* 270: 985–988

74 Takahashi T, Tagami T, Yamazaki S, Uede T, Shimuzu J, Sakaguchi N, Mak T, Sakaguchi S (2000) Immunologic self tolerance is maintained by $CD25^+CD4^+$ regulatory T cells constitutively expressing cytotoxic T lymphocyte associated antigen 4. *J Exp Med* 192: 303–310

75 Read S, Malmstrom V, Powrie F (2000) Cytotoxic T lymphocyte associated antigen 4 plays an essential role in the function of $CD25^+CD4^+$ regulatory cells that control intestinal inflammation. *J Exp Med* 192: 295–302

76 Manzotti CN, Tipping H, Perry LC, Mead KI, Blair PJ, Zheng Y, Sansom DM (2002) Inhibition of human T cell proliferation by CTLA-4 utilizes CD80 and requires $CD25^+$ regulatory T cells. *Eur J Immunol* 32(10): 2888–2896

77 Perez V, van Parjis L, Biuckians A, Zheng X, Strom T, Abbas A (1997) Induction of peripheral T cell tolerance *in vivo* requires CTLA-4 engagement. *Immunity* 6: 411–417

78 Baecher-Allan C, Brown J, Freeman G, Hafler D (2003) $CD4^+CD25^+$ regulatory cells from human peripheral blood express very high levels of CD25 *ex vivo*. *Novartis Foundation Symp* 252: 67–88

79 Yamagiwa S, Gray J, Hashimoto S, Horwitz D (2001) A role of TGF-beta in the generation and expansion of $CD4^+CD25^+$ regulatory T cells from human peripheral blood. *J Immunol* 166: 7282–7289

80 Barrat FJ, Cua DJ, Boonstra A, Richards DF, Crain C, Savelkoul HF, de Waal-Malefyt R, Coffman RL, Hawrylowicz CM, O'Garra A (2002) *In vitro* generation of interleukin 10-producing regulatory $CD4^+$ T cells is induced by immunosuppressive drugs and inhibited by T helper type 1 (Th1)- and Th2-inducing cytokines. *J Exp Med* 195: 603–616

81 Taylor A, Namba K (2001) *In vitro* induction of $CD25^+CD4^+$ regulatory T cells by the

neuropeptide alpha-melanocyte stimulating hormone (alpha-MSH). *Immunol Cell Biol* 79: 358–367

82 Taylor PA, Lees CJ, Blazar BR (2002) The infusion of *ex vivo* activated and expanded $CD4^+CD25^+$ immune regulatory cells inhibits graft-*versus*-host disease lethality. *Blood* 99: 3493–3499

83 Hoffmann P, Eder R, Kunz-Schughart LA, Andreesen R, Edinger M (2004) Large-scale *in vitro* expansion of polyclonal human CD4(+)CD25high regulatory T cells. *Blood* 104: 895–903

84 Iellem A, Mariani M, Lang R, Recalde H, Panina-Bordignon P, Sinigaglia F, D'Ambrosio D (2001) Unique chemotactic response profile and specific expression of chemokine receptors CCR4 and CCR8 by $CD4^+CD25^+$ regulatory T cells. *J Exp Med* 194: 847–854

85 Bystry R, Aluvihare V, Welch K, Kallikourdis M, Betz A (2001) B cells and professional APCs recruit regulatory T cells. *Nat Immunol* 2: 1126–1132

86 Herbelin A, Gombert J, Lepault F, Bach J, Chatenoud L (1998) Mature mainstream TCR alpha $beta^+$ $CD4^+$ thymocytes expressing L-selectin mediate "active tolerance" in the nonobese diabetic mouse. *J Immunol* 161: 2620–2628

87 Chatenoud L, Salomon B, Bluestone J (2001) Suppressor T cells – they're back and critical for regulation of autoimmunity. *Immunol Rev* 182: 149–163

88 Stassen M, Fondel S, Bopp T, Richter C, Muller C, Kubach J, Becker C, Knop J, Enk AH, Schmitt S et al (2004) Human $CD25^+$ regulatory T cells: two subsets defined by the integrins alpha 4 beta 7 or alpha 4 beta 1 confer distinct suppressive properties upon $CD4^+$ T helper cells. *J Immunol* 34: 1303–1311

89 Konieczny B, Dai Z, Elwood E, Saleem S, Linsley P, Baddoura F, Larsen C, Pearson T, Lakkis F (1998) IFN-γ is critical for long-term allograft survival indcued by blocking the CD28 and CD40 ligand T cell costimulation pathways. *J Immunol* 160: 2059–2064

90 Furtado GC, de Lafaille MAC, Kutchukhidze N, Lafaille JJ (2002) Interleukin 2 signaling is required for $CD4^+$ regulatory T cell function. *J Exp Med* 196: 851–857

91 Larsen P, Elwood E, Alexander D, Ritchie S, Hendrix R, Tucker-Burden C, Cho H, Aruffo A, Hollenbaugh D, Insley PL et al (1996) Long-term acceptance of skin and cardiac allografts after blocking CD40 and CD28 pathways. *Nature* 381: 434–438

92 Kirk A, Burkly L, Batty D, Baumgartner R, Berning J, Buchanan K, Fechner J, Germond R, Kampen R, Patterson N et al (1999) Treatment with humanised monoclonal antibody against CD154 prevents acute renal allograft rejection in nonhuman primates. *Nat Med* 5: 686–693

93 Opelz G (1995) Efficacy of rejection prophylaxis with OKT3 in renal transplantation. Collaborative Transplant Study. *Transplantation* 60: 1220–1224

94 Takatsuki M, Uemoto S, Inomata Y, Egawa H, Kiuchi T, Fujita S, Hayashi M, Kanematsu T, Tanaka K (2001) Weaning of immunosuppression in living donor liver transplant recipients. *Transplantation* 72: 449–454

95 Kupiec-Weglinski JW, Filho M, Strom T, Tilney N (1984) Sparing of suppressor cells: A critical action of cyclosporine. *Transplantation* 38: 97–101

96 Gregori S, Casorati M, Amuchastegui S, Smiroldo S, Davalli AM, Adorini L (2001) Reg-

ulatory T cells induced by 1{alpha},25-dihydroxyvitamin D3 and mycophenolate mofetil treatment mediate transplantation tolerance. *J Immunol* 167: 1945–1953

97 Vincenti F (2001) Interleukin-2 receptor monoclonal antibodies in renal transplantation: current use and emerging regimens. *Transpl Proc* 33: 3169–3171

98 Jones TR, Ha J, Williams MA, Adams AB, Durham MM, Rees PA, Cowan SR, Pearson TC, Larsen CP (2002) The role of the IL-2 pathway in costimulation blockade-resistant rejection of allografts. *J Immunol* 168: 1123–1130

Index